Topics in Applied Physics Volume 75

Springer-Verlag Berlin Heidelberg GmbH

Topics in Applied Physics Founded by Helmut K. V. Lotsch

49 **Laser Spectroscopy of Solids** 2nd Ed.
Editors: W. M. Yen and P. M. Selzer

50 **Light Scattering in Solids II**
Basic Concepts and Instrumentation
Editors: M. Cardona and G. Güntherodt

51 **Light Scattering in Solids III**
Recent Results
Editors: M. Cardona and G. Güntherodt

52 **Sputtering by Particle Bombardment II**
Sputtering of Alloys and Compounds, Electron and Neutron Sputtering, Surface Topography. Editor: R. Behrisch

53 **Glassy Metals II**
Atomic Structure and Dynamics, Electronic Structure, Magnetic Properties
Editors: H. Beck and H.-J. Güntherodt

54 **Light Scattering in Solids IV**
Electronic Scattering, Spin Effects, SERS, and Morphic Effects
Editors: M. Cardona and G. Güntherodt

55 **The Physics of Hydrogenated Amorphous Silicon I** Structure, Preparation, and Devices
Editors: J. D. Joannopoulus and G. Lucovsky

56 **The Physics of Hydrogenated Amorphous Silicon II**
Electronic and Vibrational Properties
Editors: J. D. Joannopoulos and G. Lucovsky

57 **Strong and Ultrastrong Magnetic Fields and Their Applications** Editor: F. Herlach

58 **Hot-Electron Transport in Semiconductors** Editor: L. Reggiani

59 **Tunable Lasers** 2nd Ed.
Editors: L. F. Mollenauer, J. C. White, and C. R. Pollock

60 **Ultrashort Laser Pulses Generation and Applications** 2nd Ed. Editor: W. Kaiser

61 **Photorefractive Materials and Their Applications I** Fundamental Phenomena
Editors: P. Günter and J.-P. Huignard

62 **Photorefractive Materials and Their Applications II** Survey of Applications
Editors: P. Günter and J.-P. Huignard

63 **Hydrogen in Intermetallic Compounds I**
Electronic, Thermodynamic and Crystallographic Properties, Preparation
Editor: L. Schlapbach

64 **Sputtering by Particle Bombardment III**
Characteristics of Sputtered Particles, Technical Applications
Editors: R. Behrisch and K. Wittmaack

65 **Laser Spectroscopy of Solids II**
Editor: W. M. Yen

66 **Light Scattering in Solids V**
Superlattices and Other Microstructures
Editors: M. Cardona and G. Güntherodt

67 **Hydrogen in Intermetallic Compounds II**
Surface and Dynamic Properties, Applications Editor: L. Schlapbach

68 **Light Scattering in Solids VI**
Recent Results, Including High-Tc Superconductivity
Editors: M. Cardona and G. Güntherodt

69 **Unoccupied Electronic States**
Editors: J. C. Fuggle and J. E. Inglesfield

70 **Dye Lasers: 25 Years** Editor: M. Stuke

71 **The Monte Carlo Method in Condensed Matter Physics** 2nd Ed. Editor: K. Binder

72 **Glassy Metals III**
Editors; H. Beck and H.-J. Güntherodt

73 **Hydrogen in Metals III**
Properties and Applications
Editor: H. Wipf

74 **Millimeter and Submillimeter Wave Spectroscopy of Solids**
Editor: G. Grüner

75 **Light Scattering in Solids VII**
Crystal-Field and Magnetic Excitations
Editors: M. Cardona and G. Güntherodt

76 **Light Scattering in Solids VIII**
Fullerenes, Semiconductor Surfaces, Coherent Phonons
Editors: M. Cardona and G. Güntherodt

Volume 1–48 are listed at the end of the book

Light Scattering in Solids VII

Crystal-Field and Magnetic Excitations

Edited by M. Cardona and G. Güntherodt

With Contributions by
M. Cardona, G. Güntherodt, B. Hillebrands,
G. Schaack

With 96 Figures and 24 Tables

Springer

Professor Dr., Dres. h. c. Manuel Cardona
Max-Planck-Institut für Festkörperphysik
Heisenbergstr. 1
D-70569 Stuttgart, Germany

Professor Dr. Gernot Güntherodt
2. Physikalisches Institut
Rheinisch-Westfälische Technische Hochschule Aachen
Templergraben 55
D-52074 Aachen, Germany

ISSN 0303-4216
ISBN 978-3-662-31109-7

Library of Congress Cataloging-in-Publication Data applied for.

Die Deutsche Bibliothek - CIP-Einheitsaufnahme
Light scattering in solids.

7. Crystal-field and magnetic excitations. - 2000
(Topics in applied physics; Vol. 75)
ISBN 978-3-662-31109-7 ISBN 978-3-540-48770-8 (eBook)
DOI 10.1007/978-3-540-48770-8

Originally published by Springer-Verlag Berlin Heidelberg New York in 2000
Softcover reprint of the hardcover 1st edition 2000

Cover concept: Studio Calamar Steinen
Cover production: *design & production* GmbH, Heidelberg
Typesetting: Data conversion by Steingraeber Satztechnik GmbH, Heidelberg

SPIN: 10658627 57/3144/mf - 5 4 3 2 1 0 – Printed on acid-free paper

Preface

This volume is the seventh of a series (Topics in Applied Physics, Vols. 8, 50, 51, 54, 66, 68, 75) devoted to inelastic light scattering by solids, both as a physical effect and as a spectroscopic technique.

The previous volume, Light Scattering in Solids VI (LSS VI) appeared in 1991, four years after the discovery of high-temperature superconductivity. By the time it appeared, inelastic (Raman) light scattering had established itself as one of the most powerful techniques for the investigation of electronic excitations, magnons, phonons, and electron–phonon interaction in the new high-temperature superconductors. Correspondingly, a chapter of LSS VI was devoted to Raman scattering in high-temperature superconductors. In the past eight years, and with the discovery of new families of high-T_c superconductors, Raman spectroscopy has continued to demonstrate its usefulness for the investigation and characterization of this class of materials. New exciting materials, such as fullerenes and carbon nanotubes, porous silicon, and the colossal magnetoresistance manganates, as well as materials exhibiting spin-Peierls transitions, have also shown to be excellent candidates for the investigation by means of inelastic light-scattering spectroscopy. Progress in instrumentation has extended the capabilites of Raman spectroscopy in the directions of spatial microsampling and time-resolved spectroscopy. Increasing commercial availability of laser-based equipment producing subpicosecond pulses has led to the technique of "coherent phonons" which can be considered equivalent to conventional spontaneous Raman scattering but in the time instead of the frequency domain. A chapter devoted to coherent phonons will appear soon in *Light Scattering in Solids VIII*, now in preparation.

This volume contains an introductory chapter with a review of the work in previous volumes, a summary of the contents of the present one, a preview of LSS VIII, and a survey of some of the progress in other aspects of Raman spectroscopy that has taken place since 1991.

Chapter 2 of this volume discusses electronic excitations between crystal-field split levels of transition-metal and rare-earth ions in crystals, among them high-T_c superconductors. Chapter 3 is concerned with a wide range of magnetic excitations that appear in superlattices containing magnetic metals.

The authors would like to thank once again Sabine Birtel for secretarial help and skillful use of modern word processing techniques. Thanks are also due to the Staff of Springer-Verlag, in particular Ms Friedhilde Meyer and Dr. Werner Skolaut for unbureaucratic and skillful production of this volume.

Stuttgart and Aachen, August 1999

Manuel Cardona
Gernot Güntherodt

Contents

Contributors

Prof. Dr. Manuel Cardona,
MPI für Festkörperforschung,
Heisenbergstr. 1,
D-70569 Stuttgart,
Germany
e-mail: cardona@cardix.mpi-stuttgart.mpg.de

Prof. Dr. Gernot Güntherodt
RWTH Aachen,
2. Physikalisches Institut,
Templergraben 55,
D-52074 Aachen,
Germany
e-mail: Gernot.Guentherodt@physik.rwth-aachen.de

Prof. Dr. Burkard Hillebrands,
Fachbereich Physik,
Universität Kaiserslautern,
Erwin-Schrödinger-Str. 56,
D-67663 Kaiserslautern,
Germany
e-mail: hilleb@physik.uni-kl.de
Internet: www.physik.uni-kl.de/w_hilleb
phone: +49 631 205 4228
fax: +49 631 205 4095

Prof. Dr. Gerhard Schaack,
Physikalisches Institut der Universität Würzburg,
Am Hubland,
D-97074 Würzburg,
Germany
e-mail: schaack@physik.uni-wuerzburg.de
phone: +049 931 86544
fax: +049 931 888 5142

1 Introduction

M. Cardona and G. Güntherodt

> – But look here, Krishnan. If this is true of X-Rays, it must be true of light too. I have always thought so. There must be an optical analogue of the Compton Effect. We must pursue it and we are on the right lines. It *must* and shall be found. The Nobel Prize must be won.
>
> *C.V. Raman, Nov. 1927 after hearing of the Nobel award to A.H. Compton. As reported by A. Jayaraman in: C.V. Raman (affiliated East-West Press, New Delhi 1989) p. 21*

This volume is the seventh in the series *Light Scattering in Solids* (LSS) which appears in the collection *Topics in Applied Physics.* The first volume was published in 1975, only five years after the death of Prof. C.V. Raman, and was originally intended to be a single treatise on the subject. A second edition was issued in 1983 [1.1]. Because of rapid developments in the field, volumes II [1.2] and III [1.3] became necessary; they appeared in 1982. Volume IV [1.4] was published in 1984, volume V in 1989 [1.5] and volume VI in 1991 [1.6]. Volume VIII is in preparation and will appear shortly after the present one [1.7]. It will contain chapters on scattering by phonons at semiconductor surfaces and interfaces [1.8], scattering by phonons in C_{60} (the so-called fullerites) [1.9] and the recently developed technique of coherent phonons, which is equivalent to Raman scattering in the time domain [1.10].

1.1 Survey of Previous Volumes (I–VI)

1.1.1 Contents of *Light Scattering in Solids I*

The first volume of this series [1.1] contains six chapters that cover the basic principles of the phenomenon of Raman scattering and the technique of Raman spectroscopy as applied to semiconductors and insulators. Scattering by phonons and by electronic excitations (in doped semiconductors) is discussed with respect to both, the spectra of the scattering excitations and their resonance when the laser (and/or the scattered) frequency is close to that of strong electronic interband transitions. One of the chapters in [1.1] is devoted to phonons in glasses and amorphous semiconductors, a topic which

Topics in Applied Physics, Vol. 75
Light Scattering in Solids VII
Eds.: M. Cardona, G. Güntherodt

had been already mentioned by Raman in his early publications [1.11]. Most of this series is concerned with spontaneous light scattering. The interested reader will also find in [1.1] an article devoted to stimulated (i.e., coherent) Raman scattering.

1.1.2 Contents of *Light Scattering in Solids II*

Volume II [1.2] contains an article on resonance Raman phenomena, as observed mainly in the scattering by phonons in semiconductors. A collection of practical rules and equations is given which allow the estimate of *absolute* scattering efficiencies (the solid-state equivalent to cross sections). These estimates are compared with the few data on absolute efficiencies available in the literature till 1982. Even today, data on absolute Raman efficiencies remain rather scarce [1.12].

Light Scattering in Solids II [1.2] also contains a chapter on multichannel detection, a technique that allows a reduction in the measurement time by a couple of orders of magnitude and has made a phenomenal progress in recent years. The article was updated in Volume V of the series [1.13]. For a more recent review see [1.14]. This volume also contains an article on the somewhat esoteric, but powerful nonlinear optical technique of hyper-Raman spectroscopy that allows the observation of some Raman and ir forbidden (i.e., silent) excitations (for recent hyper-Raman work dealing with the very topical material GaN see [1.15]).

1.1.3 Contents of *Light Scattering in Solids III*

Volume III [1.3] contains chapters on a multitude of light scattering phenomena observed in a wide range of crystals, from graphite to superionic conductors, from transition-metal compounds to direct gap semiconductors. Several of the articles in [1.3] are concerned with materials with magnetic ions and/or magnetic structures related to the work in Chaps. 2 and 3 of the present volume. We mention explicitly the chapter on magnetic excitations in transition-metal halides [1.16], the work on phonon anomalies in normal and superconducting metallic transition-metal compounds [1.17], which is complemented by the article by G. Schaack in the present volume, and the work on Brillouin spectroscopy using multi-pass tandem Fabry–Pérot interferometry [1.18], covering Brillouin (i.e. low frequency) scattering by magnetic materials, which is complemented by the article on magnetic structures by B. Hillebrands in the present volume.

1.1.4 Contents of *Light Scattering in Solids IV*

Volume IV [1.4] is devoted to electronic Raman scattering, surface enhanced Raman scattering and also to the effect of hydrostatic pressure on the Raman spectra of phonons (in semiconductors as well as in molecular solids). For a recent review of the latter topic, dealing also with the effects of uniaxial stress, the reader should consult [1.19]. Volume IV also contains work of relevance to the chapters in the present volume, namely two articles on light scattering in rare-earth magnetic semiconductors which contain magnetic ions [1.20]. Two chapters dealing with the interesting phenomenon of surface-enhanced Raman scattering are also contained in [1.4]. In the past two decades the phenomenon of surface-enhanced Raman scattering has developed into a powerful technique to investigate monomolecular organic layers.

1.1.5 Contents of *Light Scattering in Solids V*

Volume V [1.5] of the *Light Scattering in Solids* series appeared in 1989 after a decade of explosive development of Raman spectroscopy as applied to nanostructures. It thus covers most aspects of Raman scattering in superlattices and quantum wells. Quantum dots and wires made their grand appearance in the Raman field somewhat later. The interested reader should consult [1.21, 1.22].

Among the topics discussed in [1.5] we mention the formal macroscopic theory (based on the macroscopic elastic constants, magnetic susceptibilities, and dielectric functions) of excitations in periodic layer systems (i.e. superlattices), Raman and Brillouin scattering by phonons and electronic excitations in such systems, quasiperiodic superlattices (e.g., of the Fibonacci type), Raman investigations of surfaces and interfaces using highly sensitive multichannel detectors. The last chapter [1.23] can be regarded as a predecessor of Chap. 3 of the present volume.

1.1.6 Contents of *Light Scattering in Solids VI*

Volume VI of the series [1.6] contains two chapters on magnetic scattering and crystalline electric field phenomena closely related to the work by G. Schaack in the present volume [1.24, 1.25]. It also has an article on time-dependent phenomena in light scattering, a topic which has experienced enormous development since the appearance of [1.6] (see [1.26]). Another article in [1.6] also treats work on time-resolved Raman spectra as applied to scattering by phonons in AgCl and AgBr.

The next volume of this series (LSS VIII) will contain an article on the new technique of *coherent phonons*, which is equivalent to Raman spectroscopy in the time domain instead of frequency domain [1.10, 1.27]. Raman scattering in the time domain is also important for the theoretical description of the scattering cross sections by means of the highly sophisticated and powerful time correlator techniques. An article by J. Page in [1.6] treats in depth this theoretical approach to Raman scattering in molecular crystals whereas an article by Yacoby and Ehrenfreund discusses Raman scattering in conjugated polymers.

Last but not least, an article by C. Thomsen in [1.6] discusses the basic principles and the applications of Raman spectroscopy to high-T_c superconductors. For this family of materials, discovered in 1987, the capabilities of Raman spectroscopy had been already realized in 1991, the year of appearance of [1.6], electronic (e.g., pair breaking) excitations, magnons, crystal field transitions and various coupled versions of these excitations, already had been observed at the time when [1.6] appeared. In the past eight years, however, considerable activity has taken place in the field as new phenomena (e.g., the existence of pseudogaps, electronic crystal-field excitations) and new materials have been discovered. A few examples illustrating the progress will be discussed in Sect. 1.2. For a recent review of light scattering in high-T_c superconductors see [1.28].

1.2 Highlights and Recent Progress in Raman Spectroscopy

Since the appearance of Volume VI [1.6], Raman spectroscopy has continued to establish itself as one of the most effective and versatile tools for the investigation and characterization of solids and as a multidisciplinary technique with applications to materials sciences, microelectronics, chemistry, biology, and medicine, especially medical diagnostics. For a wide range of examples see the proceedings of the XVIth International Conference on Raman Spectroscopy (ICORS) held in Cape Town (South Africa) in August 1998 [1.29]. These developments have been triggered in part by advances in instrumentation and experimental methods and also by the discovery and synthesis of new materials. Among the new instruments now commercially available, large collection efficiency systems, based on holographic notch filters for the suppression of Rayleigh scattered light and also on the use of acousto-optic tunable filters as dispersive elements, are becoming generally accepted (for a review see [1.14]). The use of multichannel detectors, mostly of the Charge Coupled Device (CCD) variety [1.30] has also gained wide acceptance, except in cases where very high resolution is required. These advances in multichannel detectors have led to the reduction of the "exposure times" required to obtain spectra with a good signal-to-noise ratio, thus enabling the investigation by Raman spectroscopy of time-dependent phenomena in real time. In parallel with this time-resolution feature, spatial resolution is now also commercially available. Several manufacturers offer Raman systems equipped with confocal microscope arrangements that allow a lateral resolution of a few microns. Three-dimensional in-depth resolution is also obtained in the case of transparent materials.

Another reason for the increasing pace in the applications of Raman spectroscopy to condensed matter physics lies in the synthesis of new materials and the discovery of new phenomena. Since the publication of LSS VI, a few new high-temperature superconductors, in particular the Hg-12$(n-1)n$ compounds with T_c's up to ≈160 K (in Hg-123 under pressure), have come to the fore. Nevertheless, and in spite of a staggering amount of experimental and theoretical work, the mechanism leading to the phenomenon of high-T_c superconductivity is still not understood. A number of conjectures related to the high-T_c superconducting oxides have, however, found widespread acceptance. Among them we mention:

1. The existence of an optimal doping concentration and the different properties of underdoped (i.e., doped below optimal) and overdoped materials, the latter being closer than the former to conventional metals in their normal state [1.31]. Optimally doped and underdoped materials have rather anomalous "normal" state properties, exemplified among others by the linear T-dependence of their electrical resistivity.
2. The existence of Cooper-like pairs and a highly anisotropic pair-breaking gap. This gap, which vanishes at T_c, has $d_{x^2-y^2}$-like symmetry (irreducible representation B_{1g} in materials with tetragonal D_{4h} point-group symmetry). It can be directly observed by means of several spectroscopic techniques such as photoelectron spectroscopy [1.33], tunneling [1.34], and Raman spectroscopy [1.35].
3. All high-T_c materials known to date are either tetragonal (D_{4h} point group) or orthorhombic (D_{2h} point group) and possess a number of parallel CuO_2-planes (between one and about six) in each primitive cell. These planes are believed to support the superconducting carriers (usually holes, exceptionally electrons in the Nd_2CuO_4 materials).
4. The high-T_c superconductors become insulators in the absence of *doping* (e.g., if not enough carriers are present, e.g., in $YBa_2Cu_3O_6$). In the semiconducting phases the spins of the Cu^{2+} ions in the CuO_2 planes order antiferromagnetically with a rather high Néel temperature T_N. Traces of this order appear as antiferromagnetic fluctuations in the doped, superconducting phases. These fluctuations are often believed to be responsible for the superconducting pairing.
5. In the underdoped case a partial gap in the spectral density of electronic excitations appears in some regions of the Fermi surface and persists above T_c. It can be seen in Raman spectroscopy, most clearly in the B_{2g}-like (xy symmetry) spectra as illustrated in Fig. 1.1 [1.32]. In this figure we observe the decrease in the integrated spectral weight for electronic excitations (for $\hbar\omega < 700\ \mathrm{cm}^{-1}$) that appears in the B_{2g} Raman spectra below $T = 200$ K for two underdoped samples (Bi-2212 and Y-123). The reason why pseudogap structure appears in the B_{2g} configuration is not understood. A pseudogap can also be seen in ir-spectroscopy for electric fields polarized along the c-axis [1.37].
6. In spite of the likelihood of antiferromagnetic fluctuations being responsible for the superconducting pairing, it is of interest to investigate the effect of electron–phonon coupling which is responsible for conventional, BCS-like superconductivity. For this purpose, Raman spectroscopy is an

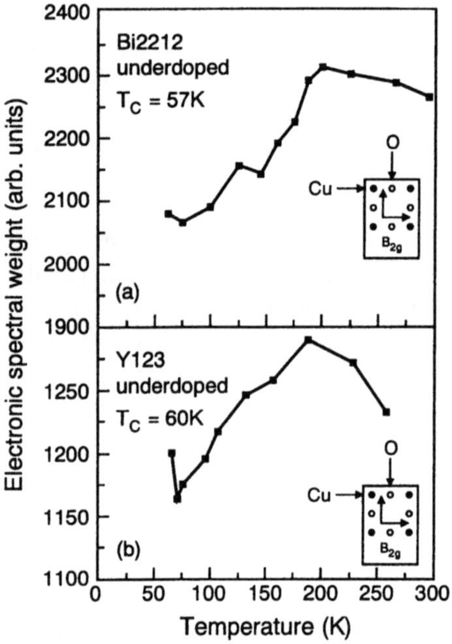

Fig. 1.1. Spectral weight of the electronic Raman spectra of optimally and underdoped Bi2212 (**a**) and underdoped Y123 (**b**) as a function of temperature. Note that the decrease below 200 K, which has been interpreted as a signature of a pseudogap, appears only in B_{2g} scattering configuration. From [1.32]

ideal technique, although its application is limited to phonons with $\boldsymbol{k} \simeq 0$. Large electron–phonon coupling reveals itself in changes of the phonon Raman efficiencies and self-energy anomalies when crossing T_c. Such effects have been observed in many high-T_c materials. We show in Fig. 1.2 one of the most spectacular examples, involving A_{1g} vibrations at 220 and 375 cm^{-1} of the oxygens in the CuO_2 planes of Hg-1234 with T_c close to that which corresponds to optimal doping, a material with four CuO_2 planes per primitive cell [1.28,1.36]. Three clear and strong anomalies appear in the spectra of these phonons when lowering T below $T_c = 123$ K: (1) A strong increase in the phonon intensities, (2) a decrease in their frequencies and (3) a decrease in their linewidths. These effects can be interpreted as resulting from the interaction of the phonons with the pair breaking electronic excitations. The latter appear in Fig. 1.2 below T_c as a broad peak centered around 600 cm^{-1}.

7. The coupling of the phonons to the electrons displayed in Fig. 1.2 is compatible with a McMillan electron–phonon interaction parameter $\lambda \approx 6$. If all phonons would have the same coupling, $\lambda = 6$ would suffice to attain $T_c \approx 100$ K solely on the basis of electron–phonon interaction. However, all phonons do not couple as strongly to the pair breaking excitations as those of Fig. 1.2. For the optimally doped samples, the effects on T_c of replacing the ionic masses by different isotopes are nearly negligible [1.38], a fact that speaks against the purely vibrational origin of the superconducting paring in high-T_c materials. We should mention, however, that as the doping is decreased the isotope effect on T_c increases and reaches,

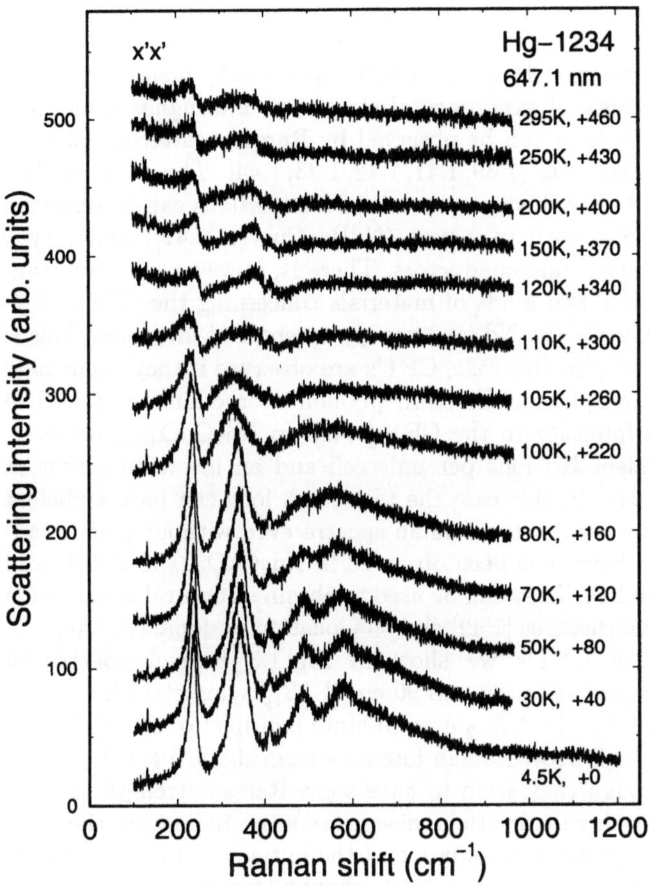

Fig. 1.2. Raman spectra of Hg-1234 (T_c = 123 K) measured at various temperatures between RT and 4.5 K in $x'x'$ polarization. The numbers in the right column give the vertical offset of the spectra with respect to that at the bottom. Note the strong changes that take place in the strength, frequency and width of the two low-frequency peaks (A_{1g} phonons) when crossing T_c. From [1.36]

for $^{16}O \rightarrow {}^{18}O$ substitution, values close to those predicted for phonon coupling ($\alpha \simeq 0.5$) at $T_c \simeq 40$ K. This effect would suggest a phononic contribution of an unconventional type to the pairing mechanism.

The contributions of Raman spectroscopy to the investigation of high-T_c superconductors are related to the following elementary excitations:

1. Phonons
2. Electrons in the normal and the superconducting state
3. Electron–phonon coupled modes
4. Magnons
5. Crystal-field transitions (CFT) in constituent rare earth ions (f-electrons)

Most of these excitations have been mentioned above. We shall next spend a few words on the Raman spectra of transitions between f-electron states of rare-earth ions. Such transitions are usually observed by inelastic neutron scattering [1.40]. Because of experimental simplicity and higher resolution, the few cases in which they can be observed by Raman scattering have received considerable attention [1.39, 1.41, 1.42, 1.43, 1.44]. The most studied rare-earth ion is Nd^{3+}. For this ion, crystal-field transitions can be observed by Raman spectroscopy both in p-type ($NdBa_2Cu_3O_7$ [1.41]) and n-type ($Nd_2CuO_{4+\delta}$ [1.42, 1.43]) superconductors. There is, however, an important difference between these two kinds of materials concerning the CFT's of f-electrons: in $NdBa_2Cu_3O_7$ the Nd-ions occupy a center of inversion. This is probably the reason why, in this case, CFT's are observed if they occur near a strongly Raman active phonon: Electron–phonon interaction transfers then some of the phonon intensity to the CFT [1.41]. In $Nd_2CuO_{4+\delta}$, however, there are two equivalent Nd ions per unit cell and an inversion center at midpoint between them. In this case the rare-earth ions can have sufficient Raman intensity to be seen in the Raman spectra even without phonon admixture. Such CFT's have thus been observed in $Nd_2CuO_{4+\delta}$ (Nd-214) and also in Pr-214 and Sm-214. They can be used to obtain structural information concerning magnetic structures [1.42] and the location of dopants [1.44].

As an example of CFT's we show in Fig. 1.3 Raman spectra of $SmBa_2Cu_3O_7$. The peaks at $\approx$ 180 and 90 cm^{-1} (B_{1g} symmetry) have been identified as CFT's within the $^6H_{5/2}$ ground-state multiplet of the Sm^{3+} ion. These electronic CFT's borrow Raman intensity from the 310 cm^{-1} phonon (see Fig. 1.3a,b) although they seem to have some Raman strength of their own [1.39]. Note that these transitions disappear when increasing the temperature, due to the equalizing population of the initial and final CF states. This can be viewed as a signature of CF vs. phonon transitions.

We close the discussion of recent progress in Raman spectroscopy of high-T_c superconductors by mentioning that isotopic substitution, as observed in the Raman spectra of phonons, is important for the characterization of samples to be used in measurements of isotope effects on T_c [1.38]. It has also been used to determine experimentally the eigenvectors of phonons [1.46]. Information on absolute efficiencies for scattering by phonons and also by electronic excitations can be found in [1.12].

Stimulated by the discovery of high-T_c superconductors, there is renewed interest in the physical properties and excitation spectra of other low-dimensional spin systems, such as one-dimensional transition-metal oxide chain or ladder compounds. Research in this field of quantum spin systems has been spurred on particularly by theoretical predictions of a spin gap and "d-wave" pairing in hole doped ladder compounds with an even number of legs [1.47]. More recently, three-leg ladders are being discussed as analogs to underdoped superconducting cuprates [1.48]. In fact, the compound $Sr_{14}Cu_{24}O_{41}$, with CuO_2 chains and Cu_2O_3 ladders as building blocks, i.e.,

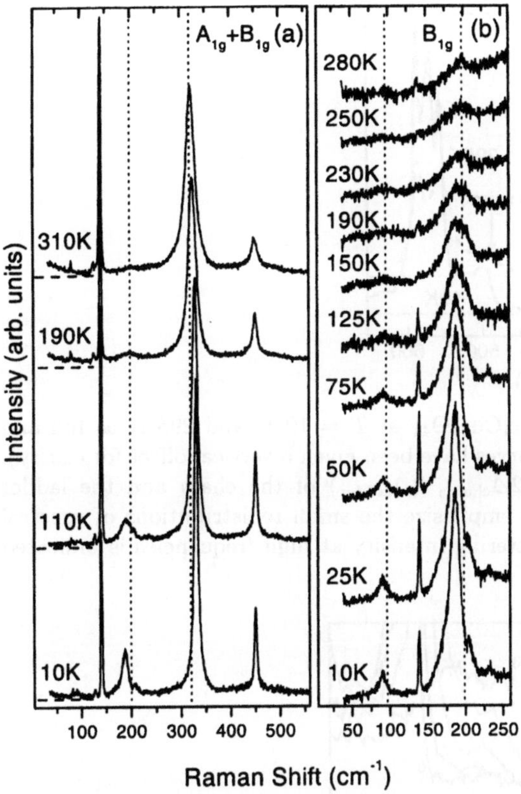

Fig. 1.3. Polarized Raman spectra in (a) $A_{1g} + B_{1g}$ and (b) B_{1g} for a nonsuperconducting $SmBa_2Cu_3O_6$ crystal at different temperatures. Crystal-field excitations are seen at 90 and 190 cm^{-1}. They disappear with increasing temperature because of the equalization of the population of the initial and the final states. From [1.39]

$[Sr_2Cu_2O_3]_7(CuO_2)_{10}$, shows upon substitution of Sr_{14} by $(Sr_{0.4}Ca_{13.6})$ a superconducting transition at $T_c = 12$ K under a pressure of 3 GPa [1.49]. Raman scattering in the undoped material shows the spin gap of the chains and the ladders [1.45] (see Fig. 1.4).

On the other hand, chain- or ladder-compounds of transition-metal oxides have been investigated quite extensively as model systems in order to identify spin–charge separation [1.50] and to characterize spin–phonon coupling and the spin-excitation spectra [1.51,1.52,1.53,1.54,1.55,1.56,1.57,1.58,1.59,1.60]. Besides neutron scattering [1.51,1.52], Raman scattering has played a significant and crucial role in the investigation of the first inorganic spin-Peierls compound $CuGeO_3$ with $T_{SP} = 14$ K [1.53,1.54,1.55,1.56,1.57,1.58]. In the high-temperature uniform phase, the Raman scattering intensity has been shown to be due to the competition of nearest and next nearest neighbor antiferromagnetic exchange interactions [1.55]. In the low temperature dimerized phase, Raman scattering has revealed transition-induced folded phonon

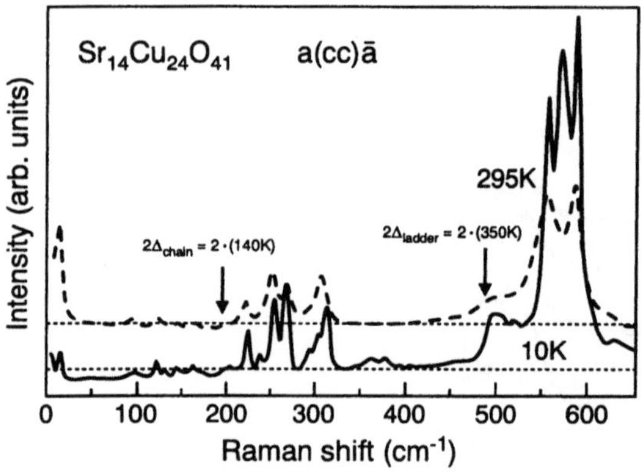

Fig. 1.4. Raman scattering in $Sr_{14}Cu_{24}O_{41}$ at $T = 10$ K and 295 K as full and dashed curves, respectively (the curves have been given a vertical offset for clarity). The double singlet-triplet gaps ($2\Delta_{\text{chain}}$, $2\Delta_{\text{ladder}}$) of the chain and the ladder system are marked by arrows. To emphasize the small redistributions of spectral weight the background of the scattering intensity at high frequencies is indicated by a dotted line. From [1.45]

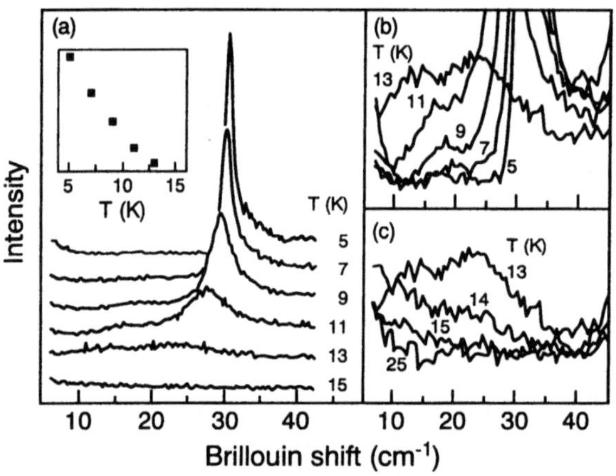

Fig. 1.5. Brillouin spectra of $CuGeO_3$ for polarizations parallel to the chain axis (cm^{-1}) for several temperatures (**a**) showing the singlet bound state at $\Delta\omega \approx 30\ cm^{-1}$ for $T < T_{SP} = 14$ K (the curves have been given an offset for clarity); (**b**) a thermally activated three-magnon scattering process near 17 cm^{-1} for $T < T_{SP} = 14$ K; and (**c**) the disappearance of the scattering intensity for $T > T_{SP}$. The inset in (**a**) shows the T-dependence of the peak intensity of the singlet bound state

modes, their Fano resonances and the occurrence of a singlet bound state [1.53, 1.54, 1.55]. The latter appears as a spin-conserving Raman excitation with twice the singlet-triplet gap energy and is displayed in Fig. 1.5a at 30 cm^{-1}. An additional triplet bound state, identified in inelastic light scattering via a three-magnon scattering process, also appears in Fig. 1.5b,c [1.56]. It

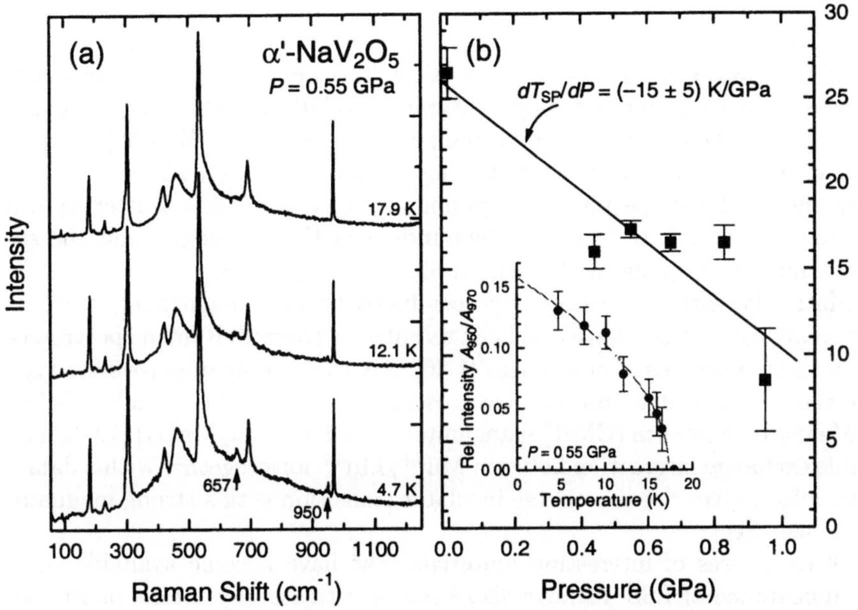

Fig. 1.6. (**a**) Raman spectra of α'-NaV_2O_5 for various temperatures at a pressure of 0.55 GPa. The transition to the dimerized phase manifests itself by the appearance of additional Raman peaks at 950 and 657 cm^{-1}. (**b**) The transition temperature T_{SP} can be determined from the temperature dependence of the intensity of the 950-cm^{-1} peak (inset). Under pressure T_{SP} decreases rapidly at a rate of 15 ± 5 K/GPa

corresponds to a transition from a thermally excited bound triplet state into the triplet continuum. In $Cu_{1-x}Zn_xGeO_3$, a bound state between a spinon and the Zn dopant was found using Brillouin light scattering [1.57, 1.58]. A second, recently discovered spin-Peierls-type compound, namely α'-NaV_2O_5, a quarter-filled spin ladder structure with $T_{SP} = 34$ K [1.61], exhibits at low temperature charge ordering, multiple magnetic bound states in Raman scattering [1.59] and strong spin–phonon coupling [1.60]. Raman spectra for this material above and below T_{SP} (at a pressure of 0.55 GPa) are shown in Fig. 1.6 together with the pressure dependence of T_{SP} [1.62]. Measurements of the spin-Peierls transition of $CuGeO_3$ vs. pressure can be found in [1.63].

Another link between the low-dimensional quantum spin systems and the cuprate high-T_c superconductors is provided by the so-called stripe phase compounds, such as the doped lanthanum nickelate , $La_{2-x}Sr_xNiO_4$, and lanthanum cuprate, $(La,Nd)_{2-x}Sr_xCuO_4$ [1.64]. These compounds exhibit a new type of real-space charge and spin ordering in topological stripe-type phases. The quasi-two-dimensional commensurate charge and spin stripe ordering in the NiO_2 planes of $La_{1.67}Sr_{0.33}NiO_4$ has been investigated by Raman scattering [1.64]. Below the charge-ordering transition $T_{CO} = 240$ K a superstructure and the opening of a pseudogap in the electron–hole excitation spectra is observed. The spin ordering is found to be driven by the charge ordering. The emergence of two peaks from the two-magnon continuum in Raman scattering of $La_{1.67}Sr_{0.33}NiO_4$ upon cooling below T_{CO} has been taken as evidence of two-dimensional antiferromagnetic spin correlation in the charge-ordered state even above the spin-ordering temperature $T_{SO} = 180$ K [1.65].

The ongoing investigation of the low-dimensional model-type spin systems and of the stripe phase compounds by means of light scattering will presumably contribute to our understanding of the pseudogap and the superconducting coupling mechanism in high-T_c compounds.

Materials with coexisting superconductivity and magnetism, such as $RuSr_2GdCu_2O_8$ [1.66], have also been recently discovered. Raman spectroscopy on these materials is in progress [1.66]. Another family of perovskite-type materials of interest to Raman spectroscopists involves the so-called Colossal MagnetoResistance (CMR) manganites (e.g., $La_{1-x}Ca_xMnO_3$) [1.67]. The double exchange interaction between Mn^{3+}/Mn^{4+} ions together with a Jahn–Teller effect, gives rise to a metal–insulator transition with a strong magnetoresistance effect.

Other classes of interesting materials that have become available since the appearance of LSS VI have also been investigated by means of Raman spectroscopy. Among them we mention porous silicon [1.68], a material that has received a lot of attention on account of its high photoluminescence efficiency. The synthesis of C_{60} (the so-called fullerite) in powder and single crystal form has also attracted the attention of Raman spectroscopists. The vibrational properties of these and related materials (e.g., C_{70}) will be the subject of a chapter in the next volume of this series [1.9]. C_{60} crystals are known to become metallic, and even superconducting with T_c's up to 40 K, when doped with alkali metals (e.g., Rb_3C_{60}). A recent publication reports the observation of the superconducting gap in these materials by means of Raman spectroscopy [1.69].

The fall of the "Iron Curtain" had a profound influence on science policy and scientific collaboration between East and West. Among other consequences, stable isotopes of many elements became available in the West at affordable prices or even at no cost (on the basis of collaboration with Russian colleagues). Bulk single crystals, single crystalline films and nanostructures composed of parts (e.g., layers) with different isotopic compositions have been grown at several laboratories. Raman spectroscopy has become a powerful technique for characterizing these materials, especially the isotopic nanostructures, whose structural properties cannot be investigated by x-rays techniques. Besides the analysis of structural properties, Raman spectroscopy, as applied to crystals with different isotopic compositions, has generated a great wealth of information concerning [1.70]:

- Phonon eigenvectors
- Anharmonic decay of phonons

- Contribution of isotopic disorder to phonon linewidths and energy renormalization
- First order scattering induced by isotopic disorder.

We close this section by mentioning recent developments in the field of Raman scattering using as a "light source" highly monochromatized synchrotron radiation [1.71] the scattering photons are in the x-ray region, a fact that enables one to scan with the $\boldsymbol{k}$-vector the full Brillouin zone, thus providing an alternative to phonon spectroscopy by means of inelastic neutron scattering. The attainable resolution, however, is still one to two orders of magnitude worse than obtainable with laser sources. Going back to Raman's conversation with his collaborator Krishnan quoted at the beginning, it seems that phonon spectroscopy has gone back to the original idea that led to Raman's discovery: the inelastic scattering of x-rays.

1.3 Contents of This Volume

1.3.1 Chapter 2

This chapter by Gerhard Schaack presents a review of the area of Crystal-Field Transitions (CFT) with emphasis on observations by means of Raman spectroscopy. CFT's are understood as electronic excitations from the ground state of a partially filled $3d$ or $4f$ shell to an intrashell excited state. These excitations have even parity (which is a good quantum number when the corresponding atomic site is an inversion center). Crystal-field excitations are thus Raman allowed from the point of view of parity. Nevertheless, the Raman cross sections are often weak since they involve intermediate virtual states that must lie outside the crystal-field manifold and are often too high in energy to lead to strong Raman efficiencies. As discussed in connection with Fig. 1.3, CFT's often require admixture of Raman-active phonons (via electron–phonon interaction) in order to be observable in the Raman spectra. The most versatile type of spectroscopy for the investigation of CFT's is Inelastic Neutron Scattering (INS): CFT's are usually associated with a magnetic dipole moment and can couple strongly to thermal neutrons.

Nevertheless, the field of Raman spectroscopy as applied to CFT's in solids has a long and illustrious 35-year history [2.10] and has often led to results that are complementary to those obtained by neutron spectroscopy. The discovery, more than one decade ago, of high-T_c superconductors, many of which contain rare-earth atoms, has given a big boost to CFT studies by means of Raman spectroscopy (see [1.41] and Fig. 1.3).

After the introductory Sect. 2.1, Sect. 2.2 of Schaack's article presents a detailed review of the theory of inelastic light scattering by electronic transitions, in particular CFT's. The theory of the frequencies of possible Raman transitions and their intensities, in particular the phenomenon of intraconfigurational Raman resonances, is discussed. Section 2.2 also contains an introduction to CFT's in magnetically ordered crystals and the required group-theoretical formalism (magnetic groups). Section 2.2 ends with a brief introduction to time-resolved Raman spectroscopy of CFT's.

Section 2.3 discusses the effect of electron–phonon interaction on CFT's. The transition-metal and rare-earth ions are no longer treated as static, as in Sect. 2.2, but are allowed to vibrate. Among the resulting phenomena discussed in Sect. 2.3 we mention the Jahn–Teller effects. The symmetry considerations applicable to the treatment of vibronic interaction are discussed in considerable detail. Section 2.3 ends with a discussion of resonant electron–phonon interaction effects that take place when the "undressed" frequencies of CFT's are close to those of phonons of the same symmetry.

Section 2.4 is concerned with the applications of Raman spectroscopy of CFT's to the investigation and characterization of materials that contain rare-earth ions. Not surprisingly, it starts with a discussion of high-T_c superconductors containing rare-earth ions. Raman spectra of CFT's have been observed for high-T_c superconductors (and their insufficiently doped semiconducting counterparts) containing Nd^{3+}, Pr^{3+}, and Sm^{3+}; CTF's in most other rare-earth ions should also be accessible to Raman spectroscopy by using appropriate laser frequencies and spectroscopic equipment. Both, hole conductors (e.g., $NdBa_2Cu_3O_7$ [1.41], Pr_2Sr_2Nd [2.187], Cu_3O_8) and electron conductors (Nd_2CuO_4 [1.42],[2.162]) have been investigated.

The above-mentioned studies of CFT's in high-T_c superconductors and their semiconducting counterparts have been performed not only *per se* but also as a means of characterizing structural properties of these materials. We mention, as an example, the investigation of the complicated antiferromagnetic structures the Cu^{2+} ions (with several magnetic phase transitions) in Nd_2CuO_4 [1.42].

As another family of materials to which Raman spectroscopy can be fruitfully applied Schaack discusses next, in Sect. 2.3, the cubic Rare-Earth (RE) garnets with general formula $3(RE)_2O_3{\cdot}5(ME)_2O_3$ (ME = Al, Ga, or Fe). These crystals have 6 RE atoms in equivalent sites of the crystallographic unit cell which, however, are not magnetically equivalent, a fact that leads to a large number of phenomena involving crystal-field transitions. After the RE-garnets, the applications of Raman spectroscopy of CFT's to rare-earth exachloro-elpasolites (general formula: $A_2B(RE)Cl_6$, where A and B are alkali metals) are presented. A brief discussion of the "perovskite" $NdAlO_3$ follows. This crystal undergoes a trigonal-to-cubic transition at low temperatures. Because of its relative simplicity, calculations of the Raman intensities of their CFT's habe been performed using the Judd–Ofelt formalism [2.60, 61]. Agreement with experimental data, however, is not very satisfactory.

The next family of materials discussed from the point of view of applications of Raman spectroscopy is that of organometallic complexes containing rare-earth ions and actinides. We mention among them a molecular crystal by the name of uranocene (see Fig. 2.44).

Section 2.4 ends with a brief discussion of the Raman spectra of crystal-field transitions of $3d$ electrons in semimagnetic semiconductors. The canonical materials of these family have an average zincblende structure with the formula $Cd_{1-x}Mn_xTe$; the cation sites are occupied at random by either Cd or Mn^{2+}. Raman investigations have been performed not only for bulk crystals but also for narrow quantum wells embedded in a material with a different Mn concentration. The Mn^{2+} ion has a half-filled $3d$ shell, with parallel spins of all 5 electrons in the ground state, according to Hund's rule.

The total orbital angular momentum is $L = 0$, hence the corresponding ground state multiplet is $^6S_{5/2}$ (where 6 denotes the spin multiplicity, S the orbital angular momentum $L = 0$, and the resulting total angular momentum is $J = 5/2$). Since $L = 0$, the cubic crystal field does not act on the ground state multiplet. Zeman transitions can, however, be observed in a magnetic field, especially for narrow quantum wells (see Fig. 2.47). This part of Sect. 2.4 complements the article of Ramdas and Rodriguez that appeared in [1.6] and covered not only crystal-field transitions but also a number of typical semiconductor phenomena such as spin flip scattering, magnetic polarons and scattering by phonons. We would like to mention, at this point, a few recent advances in these topics, as applied to the $Cd_{1-x}Mn_xTe$ system, for which Schaack and coworkers has been mainly responsible. They involve magnetic polaron mediated multiple spin-flip Raman scattering in quantum wells in the presence of high magnetic fields [1.72], spin flip scatterings from donor-bound electrons in $Cd_{1-x}Mn_xTe/Cd_{1-y}Mn_yTe$ quantum wells and, last but not least, coherent Raman spectroscopies [Stokes (so-called CSRS) and anti-Stokes (CARS)] of magnetic excitations involving conduction electrons and bound magnetic polarons [1.73].

Section 2.5 summarizes the chapter and presents an outlook of possible future work involving Raman spectroscopy of CFT's. As materials of intrest for future work, the RE-endofullerenes (C_{60} or even C_{82} containing rare-earth atom inside the "bucky ball") are suggested. From the point of view of practical applications, phosphors involving rare earths are mentioned (e.g., $Y_3Al_5O_{12}$:Cd^{3+}, a garnet). In this crystalline material, the Ce^{3+}-ion contains only one 4f electron in a $^2F_{5/2}$ ground state. A schematic diagram of the crystal field states of the Ce^{3+} ions, including the excited states in which the 4f-electron has been promoted to a $5d$ state, is shown in Fig. 1.7 [2.222]. Notice that photons corresponding to three visible wavelengths (460, 520, and 580 nm, i.e., blue, green, and orange) should appear in the deexcitation of the $5d^1$ excited electrons. Hence nearly white light is produced in this rather efficient luminescence process that can be excited by a SiC or GaN electroluminescent diode (LED) emitting in the blue. The composite device containing such LED and an $Y_3Al_5O_{15}$:Ce^{3+} phosphor has been named a LUCOLED [2.222]. If such devices reach widespread use, Raman spectroscopy would be expected to play a role in the development and characterization of the required RE phosphor materials.

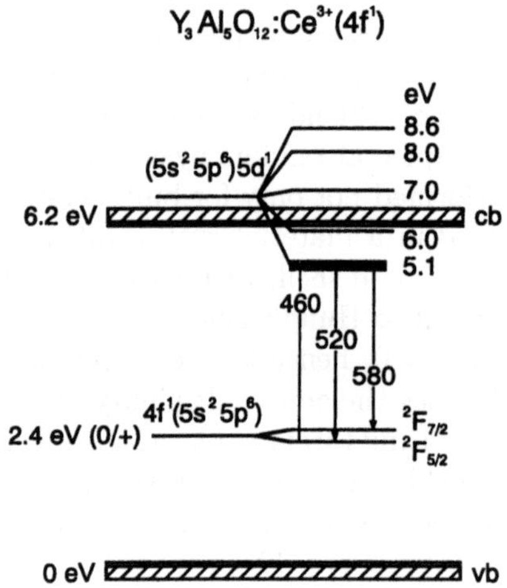

Fig. 1.7. Crystal-field levels of Ce^{3+} in the luminescent garnet $Y_3Al_5O_{15}$:Ce^{3+}. The levels include the $4f^1$ ground state and the $3d^1$ excited states of the Ce^{3+} ion. Notice the orange (580 nm), green (520 nm) and blue (460 nm) luminescent transitions leading to the emission of "white light" in LUCOLED light emitting diodes. From [2.222]

Chapter 2 is complemented by an Appendix containing a discussion of the symmetry properties of the scattering tensor and the associated selection rules, and 14 tables with selection rules for intraconfigurational transitions in ions with unfilled shells and a few other related properties (e.g., magnetic dipole, Table 2.4, and electric quadruple transitions, Table 2.5). Tables 2.7–10 contain observed and calculated relative Raman intensities while Tables 2.12–14 contain crystal-field parameters (in Wybourne notation) for a number of $4f$ materials, mainly high-T_c superconductors.

1.3.2 Chapter 3

Chapter 3 of this volume, by B. Hillebrands, on Brillouin light scattering in layered magnetic structures, gives an overview of the versatility and usefulness of this technique in the investigation of ultrathin metallic magnetic layers, multilayers, and superlattices. Besides the experimental verification of theoretically predicted new collective excitations in magnetic multilayers and superlattices, the advantages of Brillouin light scattering in the above systems go well beyond magnetic resonance studies, mainly because of measuring excitations at finite values of the wavevector ($q \neq 0$). The chapter reviews the progress in the field of magnetic excitation phenomena, observed by means of inelastic light scattering, that has taken place since the publication of P. Grünberg's article [1.23]. Particularly superior examples are the determination of magnetic anisotropies in layered magnetic systems and the measurements of the interlayer exchange coupling in the regime of ferromagnetic interaction. Another important advantage of the light scattering technique is shown to be its applications in a UHV environment, i.e., in studying, e.g., magnetic interface anisotropies as a function of layer coverage. This chapter ends with a discussion of recent developments in studying the exchange coupling at ferro-/antiferromagnetic layer interfaces, i.e., the so-called exchange bias effect, which is of importance for magnetic sensors .

Magnetic excitations observed by means of light scattering have much lower energies than typical phonons (typical frequency shifts in magnetic scattering are 10 GHz $\simeq 0.3\ \mathrm{cm}^{-1} \simeq 40\ \mu\mathrm{eV}$). These energies are comparable to those encountered in light scattering by acoustic phonons; the phenomenon is therefore usually called Brillouin Light Scattering (BLS) instead of Raman Scattering (RS). Since the difference in frequency shifts between BLS and RS is merely quantitative, the classification of light scattering spectra into either Brillouin or Raman on the basis of such small shifts is somewhat arbitrary. From the operational point of view it is, however, possible to make such classification on the basis of the type of spectrometer used. RS is normaly observed by means of grating monochromators [1.14]. Nevertheless, BLS requires better resolution and rejection of elastically scattered light (Rayleigh scattering). This is nowadays accomplished by means of multiple pass Fabry–Pérot interferometers. Since these devices are mechanically scanned perpendicular to the interferometer plates, the various spectral orders effect a periodic repetition of the measured spectra that limits the *Free Spectral Range* (FSR). In present day Brillouin spectrometers the FSR is enhanced by a large factor when using two synchronized (tandem) spectrometers with slightly different FSR. An article by John Sandercock, the inventor of these devices, in Vol. III of this series [1.18] discusses the basic instruments and some of the early applications.

As opposed to other spectroscopies (in particular inelastic neutron scattering) BLS and RS have the great advantage of probing very small sample volumes. The incident laser beams can be focused down to a diameter of 1 μm, whereas the sampling depth can be as small as 100 Å in highly absorbing samples. This feature offers a number of attractive possibilities:

1. Spectra can be obtained for very small, microscopic samples.
2. The surfaces of larger samples can be scanned and their topography investigated.
3. Size and shape effects on excitation frequencies can be studied. This feature is particularly attractive for the investigation of thin films, superlattices, quantum wires and quantum dots. The capabilities of RS for the investigation of two-dimensional electron gases in doped semiconductor layers were early pointed out by *Burstein* et al. [1.74].

BLS by spin waves in solids (magnons) was first detected in 1966 in three-dimensional (bulk) samples [1.75]. The then observed phenomenon develops into a rich range of effects when thin films, superlattices and other micro- (and nano-) structures are measured The static magnetic structures of such microstructures can differ considerably from those of the bulk due to readjustments in the Fermi levels, dependence of interlayer exchange and dipolar coupling on layer spacing, magnetic anisotropy effects and magnetoelastic phenomena. The resulting changes in the magnetic properties of the ground states of such samples must also manifest themselves in the elementary excitations that participate in BLS. The latter can become rather different from those in the corresponding bulk materials. Moreover, reduction in the possible translational symmetry operations (e.g., thin films are not invariant under translational operations perpendicular to the film) replaces part of the continuum of excitations by space quantized, so-called confined modes.

The chapter under discussion contains an introduction to the theory of BLS by thin films and their superlattices based on macroscopic response functions, e.g., magnetic susceptibilities, that complements the work by D.L. Mills which appeared in the second chapter of [1.5]. The shape effects in this theory arise mainly from the magnetostatic boundary conditions combined with magnetocrystalline and magnetoelastic anisotropies. Boundary conditions at surfaces are responsible for the appearance of spin-wave modes localized near the surface, the so-called Damon–Eshback modes [3.79]. Formally, these modes correspond to the electrostatic surface and interface modes observed for phonons and plasmons [1.5]. While they reflect the loss of translational invariance perpendicular to the film, periodic repetition of films along that direction introduces a new period, albeit much larger than that of the bulk constituents: Mini-Brillouin-zones appear.

After a brief introduction in Sect. 3.3 to the macroscopic theory of the scattering cross sections (i.e., efficiencies) based on fluctuations of the magnetic susceptibility, the state of the art concerning instrumentation for BLS is reviewed in Sect. 3.4. In Sect. 3.5 selected applications to films on bulk crystals, such as Fe on W (F/W), Co/Cu, Fe/Pd, Co/Pd, and Co/Au, are discussed. Section 3.5 concludes with several examples of BLS in superlattices (alternating films of Co and Pd, likewise for Co and Au).

Section 3.5.4 discusses BLS spectra of trilayers (e.g., Fe/Au/Fe) with the purpose of elucidating effects of oscillatory interlayer exchange coupling between the magnetic layers.

Section 3.5.5. discusses the effects of spatial inhomogeneities, such as fluctuations in layer thicknesses on the spin wave spectra. Such fluctuations are often responsible for the observed widths of the spectral peak observed in BLS. As an example, the Co/Pt and Co/Au multilayer systems are presented.

In Sect. 3.5.6 the author discusses nonlinear phenomena which are observed in BLS when spin waves are externally excited by means of microwaves. In this connection we should mention the very recent work of Hillebrands and coworkers [1.77] in which the formation of solitons and their mutual interaction (actually the lack of it) is described.

Most of the investigations discussed in Chap. 3 refers to two-dimensional films and periodic multifilm structures. The natural extension of such work

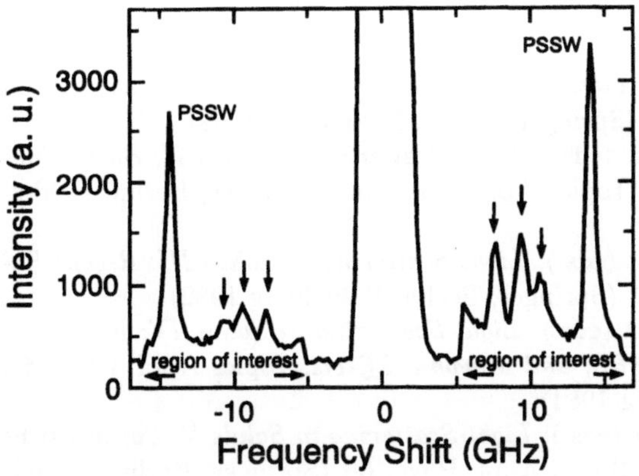

Fig. 1.8. BLS spectrum of a wire array with a width of 1.7 μm, a wire thickness 40 nm and a wire separation of 0.8 μm. The applied field is 500 Oe. The transferred in-plane wavevector $q_{\|}$ is 0.3×10^5 cm^{-1}. In the "regions of interest" [$\pm(5-17)$ GHz] the scan speed was reduced by a factor of three [1.76]. The peaks labeled PSSW correspond to "Perpendicular (to the film) Standing Spin Waves"

leads to one-dimensional wires and zero-dimensional dots. Like in the case of superlattices, wires and dots can also be arranged periodically. BLS on such structures is briefly discussed in Sect. 3.5.7. While the theoretical foundations are firmly laid, experimental work is just beginning to appear [1.76].

Most of the work related to wires and dots has been performed on patterned films. As an example we show in Fig. 1.8 the BLS spectrum of such a periodic array of wires with widths of 1.7 μm and 40 nm thicknesses, regularly spaced from each other by 0.8 μm. The array was fabricated by ion milling of a Ni–Fe film on Si. The scattering plane corresponding to Fig. 1.8 is perpendicular to the wires [1.76]. the in-plane scattering wavevector, also perpendicular to the wires, amounts to 3×10^4 cm^{-1}. Note the presence of a Stokes and an anti-Stokes component to the spectrum. The three peaks marked by arrows in each of these spectra correspond to Damon–Eshbach modes [3.79] with the frequency modified by the lateral confinement of the spin-wave excitations to the wires. Their positions in the frequency scale carry information on the dispersion relation of such modes. The large peaks labeled PSSW correspond to standing spin waves perpendicular to the film.

The Appendix at the end of Chap. 3 contains extensive reference to all BLS spectra of magnetic structures published up to the end of 1996.

References

1.1 M. Cardona (ed.): *Light Scattering in Solids I: Introductory Concepts*, 2nd edn., Topics Appl. Phys. **8** (Springer, Berlin, Heidelberg 1982)
1.2 M. Cardona, G. Güntherodt (eds.): *Light Scattering in Solids II: Basic Concepts and Instrumentation*, Topics Appl. Phys. **50** (Springer, Berlin, Heidelberg 1982)
1.3 M. Cardona, G. Güntherodt (eds.): *Light Scattering in Solids III: Recent Results*, Topics Appl. Phys. **51** (Springer, Berlin, Heidelberg 1982)
1.4 M. Cardona, G. Güntherodt (eds.): *Light Scattering in Solids IV: Electronic Scattering, Spin Effects, SERS, and Morphic Effects*, Topics Appl. Phys. **54** (Springer, Berlin, Heidelberg 1984)
1.5 M. Cardona, G. Güntherodt (eds.): *Light Scattering in Solids V: Superlattices and other Microstructures*, Topics Appl. Phys. **66** (Springer, Berlin, Heidelberg 1989)
1.6 M. Cardona, G. Güntherodt (eds.): *Light Scattering in Solids VI: Recent Results including High-T_c Superconductors* (Topics Appl. Phys. **68**) (Springer, Berlin, Heidelberg 1991)
1.7 M. Cardona, G. Güntherodt (eds.): *Light Scattering in Solids VIII*, Topics Appl. Phys. **76** (Springer, Berlin, Heidelberg 1999)
1.8 N. Esser, W. Richter: Raman Scattering by Phonons at Surfaces and Interfaces, in [1.7].
1.9 J. Menéndez, J. Page: Vibrational Spectroscopy of C_{60}, in [1.7].
1.10 T. Dekorsy, G.C. Cho, H. Kurz: Coherent Phonons in Condensed Media, in [1.7].
1.11 C.V. Raman: Scientific Papers (Ind. Acad. of Sci., Bangalore 1978) p. 478
1.12 For recent work on absolute scattering efficiencies in high-T_c superconductors see E.T. Heyen, S.N. Rashkeev, I.I. Mazin, O.K. Andersen, R. Liu, M. Cardona, O. Jepsen: Phys. Rev. Lett. **65**, 3048 (1990);
T. Strohm, M. Cardona: Phys. Rev. B **55**, 12 725 (1997);
B. Lederle: Theoretische und experimentelle Bestimmung der phononischen Ramanstreuwirkungsgrade von Hochtemperatursupraleiter, Doctoral Dissertation, University of Stuttgart (1997)
1.13 J.C. Tsang: Multichannel Detection and Raman Spectroscopy, in [1.5] p. 233
1.14 T. Ruf: Phonon Raman Scattering in Semiconductors, Quantum Wells, and Superlattices, Springer Tracts Mod. Phys. **142** (Springer, Berlin, Heidelberg 1998) p. 219
1.15 L. Philipidis, H. Siegle, A. Hofmann, C. Thomsen: Phys. Stat. Sol. (b) **2112**, R1 (1999)
1.16 D.J. Lockwood: Light Scattering by Electronic and Magnetic Excitations in Transition Metal Halides, in [1.3] p. 59
1.17 M.V. Klein: Raman Studies of Phonon Anomalies in Transition Metal Compounds, in [1.3] p. 121
1.18 J.R. Sandercock: Trends in Brillouin Scattering: Studies of Opaque Materials, Supported Films, and Central Modes, in [1.3] p. 207
1.19 E. Anastassakis, M. Cardona: Phonons, Strains and Pressure in Semiconductors, Semicond. and Semimetals **55**, 117 (1998)
1.20 G. Güntherodt, R. Zeyher: Spin Dependent Raman Scattering in Magnetic Semiconductors, G. Güntherodt, R. Merlin: Raman Scattering in Rare Earth Chalcogenides, in [1.4] p. 203, p. 343

1.21 L.E. Brus, A.L. Efros, T. Itoh (eds.): J. Lumin. **70**, 1–484 (1996)
1.22 C. Trallero-Giner, A. Debernardi, M. Cardona, E. Menéndez-Proupın, A.I. Ekimov: Phys. Rev. B **57**, 4664 -4669 (1998)
1.23 P. Grünberg: Light Scattering from Spin Waves in Thin Films and Layered Magnetic Structures, in [1.5] p. 285
1.24 E. Zirngiebl, G. Güntherodt: Light Scattering in Rare Earth and Actiuide Compounds, in [1.6] p. 207
1.25 A.K. Ramdas, S. Rodriguez: Raman Scattering in Diluted Magnetic Semiconductors, in [1.6] p. 137
1.26 J. Shah: Ultrafast Spectroscopy of Semiconductors and Semiconductor Nanostructures, second edition (Springer, Heidelberg 1999)
1.27 R. Merlin: Solid State Commun. **102**, 207 (1997)
1.28 M. Cardona: Raman Scattering in High-T_c Superconductors: Phonons, Electrons and Electron-Phonon Interaction, Physica C, in press.
1.29 Proceedings of the XVIth Int. Conf. on Raman Spectroscopy (ICORS) ed. by A.M. Heyns (J. Wiley, New York 1998)
1.30 see [1.2], Chap. 3, by R.K. Chang, M.B. Long, and [1.5] Chap. 6, by J.C. Tsang.
1.31 C. Bernhard, R. Henn, A. Wittlin, M. Kläser, Th. Wolf, G. Müller-Vogt, C.T. Lin, M. Cardona: Phys. Rev. Lett. **80**, 1762 (1998)
1.32 R. Nemetschek, M. Opel, C. Hoffmann, P.F. Müller, R. Hackl, H. Berger, L. Forro, A. Erb, E. Walker: Phys. Rev. Lett. **78**, 4837 (1997)
1.33 Z.-X. Shen, D.S. Dessau: Phys. Rep. **253**, 1–162 (1995)
1.34 O. Fischer, C. Renner, I. Maggio-Aprile: In *the Gap Symmetry and Fluctuations in High-T_c Superconductors*, ed. by J. Bok, G. Deutscher, D. Pavune, S.A. Wolf: NATO ASI, Ser. B Phys. **371**, 487 (1998)
1.35 T. Strohm, M. Cardona, A.A. Martin: Proc. of the Workshop on High-T_c Superconducitivity, Miami 1999, Am. Inst. of Physics (in press)
1.36 V.G. Hadjiev, Xingjiang Zhou, T. Strohm, M. Cardona, Q.M. Lin, C.W. Chu: Phys. Rev. B **58**, 1043 (1998)
1.37 C. Bernhard, D. Munzar, A. Wittlin, W. König, A. Golnik, C.T. Lin, M. Kläser, Th. Wolf, G. Müller-Vogt, M. Cardona: Phys. Rev. B, March 1, Rapid Communication (1999)
1.38 J.P. Franck: Experimental Studies of the Isotope Effect in High T_c Superconductors, in *Physical Properties of High T_c Superconductors IV*, ed. by D.M. Ginsberg (World Scientific, Singapore 1996) p. 189.
1.39 A.A. Martin, V.G. Hadjiev, T. Ruf, M. Cardona, T. Wolf: Phys. Rev. B **58**, 14211 (1998)
1.40 J. Mesot, A. Furrer: The crystal field as a local probe in rare-earth based high-T_c superconductors, in *Neutron Scattering in High T_c Superconductors*, ed. by A. Furrer (Kluwer, Dordrecht 1998) p. 335;
J. Mesot, A. Furrer: J. Supercond. **10**, 623 (1997)
1.41 T. Ruf, R. Wegerer, E.T. Heyen, M. Cardona, A. Furrer: Solid State Commun. **85**, 297 (1993)
1.42 P. Dufour, S. Jandl, C. Thomsen, M. Cardona, B.M. Wanklyn, C. Changkang: Phys. Rev. B **51**, 1053 (1995)
1.43 J.A. Sanjurjo, G.B. Martins, P.G. Pagliuso, E. Granado, I. Torriani, C. Rettori, S. Oseroff: Z. Fisk, Phys. Rev. B **51**, 1185 (1995)
1.44 V. Nekvasil, S. Jandl, T. Strach, T. Ruf, M. Cardona: J. Magnetism Mag. Mater. **177–181**, 535 (1998)

1.45 P. Lemmens, M. Fischer, M. Grove, P.H.M. v. Loosdrecht, G. Els, E. Sherman, C. Pinettes, G. Güntherodt: Quantum Spin Systems: From Spin Gaps to Pseudo Gaps, Festkoerperprobleme – Adv. in Solid State Physics, Vol. 39 (Vieweg, Braunschweig/ Wiesbaden 2000), in press.
1.46 R. Henn, T. Strach, E. Schönherr, M. Cardona: Phys. Rev. B **55**, 3285 (1997)
1.47 E. Dagotto, T.M. Rice: Science **271**, 618 (1996)
1.48 T.M. Rice, S. Haas, M. Sigrist, F.-Ch. Zhang: Phys. Rev. B **56**, 14655 (1997)
1.49 M. Uehara, T. Nagata, J. Akimitsu, H. Takahashi, N. Mori, K. Kinoshita: J. Phys. Soc. Jpn. **65**, 2764 (1996)
1.50 C. Kim, A.Y. Matsuura, Z.-X. Shen, N. Motoyama, H. Eisaki, S. Uchida, T. Tohyama, S. Maekawa: Phys. Rev. Lett. **77**, 4054 (1996)
1.51 J. Boucher, L. Regnault: J. Phys. I (Paris) **6** 1936 (1996)
1.52 M. Nishi, O. Fujita, J. Akimitsu: Phys. Rev. B **50**, 6508 (1994)
1.53 H. Kuroe, T. Sekine, M. Hase, Y. Sasago, K. Uchinokura, H. Kojima, I. Tanaka, Y. Shibuya: Phys. Rev. B **50**, 16468 (1996)
1.54 P. van Loosdrecht, J. Boucher, G. Martinez, G. Dhalenne, A. Revcolevschi: Phys. Rev. Lett. **76**, 311 (1996)
1.55 V. Muthukumar, C. Gros, W. Wenzel, R. Valenti, P. Lemmens, B. Eisener, G. Güntherodt, M. Weiden, C. Geibel, F. Steglich: Phys. Rev. B **54**, R 9635 (1996)
1.56 G. Els, P. van Loosdrecht, P. Lemmens, H. Vonberg, G. Güntherodt, G. Uhrig, O. Fujita, J. Akimitsu, G. Dhalenne, A. Revcolevschi: Phys. Rev. Lett. **79**, 138 (1997)
1.57 G. Els, G. Uhrig, P. Lemmens, H. Vonberg, P. van Loosdrecht, G. Güntherodt, O. Fujita, J. Akimitsu, G. Dhalenne, A. Revcolevschi: Europhys. Lett. **43**, 463 (1998)
1.58 P.H.M. van Loosdrecht: In *Contemporary studies in condensed matter physics*, ed. by M. Davidovic, Z. Ihonic: Solid State Phenom. **61–62** (Scitec, Switzerland 1998) p. 19
1.59 P. Lemmens, M. Fischer, G. Güntherodt, A. Mishchenko, M. Weiden, R. Hauptmann, C. Geibel, F. Steglich: Phys. Rev. B **58**, 14159 (1998)
1.60 E.Ya. Sherman, M. Fischer, P. Lemmens, P. van Loosdrecht, G. Güntherodt: unpublished
1.61 M. Isobe, Y. Ueda, J. Phys. Soc. Jpn. **65**, 1178 (1996)
1.62 I. Loa, K. Syassen: private communication
1.63 P.H.M. van Loosdrecht, J. Zeman, G. Martínez, G. Dhalenne, A. Revcolevschi: Phys. Rev. Lett. **78**, 487 (1997)
1.64 G. Blumberg, M.V. Klein, S.-W. Cheong: Phys. Rev. Lett. **80**, 564 (1998) and references therein
1.65 K. Yamamoto, T. Katsufuji, T. Tanaba, Y. Tokura: Phys. Rev. Lett. **80**, 1493 (1998)
1.66 V.G. Hadjiev, A. Fainstein, P. Etchegoin, H.J. Trodahl, C. Bernhard, M. Cardona, J.L. Tallon: Phys. Stat. Solidi (b) **211**, R5 (1999)
1.67 V.B. Podobedov, A. Weber, D.B. Romero, J.P. Rice, H.D. Drew: Phys. Rev. B **58**, 43 (1998);
E. Granado, N.O. Moreno, A. Garcia, J.A. Sanjurjo, C. Rettori, I. Torriani, S.B. Oseroff, J.J. Neumeier, K.J. McClellan, S.-W. Cheong, Y. Tokura: Phys. Rev. B **58**, 11 435 (1998)

1.68 F. Kozlowski, P. Steiner, W. Lang: In *Porous Silicon*, ed. by Zhe Chuan Feng, R. Tsu (World Scientific 1994) p. 149;
H.D. Fuchs, M. Stutzmann, M.S. Brandt, M. Rosenbauer, J. Weber, A. Breitschwerdt, M. Cardona: Phys. Rev. B **48**, 8172 (1993)
1.69 G. Els, P. Lemmens, P.H.M. van Loosdrecht, G. Güntherodt, H.P.Lang, V. Thommen-Geiser, H.-J. Güntherodt: Physica C **307**, 79 (1998)
1.70 M. Cardona: Physica B (Amsterdam) **263–64B**, 376 (1999)
1.71 M. Schwoerer-Böhning, A.T. Macrander: Phys. Rev. Lett. **80** (1998)
1.72 J. Stühler, G. Schaack, M. Dahl, A. Waag, G. Landwehr, K.V. Kavokin, I.A. Merkulov: Phys. Rev. Lett. **74**, 2567 (1995)
1.73 R. Rupprecht, B. Müller, H. Pascher, I. Miotkowski, A.K. Ramdas: Phys. Rev. B **58**, 16 123 (1998)
1.74 E. Burstein, A. Pinczuk, S. Buchner: In *Physics of Semiconductors 1958*, ed. by B.L.H. Wilson (Inst. of Physics, London 1979) p. 123
1.75 P.A. Fleury, S.P.S. Porto, L.E. Cheesman, H.J. Guggenheim: Phys. Rev. Lett. **17**, 84 (1966)
1.76 S.O. Demokritov, B. Hillebrands: J. Mag. Magn. Mater (in press)
1.77 O. Büttner, M. Bauer, S.O. Demokritov, B. Hillebrands, M.P. Kostylev, B.A. Kalinikos, A.N. Slavin: Phys. Rev. Lett. **82**, 4320 (1999)

2 Raman Scattering by Crystal-Field Excitations

G. Schaack

Abstract. In this chapter we review the studies, via Raman scattering, of low lying electronic states of ions which have unfilled electronic shells, in inorganic insulators, high-T_c superconductors, (semimagnetic) semiconductors, and organic complexes. Mostly rare earths ($4f$), but also transition elements ($3d$), and actinides ($5f$) are considered. We concentrate on the basic features of electronic Raman scattering from an experimental point of view rather than striving for a complete overview of experimental data.

2.1 Introduction

The study of excited electronic states in insulating crystalline compounds is usually considered as the domain of optical spectroscopy, in particular of one-photon absorption and fluorescence spectroscopy. Low-lying electronic levels, e.g. those belonging to the ground state multiplet, having excitation energies E_0 comparable to optical phonons or other collective excitations in solids ($0.1\,\mathrm{meV} \leq E_0 \leq 0.5\,\mathrm{eV}$, $\approx 1\,\mathrm{cm}^{-1} \leq E_0 \leq 4000\,\mathrm{cm}^{-1}$) can, however, often be studied only with some difficulties by these methods because the absorption or emission cross sections of competing excitations are much stronger. Raman spectroscopy is a method which offers great advantages in this situation, both because of the flexibility in symmetry-based selection rules inherent to two-photon process, and also the ease of detecting even very weak signals with modern equipment. For this purpose tuneable laser sources with a narrow bandwidth ($\Delta\bar{\nu} \leq 1\,\mathrm{cm}^{-1}$), low (helium-) temperatures of the sample in optical cryostats with high-field superconducting magnets, triple-path monochromators with high spectral resolution and luminosity, liquid-nitrogen cooled CCD-type detectors with extremely low electronic noise are in use today. The weakness of electronic Raman signals is often characterized by its relation to strong Raman scatterers such as liquid benzene: The total Raman cross section of the 992-cm^{-1} vibrational mode (labelled ω_1, A_{1g}-symmetry, corresponding to a C–C bond stretch) amounts to $2.7 \times 10^{-28}\,\mathrm{cm}^2$ per molecule, while electronic Raman cross sections for rare earth ions in crystals range from $\approx 5 \times 10^{-29}\,\mathrm{cm}^2$ to $1 \times 10^{-33}\,\mathrm{cm}^2$ per atom. Crystal

Topics in Applied Physics, Vol. 75
Light Scattering in Solids VII
Eds.: M. Cardona, G. Güntherodt

field excitations have also been studied by inelastic neutron scattering [2.1, 2.2], see also Sect. 2.4.1. Since this technique requires extensive instrumentation, its application has been limited mostly to metals, semiconducting or superconducting opaque materials.

In cases of centrosymmetric sites of the metal ion with an unfilled shell, parity is strictly defined (Sect. 2.2.3) and intra-configurational transitions (e.g. $4f \rightarrow 4f$) cannot be observed by one-photon electric dipole transitions but are only allowed for parity conserving two-photon processes (Raman and two-photon absorption spectroscopy).

Ions with unfilled electronic shells, especially rare earth ions with their extremely sharp $4f^n \rightarrow 4f^n$ transitions, are sensitive probes of internal fields in the crystal and (or) external morphic effects (mechanical strains, electric and magnetic fields). Ground-state multiplets usually have a simpler electronic structure than excited levels, which are often subject to strong interconfiguration mixing by the crystal field, and observed effects are thus easier to interpret theoretically. Raman spectroscopy offers many ways of access to the exploration of this area.

Although Raman spectroscopy is restricted to a smaller interval of detectable excitation energies than e.g. an absorption or fluorescence experiment, the crystal field levels derived from the free-ion ground state multiplet are often the most important ones to characterize a material, since they govern the magnetic, low-temperature thermal, and semi- or superconducting (transport) properties. The optical properties of many solids and their potential as fluorescent or laser materials are to a large degree determined by the low-lying electronic excitations.

Low-lying excitations of different origin but with the same types of symmetry will, in general, interact resonantly and hybridize. Energy shifts caused by this interaction (anticrossing phenomena), asymmetric line shapes of Fano-type, changes of spectral strengths and selection rules are the consequences; they may hamper a straightforward interpretation of the data. On the other hand, these effects are of special scientific interest as they offer a deeper insight into the nature of the mixing excitations and their various and often complex interaction mechanisms. The energy shifts induced by mode coupling have to be taken properly into account when adjusting simple theoretical models for the uncoupled excitations to the experimental data. These topics will be covered in Sects. 2.3 and 2.4.

Cases of degenerate electronic ground states, and of excitations which are close to resonance with the incident or scattered light quanta, will influence the Raman spectra of phonons in such compounds: Placzek's polarizability theory [2.3], which assumes a nondegenerate ground state and off-resonant intermediate states in the scattering process, may then be invalid. Under such conditions, asymmetric scattering tensors may be encountered even for

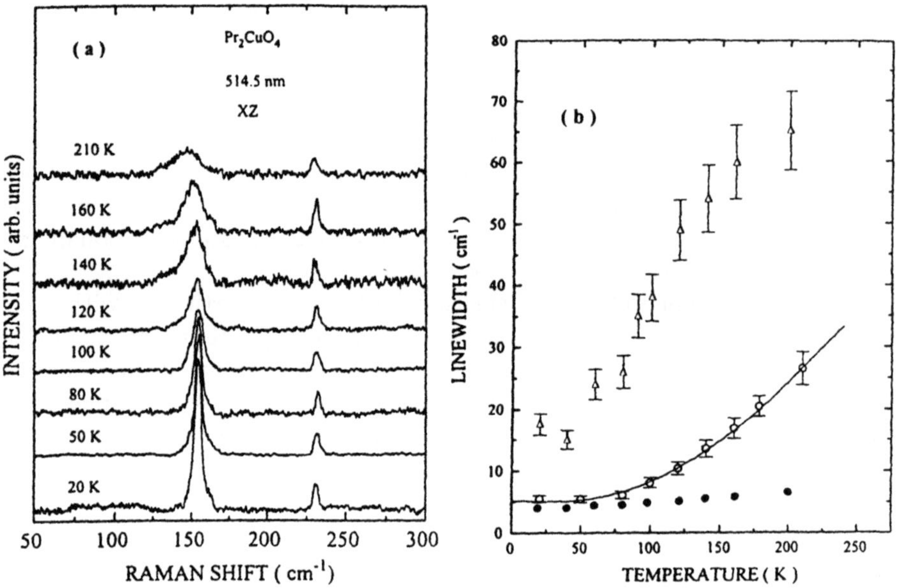

Fig. 2.1. (**a**) Temperature dependence of the electronic Raman transition at 156 cm^{-1} in Pr_2CuO_4 (D_{4h}^{17}, site symmetry: C_{4v}, transition $\Gamma_4 \rightarrow \Gamma_5$), from [2.18]. Scattering geometry y(xz)$\bar{y}$ (backscattering), incident wavelength: 514.5 nm. The peak at 233 cm^{-1} is due to polarization spillover from the A_{1g} phonon (zz), spectral resolution $\leq$ 4 cm^{-1}. (**b**) Temperature dependence of the linewidths of two crystal-field excitations in Pr_2CuO_4, [2.18]: $\circ$: 156 cm^{-1}, $\triangle$: 675 cm^{-1}; $\bullet$: A_{1g} phonon at 233 cm^{-1} for comparison. The solid line is the best fit of the data considering two contributions to the linewidth: Nonradiative decay between CF levels with emission or absorption of acoustical phonons ("direct process"); and transitions between CF levels due to inelastic scattering of phonons ("Raman process"), see e.g. [2.29]

phonon transitions.[1] This effect will be enhanced, if phonon–electron interaction and vibronic coupling are large.

Since practically all low-energy excitations of any type in a crystal are by some mechanism Raman-active, it requires particular care to attribute a certain Raman line to a specific excitation. In crystals with a multi-atom unit cell phonons usually swamp a Raman spectrum with many transitions. Raman measurements at different polarizations (different components of the Raman polarizability tensor), various temperatures, under external magnetic fields or changing the type and percentage of doping with magnetic ions in an isomorphous host, will help to clarify the situation.

Some practical remarks concerning the discrimination of electronic Raman transitions against other transitions in a scattering spectrum appear appro-

[1] Within Placzek's theory Raman tensors are symmetric as expected for the derivative of a symmetric polarizability with respect to the amplitude of the excitation.

priate. In most cases electronic Raman transitions in crystals with transitions metal ions, lanthanides (RE), or actinides have to compete both with strong fluorescence and with phonon Raman transitions. An annoying fluorescence can often be avoided by choosing a longer wavelength of the laser excitation: fluorescence transitions stay at a fixed frequency while Raman transitions display a constant frequency shift, independent of the exciting wavelength. The strong temperature dependence of intensities and halfwidths of electronic transitions (Fig. 2.1) is another characteristic quality of these transitions. It is due to changes in the electron populations and the enhanced (phonon induced) decay of the excited electronic states to the phonon system with raising temperature. In special cases, a strongly temperature dependent anti-Stokes companion of the Raman line may help in the identification. Higher-order coherent spectroscopy techniques, e.g. Coherent Anti-Stokes Raman Scattering (CARS) may be used; they allow a very effective suppression of background fluorescence. Using a pulsed laser (pulsewidth 10 ns) and fast time discriminating electronics, *Koningstein* and *Myslinski* [2.4] have also been able to separate the instantaneous off-resonance Raman signal from fluorescence with a finite decay time. This technique might be upgraded by applying shorter laser pulses in the ps-region, if bandwidth is not a problem, and new faster detection electronics [2.5]. For some technical details see also [2.6].

The problem of the often-dominant elastic Rayleigh scattering can be overcome by applying rejection filters such as holographic notch filters and crystalline colloidal array filters, which make it possible to use single uv-spectrographs with large throughput instead of the costly double- and triple-instruments [2.7]. Electronic scattering will certainly profit from these developments: a larger frequency span will be accessible for the scattered light and scattering under resonance conditions will play a major role than in the past. On the other hand, the use of long wavelength excitation, e.g. $\lambda = 1.06\,\mu\mathrm{m}$ of a Nd^{3+}-laser, often avoids the occurrence of Rayleigh scattering from the beginning: Fourier-transform (FT) Raman spectroscopy has experienced considerable technological progress and will be an attractive alternative as soon as low-noise detectors will become available for the near infrared [2.8].

Phonon Raman transitions usually exhibit a weaker temperature dependence of their intensity than electronic transitions. At room temperature the phonon spectrum of a crystal can be completely observed in most cases, while electronic transitions usually require temperatures in the liquid helium range for their clear appearance. In many cases electronic transitions obey selection rules different from those of phonons, especially when the number of electrons in the shell under study is odd (Kramers ion) ([2.9], see Table 2.A.1). In addition, electronic transitions may display asymmetric scattering intensities, when interchanging the polarizations of incident and scattered light and a Zeeman effect in an external magnetic field. The asymmetry of the anti-Stokes transitions is reversed compared to that for the corresponding Stokes transition. In general, and whenever possible, Raman experiments should be combined with and checked against one-photon absorption and fluorescence experiments.

Raman scattering by electronic excitations in transition metal and rare-earth ions has had its "golden age" in the two decades following the first observation of an electronic Raman spectrum for Pr^{3+}-ions ($4f^2$) in a $PrCl_3$ single crystal by *Hougen* and *Singh* in 1963 [2.10]. In the following years Koningstein and his coworkers have pioneered many aspects of the field which have been subsequently widely investigated.

At present the Raman community is experiencing another revolution due to the recent development of reliable ultraviolet, narrow-line, low-noise, cw laser sources: frequency-doubled argon- and krypton-ion lasers, frequency-multiplied Nd:YAG-lasers and Ti–sapphire lasers, excimer lasers, combinations with dye-lasers for the uv etc., may soon be standard equipment in a Raman laboratory. The recent development of confocal microscopy improves the lateral resolution of optical spectroscopies to less than 200 nm, and less than 500 nm along the optical axis, adding new potential to the classical role of *d*- or *f*-ions as probes of the local environment in solids.

Interest in the Raman technique and its potentials is growing again at present, due to some rapidly developing fields of endeavour, both in basic research and in applications:

1. Solid state lasers based on $3d^n$- or $4f^n$ ions in crystalline or glass hosts are being developed in many laboratories. In this field, a deep knowledge of the energies of the low lying crystal field levels and their relaxation mechanisms is required. Raman spectroscopy also offers the opportunity to study excited metastable states, as demonstrated e.g. for the excited ^2E states of ruby [2.11]. Such metastable states can generally give rise to parasitic absorption of pump radiation resulting from transitions between excited states.
2. High-T_c superconductors of the $LaCuO_4$ and $YBa_2Cu_3O_{7-\delta}$-type accept RE^{3+}-ions on the La- or Y-sites as dopants. Here electronic Raman scattering is used, competing with inelastic neutron scattering, to study e.g. effects of oxygen doping on structure etc.
3. There exists at present a growing interest in new magnetic materials incorporating RE ions and in magnetic phenomena which can be studied in such systems.

   In the field of basic research the following topics have attracted attention:
4. The investigation of intensities of electronic Raman scattering processes and their resonance behavior.
5. The study of the various coupling phenomena between localized electronic states, especially the $4f^n$-states, and other elementary excitations, such as phonons.
6. In diluted magnetic semiconductors ($A^{II}-B^{VI}$ compounds, where A^{II}-ions are replaced by magnetic ions), e.g. $(Cd_{1-x}Mn_x)Te$, of cubic zincblende

structure for $x \leq 0.7$, the spin-flip excitation of free carriers, the optically detected paramagnetic resonance signal of the $3d^n$-configuration, and the excitations of exchange-coupled clusters or magnetic clusters are easily observed in the low-temperature Raman spectra.

There are numerous previous reviews on electronic Raman scattering which cover different aspects of the subject. *Koningstein* and *Mortensen* [2.12] have concentrated on the $4f$-states in insulators, *Clark* and *Dines* [2.13] have covered the full subject including molecular species ($5d^n$ and metallocene complexes). *Klein* [2.14] has treated electronic scattering processes in semiconductors. Some topics which might fall under the title of the present review, have actually been treated comprehensively before in this series. Among them we mention the investigation of crystal fields effects in intermetallic compounds [2.15] and of localized electronic excitations in semimagnetic semiconductors [2.16], as well as light scattering by electronic and magnetic fluctuation processes in high-T_c superconductors [2.17]. We shall touch upon these topics only for the purpose of compiling some recent new results.

2.2 Theory of Inelastic Light Scattering by Electronic Transitions

2.2.1 Basic Relations in Electronic Raman Scattering

We consider the standard experimental situation in a Raman scattering experiment: Monochromatic (laser) light of angular frequency ω_L is incident on the scattering medium containing ions with excited electronic levels at low energies (crystal-field (CF) states). The electric field vector $\boldsymbol{E}_L$ is polarized in the direction of the unit vector $\boldsymbol{e}_L$ and the direction of the incident beam is parallel to the wavevector $\boldsymbol{k}_L$ in the medium,

$$k_L = \frac{2\pi n_L}{\lambda_L}, \tag{2.1}$$

where λ_L is the vacuum wavelength of the incident laser radiation and n_L is the refractive index of the medium at λ_L. $\boldsymbol{E}_L$ induces a polarization $\boldsymbol{P}$ in the medium with cartesian components P^ν given by:

$$P^\nu(\boldsymbol{r},t) = \epsilon_0 \sum_\mu \chi^{\nu\mu}(\boldsymbol{r},t) E_L^\mu(\boldsymbol{r},t), \qquad (\mu,\nu = x,y,z). \tag{2.2}$$

(χ) is the linear susceptibility tensor, ϵ_0 is the free space permittivity. The excitations in the medium at energies $\hbar\omega_0$ will modulate χ and hence produce fluctuating terms in $\boldsymbol{P}$ with a Fourier component at $\omega_S = \omega_L \pm \omega_0$ (sidebands), which in turn will give rise to a scattered electromagnetic wave, polarised in the ν-direction, with electric field strength $\boldsymbol{E}_S$ and frequency

ω_S.[2] The susceptibility $\chi(\omega_0)$ usually has to be calculated in third order perturbation theory, except for localized electronic excitations where already the second order produces the required frequency dependence [2.19].

Inelastic scattering will occur for $\omega_S \neq \omega_L$. The experiment will detect the scattered radiation with wavevector $\boldsymbol{k}_S$ and unit polarization vector $\boldsymbol{e}_S$ of its electric field, $k_S = \frac{2\pi n_S}{\lambda_S}$ and thus measure the differential scattering cross section $\mathrm{d}\sigma/\mathrm{d}\Omega$, where $\mathrm{d}\Omega$ is the element of solid angle $\mathrm{d}\Omega = \Delta A^2/R^2$, and the scattered radiation is detected at the center frequency ω_S at a distance R from the sample by the detector area (ΔA^2), [2.19, 2.20, 2.23, 2.24].

Using Maxwell's equations and the approximation of a macroscopic sample of linear dimension l and volume $\bar{V}$, $(l \approx \bar{V}^{\frac{1}{3}}, l \gg \lambda)$ and $R \gg l$, the scattered power $\bar{I}_S$ is obtained:

$$\bar{I}_S = 2\epsilon_0 c n_S \bar{V} \left(\frac{\omega_S^2}{4\pi\epsilon_0 c^2}\right)^2 \int_{(\bar{V})} \mathrm{d}^3(\boldsymbol{r}_1 - \boldsymbol{r}_2) \exp[\mathrm{i}\boldsymbol{k}_S \cdot (\boldsymbol{r}_1 - \boldsymbol{r}_2)] \\ \times \langle [\boldsymbol{e}_S \cdot \boldsymbol{P}_{\omega_0}{}^*(\boldsymbol{r}_1, \omega_S)][\boldsymbol{e}_S \cdot \boldsymbol{P}_{\omega_0}(\boldsymbol{r}_2, \omega_S)] \rangle. \tag{2.3}$$

$\boldsymbol{P}_{\omega_0}$ is the polarization due to the excitation ω_0.

The scattering cross section (s.c.s.) is defined as:

$$\frac{\mathrm{d}\sigma}{\mathrm{d}\Omega} = \left(\frac{\omega_L}{\omega_S}\right) \frac{\bar{I}_S}{I_L}, \tag{2.4}$$

where the intensity I_L of the incident beam is $I_L = 2\epsilon_0 c n_L |\boldsymbol{E}_L|^2$. According to this definition, the s.c.s. gives the number of light quanta scattered by the sample into the solid angle $\mathrm{d}\Omega$ relative to the number of incident laser quanta per unit area. Hence the well known formula for the s.c.s. is found [2.20], [2.23]:

$$\frac{\mathrm{d}\sigma}{\mathrm{d}\Omega} = \frac{\omega_L \omega_S^3 n_S \bar{V}}{(4\pi\epsilon_0 c^2)^2 n_L |E_L|^2} \int_{(\bar{V})} \mathrm{d}^3(\boldsymbol{r}_1 - \boldsymbol{r}_2) \exp[\mathrm{i}\boldsymbol{k}_S \cdot (\boldsymbol{r}_1 - \boldsymbol{r}_2)] \\ \times \langle [\boldsymbol{e}_S \cdot \boldsymbol{P}_{\omega_0}{}^*(\boldsymbol{r}_1, \omega_S)][\boldsymbol{e}_S \cdot \boldsymbol{P}_{\omega_0}(\boldsymbol{r}_2, \omega_S)] \rangle. \tag{2.5}$$

[2] This modulation is a phenomenological interpretation of the fact that the electronic dipole transition moments in χ (see (2.13)) may e.g. depend parametrically on the nuclear vibrational coordinates via the electron–phonon interaction (see [2.3] for details). In the case of magnetic excitations, χ is expanded in terms of the variables that describe the excitations propagating within the solid [2.20]. This expansion can either be carried out phenomenologically or can be deduced from microscopic considerations of the magneto-optical coupling mechanisms. Equation (2.2) is in fact the first (linear) term of an expansion of χ with respect to electric-field amplitudes of incident and scattered light in increasing powers, depending on the process under consideration. Temperature dependent nonlinear susceptibilities which may again depend on other excitations in solids, have been calculated using a density matrix formalism [2.21], [2.22], e.g. for two-photon absorption, stimulated Raman effect, or CARS (Sect. 2.2.6).

Apart from its variation with $\bar{V}$ and the factor $\approx \omega^4$, the s.c.s. depends on the spatial correlation function of the fluctuating polarization $\boldsymbol{P}$. Introducing the susceptibilities (2.2) and $\boldsymbol{E}_{\mathrm{L}}(\boldsymbol{r},t) = \boldsymbol{e}_{\mathrm{L}} E_{\mathrm{L}} \exp[\mathrm{i}(\boldsymbol{k}_{\mathrm{L}} \cdot \boldsymbol{r} - \omega_{\mathrm{L}} t)]$, equation (2.5) can be recast for the differential s.c.s. (per spectral element $\mathrm{d}\omega_{\mathrm{S}}$) as a correlation function of the Fourier transformed susceptibilities:

$$\frac{\mathrm{d}^2\sigma}{\mathrm{d}\Omega \mathrm{d}\omega_{\mathrm{S}}} = \frac{\omega_{\mathrm{L}} \omega_{\mathrm{S}}^3 n_{\mathrm{S}} \bar{V}}{(4\pi c^2)^2 n_{\mathrm{L}}} \sum_{\alpha,\beta,\mu,\nu} \langle (\chi_{\boldsymbol{k}}^{\alpha\beta})^* (\chi_{\boldsymbol{k}}^{\mu\nu}) \rangle_{\omega_0} \cdot e_{\mathrm{L}}^{\alpha} e_{\mathrm{S}}^{\beta} e_{\mathrm{L}}^{\nu} e_{\mathrm{S}}^{\mu}, \tag{2.6}$$

where the Fourier transformation

$$\begin{aligned} \langle (\chi_{\boldsymbol{k}}^{\alpha\beta})^* (\chi_{\boldsymbol{k}}^{\mu\nu}) \rangle_{\omega_0} = & \int\int \mathrm{d}t \mathrm{d}^3(\boldsymbol{r}_1 - \boldsymbol{r}_2) \langle [\chi^{\alpha\beta}(\boldsymbol{r}_1,t)]^* \chi^{\mu\nu}(\boldsymbol{r}_2,t) \rangle \\ & \times \exp[\mathrm{i}\omega_0 t - \mathrm{i}\boldsymbol{k} \cdot (\boldsymbol{r}_1 - \boldsymbol{r}_2)] \end{aligned} \tag{2.7}$$

has been applied; $e_{\mathrm{L}}^{\alpha}, \ldots$ are unit vectors of polarization.[3] The wavevector $\boldsymbol{k}$, (derived from (2.5)) in the case of a crystalline sample, and frequency $\omega_0, (E_0 = \hbar\omega_0)$ appearing in (2.6) refer to the excitation under study; for a Stokes process they are defined as:

$$\boldsymbol{k} = \boldsymbol{k}_{\mathrm{L}} - \boldsymbol{k}_{\mathrm{S}}, \tag{2.8}$$

$$\omega_0 = \omega_{\mathrm{L}} - \omega_{\mathrm{S}}. \tag{2.9}$$

Equations(2.8),(2.9) guarantee the conservation of crystal momentum and energy. Because $|\boldsymbol{k}_{\mathrm{L}}| \approx |\boldsymbol{k}_{\mathrm{S}}| \ll |\boldsymbol{k}_B|$, the Brillouin zone boundary, in first-order Raman scattering only collective excitations near the Γ-point can be observed. In disordered crystals or amorphous solids, translational symmetry is lost and rule (2.8) does not apply.

Equation (2.6) is the starting point for the analysis of the scattering process by the fluctuation–dissipation theorem [2.25]. This approach is useful in nonlinear optics and near phase transitions; for other applications we continue with (2.5) and introduce here the quantum mechanical second-order perturbation expression for the differential s.c.s. due to electronic transitions [2.19]:

$$\begin{aligned} \frac{\mathrm{d}^2\sigma}{\mathrm{d}\Omega \mathrm{d}\omega_{\mathrm{S}}} = & \bar{n} g_{\omega_0}(\omega) \frac{e^4 \omega_{\mathrm{L}} \omega_{\mathrm{S}}^3}{(4\pi\epsilon_0 c^2)^2} \\ & \times \left| \sum_l \left\{ \frac{\boldsymbol{e}_{\mathrm{S}} \cdot \mathbf{D}_{fl} \; \boldsymbol{e}_{\mathrm{L}} \cdot \boldsymbol{D}_{li}}{E_l - \hbar\omega_{\mathrm{L}} - \mathrm{i}\gamma_l} + \frac{\boldsymbol{e}_{\mathrm{L}} \cdot \boldsymbol{D}_{fl} \; \boldsymbol{e}_{\mathrm{S}} \cdot \boldsymbol{D}_{li}}{E_l + \hbar\omega_{\mathrm{S}} - \mathrm{i}\gamma_l} \right\} \right|^2, \end{aligned} \tag{2.10}$$

where $g_{\omega_0}(\omega)$ is an empirical lineshape function of the excitation at $\omega_0 = \omega_{\mathrm{L}} - \omega_{\mathrm{S}}$ (2.9), $\int g_{\omega_0}(\omega)\,\mathrm{d}\omega = 1$, and $\bar{n}$ is the number of active ions in the scattering volume $\bar{V}$. In (2.10) the summation runs over all (intermediate) electronic

[3] Another frequently used concept is the Raman scattering efficiency S, i.e. a reciprocal length [2.24], obtained by dividing(2.3–2.6) by $\bar{V}$. Multiplying S by the scattering length (sample length) and the incident power results in the scattered power per unit solid angle.

states $|l\rangle$ with energies E_l and lifetime broadening γ_l of the excited metal ions. Equation (2.10) is the Kramers–Heisenberg formula for the differential s.c.s. in dipole approximation. The total electric dipole moment of all N electrons in an ion is

$$\boldsymbol{D} = \sum_{j=1}^{N} \boldsymbol{r}_j, \tag{2.11}$$

and its matrix elements between the initial $|i\rangle$ or final $|f\rangle$ and intermediate $|l\rangle$ states are:

$$\boldsymbol{D}_{fl} = \langle f|\boldsymbol{D}|l\rangle. \tag{2.12}$$

The use of $\bar{n}$ in (2.10) and $\boldsymbol{D}$ in (2.11) (dipole approximation) is particularly useful for low concentrations of the scattering ions or states well localized in the ions. When ion–ion interactions become important, i.e. for propagating states $|i\rangle, |f\rangle$, the $\boldsymbol{A}\cdot\boldsymbol{p}$ representation will be more appropriate ($\boldsymbol{A}$: vector potential, $\boldsymbol{p}$: electron momentum).

For the following discussion of the tensor properties of the Raman polarizability we introduce the tensor of the single-ion Raman polarizability $(\alpha^{\text{s.i.}})$:

$$(\alpha^{\text{s.i.}}) = e^2 \sum_l \left\{ \frac{\boldsymbol{e}_\text{S}\cdot\boldsymbol{D}_{fl}\ \boldsymbol{e}_\text{L}\cdot\boldsymbol{D}_{li}}{E_l - \hbar\omega_\text{L} - \text{i}\gamma_l} + \frac{\boldsymbol{e}_\text{L}\cdot\boldsymbol{D}_{fl}\ \boldsymbol{e}_\text{S}\cdot\boldsymbol{D}_{li}}{E_l + \hbar\omega_\text{S} - \text{i}\gamma_l} \right\}, \tag{2.13}$$

Assuming localized excitations for simplicity, the electronic susceptibility of the crystal is $(\chi) = \bar{n}(\alpha^{\text{s.i.}})/\epsilon_0\bar{V}$. The tensors (α) or (χ) are asymmetric. They can be decomposed into their symmetric (s, $\chi^{\text{s}}_{\alpha\beta} = \chi^{\text{s}}_{\beta\alpha}$) and antisymmetric (as, $\chi^{\text{as}}_{\alpha\beta} = -\chi^{\text{as}}_{\beta\alpha}$) parts with 6 (3 diagonal plus 3 off-diagonal) and 3 off-diagonal components, respectively:

$$\chi_{\alpha\beta} = \chi^{s}_{\alpha\beta} + \chi^{\text{as}}_{\alpha\beta}, \qquad (\alpha,\beta = x,y,z) \tag{2.14a}$$

$$\chi^{s}_{\alpha\beta} = \frac{e^2\bar{n}}{2\epsilon_0(\bar{V})} \sum_l \left[\{(E_l - \hbar\omega_\text{L} - \text{i}\gamma_l)^{-1} + (E_l + \hbar\omega_\text{S} - \text{i}\gamma_l)^{-1}\} \times \left\{ D^{\beta}_{fl} D^{\alpha}_{li} + D^{\alpha}_{fl} D^{\beta}_{li} \right\} \right], \tag{2.14b}$$

$$\chi^{\text{as}}_{\alpha\beta} = \frac{e^2\bar{n}}{2\epsilon_0(\bar{V})} \sum_l \left[\{(E_l - \hbar\omega_\text{L} - \text{i}\gamma_l)^{-1} - (E_l + \hbar\omega_\text{S} - \text{i}\gamma_l)^{-1}\} \times \left\{ D^{\beta}_{fl} D^{\alpha}_{li} - D^{\alpha}_{fl} D^{\beta}_{li} \right\} \right]. \tag{2.14c}$$

Raman scattering described by the symmetric tensor χ^{s} has an isotropic component that results in "trace scattering" [2.3] and involves the isotropic part of the tensor ($\chi^0 = \frac{1}{3}(\chi_{xx} + \chi_{yy} + \chi_{zz})$). It corresponds to the so-called polarized Raman scattering (xx, yy, zz). The rest of the symmetric tensor causes symmetric depolarized (quadrupolar) scattering, $\chi^{\text{s}}_{\alpha\beta} = \frac{1}{2}(\chi_{\alpha\beta} +$

$\chi_{\beta\alpha}) - \chi^0_{\alpha\beta}\delta_{\alpha\beta}$; $\alpha, \beta \equiv x, y, z$. The antisymmetric part of the tensor is equivalent to a pseudovector. It is responsible for magnetic dipole scattering, $\chi^{\mathrm{as}}_{\alpha\beta} = \frac{1}{2}(\chi_{\alpha\beta} - \chi_{\beta\alpha})$ which is also depolarized.

Equations (2.14a) have important consequences for electronic Raman scattering: Besides the symmetric scattering due to χ^{s}, as usually found for phonons, there exists the contribution of pseudovector scattering (χ^{as}). Expanding the energy denominators in (2.13) and retaining the leading terms, we find for the ratio of the intensities I^{s} and I^{as} due to both contributions, not considering possible differences in the matrix elements:

$$\frac{I^{\mathrm{s}}}{I^{\mathrm{as}}} \approx \left[\frac{E_l}{E_{\mathrm{L}} - E_0/2}\right]^2 ; \qquad E_{\mathrm{L}} = \hbar\omega_{\mathrm{L}},\ E_0 = \hbar\omega_0 . \tag{2.15}$$

Equation (2.15) suggests that pseudovector scattering can be neglected, if the electronic ground state is nondegenerate and $E_l \gg E_{\mathrm{L}}, E_0$. These are also the conditions for the applicability of Placzek's polarizability theory [2.3], which is a good approximation for phonon scattering in insulators with large bandgaps ("off-resonance case"). For elastic Rayleigh scattering ($|i\rangle = |f\rangle$) : $\chi^{\mathrm{as}}_{\alpha\beta} = \chi^{as,*}_{\beta\alpha} = 0$, (*: conjugate complex), i.e. for real matrix elements the scattering tensor is symmetric.

In systems with impurities containing d- or f-electrons, intermediate states at E_{l} which give sizeable matrix elements $\boldsymbol{D}_{fl}$, usually have lower energies than the band gap energies found in the undoped host crystals, $E_{\mathrm{l}} < (\ll)$ E_{gap}. The $4f^n \rightarrow 4f^{n-1}5d^1$ transitions have an oscillator strength f close to 1: they are not forbidden by the parity selection rule for electric dipole transitions, as they would be for the intra-configurational ($4f^n \rightarrow 4f^n$) transitions for which $f \approx 10^{-6}$. As a rough estimate, the excited states of $3d^n$-systems at 10 Dq in an octahedral crystal field are found at $\approx 10\,000\,\mathrm{cm}^{-1}$ for divalent metal ions and at $\approx 20\,000\,\mathrm{cm}^{-1}$ for trivalent ions [2.26, 2.27]. In RE^{3+}-ions the $4f^n \rightarrow 4f^{n-1}5d^1$ transitions contribute: they start in the free ions between $\approx 45\,000\,\mathrm{cm}^{-1}$ (Ce^{3+}) and $\approx 100\,000\,\mathrm{cm}^{-1}$ (Yb^{3+}) [2.28, 2.29, 2.30]. In solids, this general trend is preserved, however the numbers are shifted to $\approx 30\%$ lower values.[4] For RE^{2+} ions the $4f^n \rightarrow 4f^{n-1}5d^1$ separation is even smaller, being shifted into the visible region (e.g. EuO). Thus, electronic scattering by d- and f-electrons is in many cases "close-to-resonance" scattering! This has been verified for e.g. Ce^{3+} in $LuPO_4$ [2.31], where a change in the incident laser energy from the Ar^+ green line (514.5 nm) to the frequency-tripled output of a Nd^{3+}:YAG laser (355 nm) was shown to enhance the electronic Raman scattering intensities by a factor close to 100, Rayleigh's usual ω^4 factor taken into account.

Antisymmetry in the scattering tensor can only be detected in the measured spectrum if selection rules permit a specific transition only for a pseu-

[4] In Ce^{3+}:YAG the lowest of the five crystal-field split components of the $4f^05d^1$ configuration is found at 21 500 cm^{-1} due to an extremely strong crystal field at the Ce^{3+}-site.

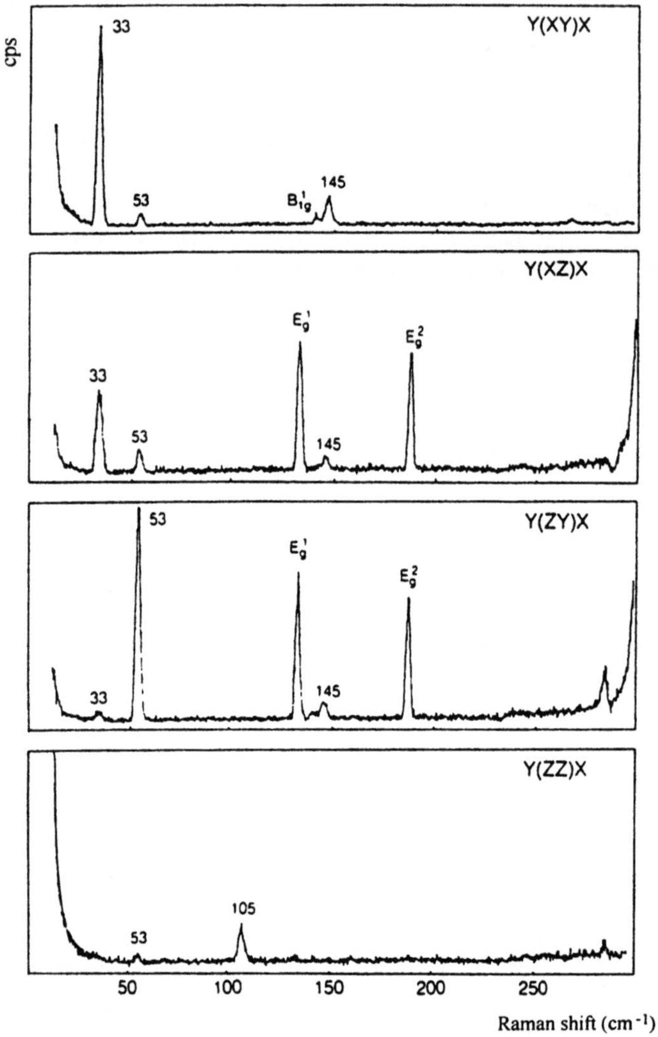

Fig. 2.2. Low-energy Raman spectrum of $ErPO_4$, $T \approx 4.2\,\mathrm{K}$, $\lambda_{\mathrm{Laser}} = 514.5\,\mathrm{nm}$; site symmetry of Er^{3+}:D_{2d}; electronic transitions within the $(4f^{11}, {}^4\mathrm{I}_{\frac{15}{2}})$ configuration. The XY-, XZ- (= YZ), ZY- (= ZX), and ZZ-polarizations are displayed. The asymmetry of the scattering (XZ versus ZY) for the 33- and 53$-$cm^{-1} lines is apparent. E^1_g, E^2_g, B^1_{1g} are phonon transitions. Full scale is 1000 cps, (counts per second). From [2.73]

dovector scattering process or if, for some excitation energy ω_0, both $\chi^{\mathrm{s}}_{\alpha\beta}$ and $\chi^{\mathrm{as}}_{\alpha\beta}$ are allowed by selection rules for a given polarization and the two amplitudes can interfere with each other. The scattered intensities after polarization reversal $\alpha \leftrightarrow \beta$ will be proportional to $|\chi^{\mathrm{s}}_{\alpha\beta} + \chi^{\mathrm{as}}_{\alpha\beta}|^2$ and $|\chi^{\mathrm{s}}_{\beta\alpha} - \chi^{\mathrm{as}}_{\beta\alpha}|^2$, respectively. Figure 2.2 displays part of the electronic Raman spectrum of $ErPO_4$, where the Er^{3+}-ions occupy sites of D_{2d} symmetry. The lines at 33 and 53 cm^{-1} clearly reveal a strong antisymmetric contribution because of the different intensities when interchanging the polarizations of incident and scattered light [2.32]. Such interference of scattered amplitudes is generally encountered if both final states are the same. If the initial and (or) the final states are degenerate, the intensities of the transitions to the various orthogonal components of the final state have to be added, and a thermal average over the initial states has to be performed. In Fig. 2.3 another example of antisymmetric Raman scattering is displayed [2.33]. In ferromagnetic TbF_3, (D_{2h}^{16}, $Z = 4$, see Sect. 2.3.3.1), the B_{1g}, B_{2g}, and B_{3g} spectra are simultaneously active in off-diagonal symmetric and in pseudo vector scattering. In the B_{1g} mode the asymmetric contribution is smaller than the symmetric one, in the B_{2g} spectra both contributions have the same size and cancel in the $z(xz)y$ spectrum completely, while in the B_{3g} spectra the symmetric contribution prevails. Typically the B_{1g}, B_{2g}, B_{3g} modes are also active as (one-photon) magnetic dipole transitions with peculiar features. They can be observed both in absorption, recognized by their polarization properties which differ from those for electric dipole transitions [2.35], and in a reflectivity experiment as "reststrahlen"bands (Fig. 2.3). These magnetic dipole bands display as a characteristic feature their reflectivity minimum on the low-frequency side of the band, while an electric-dipole transition has its reflectivity dip generally on the high-frequency side.[5]

Time-reversal symmetry of crystal field states plays a decisive role in our context. This symmetry is governed [2.37] by the time-reversal operator T, which is antilinear ($T(\alpha|\psi\rangle) = \alpha^* T|\psi\rangle$) and antiunitary, i.e. it fulfills

[5] This reversed shape [2.34] has a simple cause and some interesting consequences: Considering the reflectivity $R(\omega)$ at normal incidence along one of the symmetric directions of a cubic, uniaxial or orthorhombic crystal ($\mu \neq 1$, damping neglected): $R(\omega) = |(\sqrt{\mu_\perp(\omega)} - \sqrt{\varepsilon_\perp(\omega)})/(\sqrt{\mu_\perp(\omega)} + \sqrt{\varepsilon_\perp(\omega)}|^2$ [2.35], the reflectivity minimum occurs at ω_0 where $\mu_\perp(\omega_0) = \varepsilon_\perp(\omega_0)$. We assume that both $\mu_\perp(\omega)$ and $\varepsilon_\perp(\omega)$ follow a dispersion relation of damped oscillators or, in its factorized form, a *Kurosawa* relation [2.36] where the LO frequency of the dielectric material is replaced by the antiresonance frequency ω_A in the magnetic case [2.34] at $\mu_\perp(\omega_A) = 0$. It becomes evident that in a dielectric material ($\mu_\perp \equiv 1$), the condition $R(\omega_0) = 0$ can only be fulfilled at $\omega_0 > \omega_{TO} (= \omega_R(\text{magn.})), \omega_{LO}$, while for $\varepsilon_\perp(\omega) > 1$ and $\mu_\perp = f(\omega)$, this condition requires $\omega_0 < \omega_R$. If in a magnetic crystal ($\mu > 1$) both the electric dipole $\varepsilon_\perp(\omega)$ and the magnetic dipole (pseudo-vector) $\mu_\perp(\omega)$ transitions contribute to the reflectivity, a cancellation of the reflected intensity may occur, in formal analogy to the case of asymmetric Raman scattering. This becomes evident by comparing $R(\omega)$ with (2.36) where the contribution of quadrupolar scattering ($F^{(2)}$) substitutes for $\sqrt{\varepsilon_\perp(\omega)}$, the response of the dielectric background of the magnetic crystal. Thus both the shape of (one-photon) magnetic reststrahlen bands and asymmetric electronic Raman scattering are due the peculiar properties of a pseudo-vector.

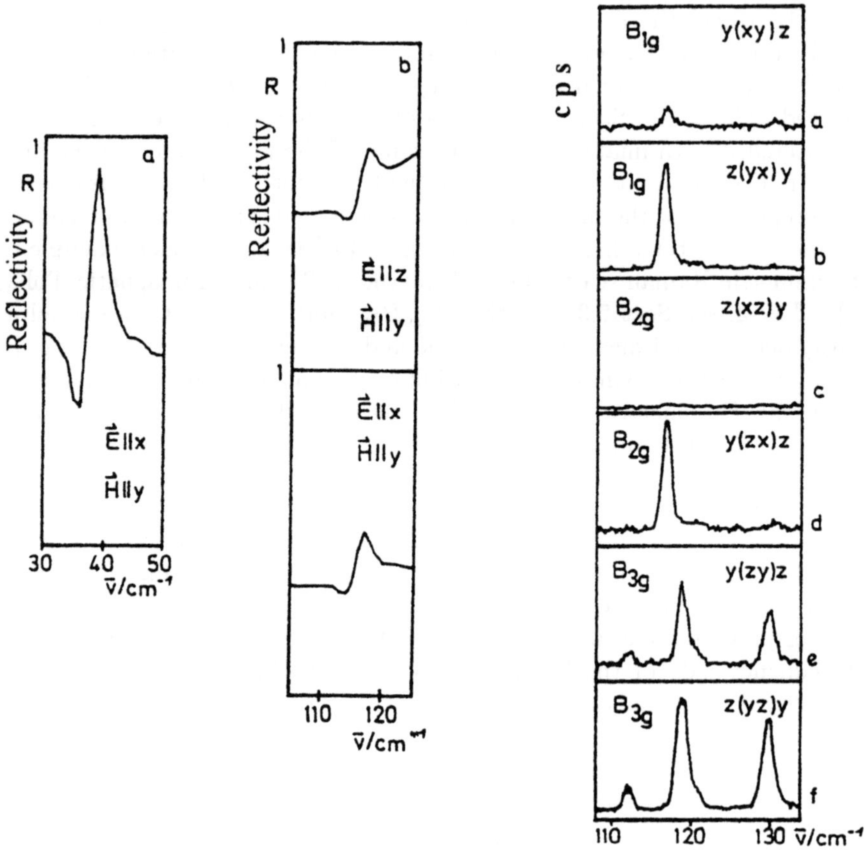

Fig. 2.3. (*Left* and *center*): Reststrahlen bands due to magnetic dipole transitions between crystal field excitations in HoF_3 (**a**) and in TbF_3 (**b**) at $T = 1.6\,K$, type of symmetry: B_{2g}. (Right): Raman spectra of TbF_3 in the ferromagnetic phase ($T \approx 2\,K$). Spectra (a, d, e) and (b, c, f), respectively, have been taken with the same instrumental adjustment. The $B_{1g} - (xy \leftrightarrow yx)$ and B_{2g} spectra ($xz \leftrightarrow zx$) display symmetric and antisymmetric (pseudovector) scattering simultaneously. The B_{2g} transition is also active as a (one-photon) magnetic dipole transition, see center. From [2.33]

$\boldsymbol{r}' \equiv T\boldsymbol{r}T^+ = \boldsymbol{r}$, $\boldsymbol{p}' \equiv T\boldsymbol{p}T^+ = -\boldsymbol{p}$, $\boldsymbol{s}' \equiv T\boldsymbol{s}T^+ = -\boldsymbol{s}$, where $\boldsymbol{r}, \boldsymbol{p}, \boldsymbol{s}$ are expectation values of variables in real, momentum, and spin space, respectively. The state vector after time-reversal

$$|\psi(t)\rangle' = T|\psi(-t)\rangle \tag{2.16}$$

obeys the same time-reversed Schrödinger equation with the same eigenvalue as $|\psi(t)\rangle$. For a single electron with spin, $T = \mathrm{i}\sigma_y K_0$, where σ_y is the y-component of the Pauli spin operator [2.38] and K_0 is the complex conjugation operator. For n electrons in a configuration, the time-reversal operator T_n becomes the product:

$$T_n = \prod_1^n T,\; T_n^2 = \begin{cases} -1 & : \text{ n= odd,} \\ +1 & : \text{ n= even.} \end{cases} \tag{2.17}$$

If $T_n^2 = -1$, then $(T_n^+)^2 = -1$ and, because of antilinearity, $\langle\psi|\bar{\psi}\rangle \equiv \langle\psi|(T_n|\psi\rangle) = (\langle T_n|\psi\rangle)|(T_n^2|\psi\rangle))^* = (T_n^2\langle\psi|)(T_n|\psi\rangle) = -\langle\psi|(T_n|\psi\rangle) = 0$, i.e. $|\psi\rangle$ and $|\bar{\psi}\rangle$ are orthogonal.

The Kramers degeneracy for n = odd is a consequence of this orthogonality: In the $|J, M_J\rangle$ representation, using eigenstates of the total angular momentum $\boldsymbol{J}$, which provide a convenient basis for expanding crystal field states:

$$|\Psi\rangle = \sum_{J,M_J} a_{J,M_J}|J, M_J\rangle, \tag{2.18}$$

a time-reversed, i.e. a Kramers conjugate state exists at the same eigenvalue that differs from $|\Psi\rangle$ for n = odd [2.38]:

$$|\bar{\Psi}\rangle = \sum_{J,M_J} a^*_{J,M_J}(-1)^{J-M_J}|J, -M_J\rangle. \tag{2.19}$$

Equation (2.19) is the result of Kramers theorem, which states that all atomic states with an odd number of electrons are at least twofold degenerate (Kramers degeneracy), provided time-reversal symmetry holds. This symmetry will be lifted by an external magnetic field but clearly not by an electric field such as the crystal field. States with an even number of electrons will all be singlets in cases of orthorhombic and lower symmetry, except for accidental degeneracy; under uniaxial (teragonal, trigonal, hexagonal) and higher (cubic) symmetries both singlets and degenerate levels due to spatial and (or) time reversal symmetry will occur.

The role of time reversal in antisymmetric light scattering has been discussed in detail in Refs. [2.39]; it is of essential importance when an external magnetic field is applied and in the case of resonance Raman spectroscopy (Sect. 2.2.5) and scattering by vibronic states in the presence of Kramers degenerate ground states (Sect. 2.3).

2.2.2 The Scattering Tensor

It has been shown in Sect. 2.2.1 that the electronic Raman scattering process is generally described by a second-rank scattering tensor $\alpha_{\rho\sigma}$, where ρ, σ are either three Cartesian or other coordinates, such as spherical or cylindrical coordinates. The element $\alpha_{\rho\sigma}$ represents the amplitude of the scattered wave of polarization ρ due to an incident wave polarized in the σ-direction. The intensity of the Raman process is proportional to $|\alpha_{\rho\sigma}|^2$. The Cartesian coordinates are preferred for describing experiments, the cylindrical coordinates for special experiments such as studying antisymmetry or, when circularly polarized radiation is applied, in the presence of a static magnetic field along a symmetry axis. The spherical coordinates are to be preferred for theoretical calculations using the methods of spherical tensor operators $\underline{\alpha}_q^k$, as in atomic theory. A spherical tensor operator α_q^k of rank k has $2k+1$ components ($q=-k,-k+1,\dots,k$) that are defined by applying the Wigner–Eckart theorem [2.40, 2.41, 2.42], in the (J,M)-representation of the l^n configuration to matrix elements of $\underline{\alpha}_q^k$:

$$(l^n\psi JM|\underline{\alpha}_q^k|l^n\psi' J'M') = (-1)^{J-M}\begin{pmatrix} J & k & J' \\ -M & q & M' \end{pmatrix} \times (l^n\psi J||\underline{\alpha}^k||l^n\psi' J'). \tag{2.20}$$

The 3-j symbol is zero unless $J+J' \geq k \geq |J-J'|$, and $M-M'=q$, the reduced matrix element $(l^n\psi J||\underline{\alpha}^k||l^n\psi' J')$ can be evaluated using standard methods [2.40, 2.42, 2.43]. The concept of Cartesian representations of spherical tensors of rank two in Raman scattering has been treated in [2.44] and, on a more tutorial basis, in [2.45].

For the ease of reference the relations for the conversion of the tensor elements between Cartesian and cylindrical coordinate systems and representations by spherical tensors have been compiled in the Appendix.

The transformations of the electric dipole moments from linear Cartesian into cylindrical coordinates, using $\bar{D}_0, \bar{D}_{\pm 1}$, are:

$$\begin{aligned} \bar{D}_0 &= eD_0^{(1)} = e\sum z, \\ \bar{D}_{\pm 1} &= eD_{\pm 1}^{(1)} = \mp e\sum (x \pm \mathrm{i}y)/\sqrt{2}. \end{aligned} \tag{2.21}$$

These relations can also be found in the Appendix.[6]

A particular scattering geometry is often described in Porto's notation [2.46], e.g. $x(yz)y$: the letters outside the brackets give the propagation directions of the incident (x) and scattered (y) linear polarized light, the bracketed letters y,z are the polarization directions of the incident and scattered light, respectively (+ or − (l or r) for left or right circularly polarized light).

[6] Care has to be taken when applying circular polarization (± 1) of the incident and (or) the scattered radiation along an axis in a uniaxial or in a cubic crystal or in Faraday orientation of a magnetic field ($B||z$). In this article we follow the usual definition, in left (σ^+) or right (σ^-) circular polarization the electric field vectors vectors rotate in counterclockwise or clockwise directions when facing the incoming beam. A left-hand circularly polarized photon (σ^+) has a helicity $+\hbar$, and vice versa for (σ^-). If the sense of rotation of the field is preserved when changing the direction of propagation ($\mathbf{k} \leftrightarrow -\mathbf{k}$), the helicity changes its sign.

2.2.3 Selection Rules for Light Scattering by Crystal-Field Excitations

Qualitative, but in many cases very helpful information on electronic Raman activity can be obtained if the quantum numbers (J, M_J) in the $4f$-case (M_L, M_S in the $3d$-case) are considered to be "good" quantum numbers (as they would be for a free ion). In this approximation (Russell–Saunders limit) the relations (2.20) tell us [2.3] that trace scattering is allowed for $\Delta J = 0$, $\Delta M_J = 0$, while magnetic dipole scattering is permitted for $\Delta J = 0, \pm 1$; $\Delta M_J = 0, \pm 1$, except for $J = 0 \rightarrow J = 0$ transitions. Quadrupolar scatteringis allowed for $\Delta J = 0, \pm 1, \pm 2$; $\Delta M_J = 0, \pm 1, \pm 2$, (except for $J = 0 \rightarrow J = 0$, $J = 0 \leftrightarrow J = 1$, and $J = \frac{1}{2} \rightarrow J = \frac{1}{2}$ transitions). $\Delta M_S = 0$ is also approximately valid. Usually Raman transitions obeying these rules will be more prominent in the spectra than other transitions. If weak L–S coupling applies, these rules can be recast into: $\Delta M_S = 0, \Delta M_L = 0, \pm 1, \pm 2$.

Another selection rule valid in the free ion and often "surviving" in the crystal for intra-configurational transitions is the parity selection rule (Laporte rule [2.3], [2.47], [2.47a]). The parity I of a n-electron state (e.g., $4f^n$) is defined as $I = (-1)^{\sum_{i=1}^{n} l_i}$, where l_i is the orbital angular momentum of a single electron ($l_i = 3$ in the f-configurations) and is constant for intra-configurational transitions. Raman scattering and two-photon absorption are parity conserving, i.e. allowed in first order, while (one-photon) electric dipole transitions are non-conserving and are forbidden in first order. Van Vleck pointed out that electric dipole crystal field transitions can only occur in crystals because of the existence of small admixtures of $4f^{n-1}ml$ configurations into the free-ion $4f^n$ states, where ml (e.g. $5d$) has to be chosen to provide parity opposite to $4f^n$. Such admixtures are caused by odd parity crystal field components and, to a weaker extent, by odd parity phonon excitations. In crystal sites with an inversion center, parity is strictly defined and electric dipole transitions among levels split by the static crystal field are forbidden also in higher order.

Contrary to the case of electronic excitations in ordinary metals and semiconductors, where the electrons are delocalized, the crystal field states in insulators are basically localized on a specific lattice site. This is especially valid for $4f$-states in rare earth compounds.[7] For symmetry-determined selection rules the localization has the consequence that the irreducible representations (reps) Γ_i of the specific crystallographic point group of the metal site in the crystal apply for deriving the selection rules.

The existence of odd-number electron states, i.e. of Kramers degeneracy, requires the application of double group representations [2.49] for analyses

[7] In $3d$- and $5f$-metals, the electrons in the d-states or the f-states are neither truly itinerant nor localized. Specific heat experiments reveal the existence of Schottky anomalies, indicating low-lying crystal-field-split states. Transport and photoelectron spectroscopy measurements, on the other hand, indicate a certain degree of delocalization of these states [2.48].

based on symmetry. The characters χ of the reps of $R^+(3)$, the group of the proper rotations, $R(\beta, \boldsymbol{n})$ in three dimensions of angle β about axis $\boldsymbol{n}$, are [2.49, 2.50, 2.51]

$$\chi^J(\beta) \equiv \chi^J[\exp(-\mathrm{i}\beta \boldsymbol{J} \cdot \boldsymbol{n})] = \sum_{M=-J}^{J} \exp(-\mathrm{i}\beta M) = \frac{\sin \beta(J+\frac{1}{2})}{\sin \frac{1}{2}\beta}. \quad (2.22)$$

Equation (2.22) applies to reps of double as well as single groups. For half integral values of $J, \chi^J(\beta \pm 2\pi) = -\chi^J(\beta)$, i.e., $\chi^J(\beta)$ is double valued. This double valuedness arises because the character of the operation $\bar{E} = R(2\pi, \boldsymbol{n})$ can take the value +1 or −1 when acting on a wavefunction of a system with an integral or half integral angular momentum, respectively. In the latter case one has to deal with spinor rather than scalar functions of position [2.50]. Double point groups G' are defined by introducing an additional operation $\bar{E}$ into the original group G and redefining the multiplication table to distinguish between $R(\beta, \boldsymbol{n})$ and $R(\beta + 2\pi, \boldsymbol{n})$. The number of group elements in G' has doubled with respect to G, while the number of classes is larger in G' than in G, but not always twice as large [2.50].

Character tables of the reps of simple and double point groups can be found, e.g., in the "Koster Tables" [2.50], and in [2.49, 2.57]. The selection rules for the electronic Raman effect have been derived and tabulated for all 32 point groups by *Kiel* and *Porto* [2.9]. They are compiled in Table 2.A.1 in the Appendix. Selection rules for the accompanying phonon spectrum can be easily derived applying the methods and tables of [2.52, 2.53, 2.54, 2.55].

The numbers of crystal field levels originating from a free-ion multiplet component (J, M_J) and their types of symmetry can be derived from (2.22) by breaking up the reps $\chi^J(\beta)$ into reps of the various point-(sub-)groups [2.49]. The results have been tabulated in the "Full Rotation Group Compatability Tables" [2.50]; see also [2.51].

The Raman polarizability tensor for a transition between two states $|i\rangle, |f\rangle$ with the reps Γ_i, Γ_f spans the reps $\Gamma_{\mathrm{R}} = (\Gamma_i \otimes \Gamma_f)$ and Raman activity is allowed if:[8]

$$\Gamma_{\mathrm{R}} \cap \Gamma_{\mathrm{S}} \neq \emptyset, \quad (2.23)$$

where Γ_{S} indicates the set of reps which comprise all components of the scattering tensor (2.13). For phonon transitions (hot bands excluded), $\Gamma_i \equiv \Gamma_1$, the totally symmetric rep, i.e. $\Gamma_{\mathrm{R}} \equiv \Gamma_f$ and (2.23) reduces to the familiar result:

$$\Gamma_f \cap \Gamma_{\mathrm{S}} \neq \emptyset. \quad (2.24)$$

For intra-configurational electric dipole transitions the Laporte rule [2.3, 2.47, 2.47a] has to be obeyed in addition. It applies strictly to lattice sites with

[8] $\Gamma_i \otimes \Gamma_f$: direct product of reps Γ_i, Γ_f; $\Gamma_{\mathrm{R}} \cap \Gamma_{\mathrm{S}}$: intersection of Γ_{R} and Γ_{S}; $\emptyset$: empty set of reps.

inversion symmetry. In Table 2.A.1 the selection rules for all 32 point groups have been compiled. The table is based on [2.9, 2.50, 2.53, 2.54, 2.55, 2.56] whereby some additions and corrections have been included. For the comfort of the user, all 32 groups have been fully tabulated despite of some isomorphisms or direct-product relations with the group C_i, which separate the reps of some groups into g- and u-classes. The one-photon selection rules have been included, because a laser-excited electronic Raman spectrum will, in most cases, be disturbed by or has to compete with one-photon spontaneous emission.

The tabulated selection rules for non-cubic groups can also be derived using *Hellwege*'s crystal quantum numbers [2.56]. These numbers are determined using the transformation properties of the solutions U of the Schrödinger equation of the atom in the crystal field; they are basically linear combinations of the free ion functions $\Psi_{\gamma J}^{M_i}$ [2.57]: $U = \sum_{\gamma,J,i} a_{\gamma J}^{M_i} \Psi_{\gamma J}^{M_i}$, where γ represents all quantum numbers not specifically mentioned, (e.g., n). Here it is sufficient to consider the crystal field states in zeroth order, i.e. for vanishing crystal field strength. In this case γ and J are fixed, $U \to u_{\gamma J}$. Investigating the transformation properties of $u_{\gamma J}$ under the elements of the cyclic groups C_p with a p-fold rotation axis, $(p = 1, 2, 3, 4, 6)$, it is evident that U includes only those functions Ψ^M which obey $M \equiv \mu \pmod{p}$. This relation defines the crystal quantum numbers μ, which can take p different values and which correspond to the group theoretical symbols for the reps. μ is integer if n is even and half-integer for n odd. In Table 2.A.2 the relations between p, μ, and the Γ_i are given as an example.

Various crystal quantum numbers have been defined to account for additional symmetry operations in other point groups: In the groups with rotation-inversion axes $(C_i, C_{3i}, C_s, S_4, C_{3h}), \mu$ is replaced by μ_I or $\bar{\mu}$: $\mu_I \equiv M_l \pm \frac{p}{2} \sum_k^n l_k \pmod{p}$ (l_k: orbital angular momentum of electron k); or $\bar{\mu} \equiv \mu + z\frac{p}{2}, z = 0$ (n even) or $z = 1$ (n odd).[9] Other quantum numbers are the parity $I = \pm 1$ in groups with inversion centers, where the reps can be separated into g- and u-types; $S = \pm 1$ in groups with vertical mirror planes (xz) and $\nu = 0, 1$ or $\pm\frac{1}{2}$ in dihedral groups (D_i, some D_{ih}, D_{id}). Selection rules based on the crystal quantum numbers have been derived in [2.56], [2.57]. These rules based on crystal quantum numbers have the advantage of an obvious relation to the familiar ΔM rules of the free atom. For example, the exchange of angular momentum $(0\hbar, 1\hbar, 2\hbar)$ between the atomic system and the radiation field in a certain scattering configuration becomes immediately evident, a fact which is especially helpful in Zeeman effect studies. They suffer from the disadvantage that their application to cubic groups is not straightforward: the group-theoretical method is more general. Recently the application of crystal quantum numbers has concentrated on the quantum number μ only [2.42], [2.58]. In this case the one-to-one correspondence of

[9] For C_i and C_{3i} (p odd) and $\sum_k^n l_k$ odd, μ_I is half-integer for n even and integer for n odd.

crystal quantum numbers and reps is lost, the reps of the point group usually form a wider class of quantum numbers for labeling the crystal field levels than the corresponding crystal field quantum numbers μ. Hence values of μ attributed to a specific rep may vary between *Hellwege*'s definition and the new classification.

The application of the selection rules, as given in Table 2.A.1, is not straightforward in cases of a low site symmetry of the metal ion incorporated in a unit cell of high symmetry, e.g. cubic garnets with a D_2 site of the $4f$-ion. Here the local axes of quantization do not coincide with any of the cubic axes. This situation is discussed in detail in Sect. 2.4.1, (2.92).

The application of an homogeneous magnetic field along a direction of symmetry usually lifts the degeneracies left over by the crystal electric field, especially the Kramers degeneracy, i.e. by raising the time reversal symmetry (see Sect. 2.2.1 and 2.2.7). If the subtleties of time reversal are not essential, only the spatial symmetry elements of the total system (crystal and field) which are common to both the point group of the site (G_s) and the group $C_{\infty h}$ of the homogeneous field have to be considered. The resulting subgroups $g_s \equiv G_s \cap C_{\infty h}$ of the site groups are abelian, i.e. they possess only one-dimensional reps. These abelian groups (other than the trivial C_1) are listed in Table 2.A.1. A linear Zeeman effect is possible in the case of an odd electron number N (Kramers degeneracy), or for even N, whenever Kramers or other symmetry degeneracy occurs. If the subgroup possesses rotation axes with $p \geq 3$, the characters of some reps are complex conjugates with their bases $\mp(x \pm \mathrm{i}y)$ describing dipoles rotating in the planes of symmetry. In this case, circularly polarized radiation propagating along the field has to be applied in order to take full advantage of the symmetry of the system (see Appendix).

In general, the Zeeman splitting is small compared with the crystal field energies. Hence the crystal field states $u_{\gamma J}$ can be used as a basis to calculate the Zeeman splitting by forming the matrix elements $\langle u_{\gamma J}|\mu_{\mathrm{B}}\boldsymbol{H} \cdot (\boldsymbol{L} + 2\boldsymbol{S})|u_{\gamma' J}\rangle$, where J mixing has been neglected. Due to the anisotropic crystal field, the Zeeman effect will also be anisotropic. This can be formally described by an anisotropic g-factor (g-tensor), subject to site symmetry. In a uniaxial crystal, i.e. in a crystal belonging to one of the trigonal, tetragonal or hexagonal crystal classes, two kinds of matrix elements have to be distinguished:

parallel Zeeman effect, see (2.18) and above:

$$\begin{aligned}&\langle \gamma, J, M_{J,i}|\mu_{\mathrm{B}} H_{\|} \cdot (L_z + 2S_z)|\gamma', J, M_{J,i}\rangle \\ &\quad = \mu_{\mathrm{B}} g_J H_{\|} \sum_i |a_i|^2 M_{J,i} = s_{\|} H_{\|}\end{aligned} \tag{2.25}$$

and perpendicular Zeeman effect

$$\begin{aligned}&\langle \gamma, J, M_{J,i}|\mu_{\mathrm{B}} H_{\perp} \cdot (L_x + 2S_x)|\gamma', J, M_{J,i} \pm 1\rangle \\ &\quad = \frac{1}{2}\mu_{\mathrm{B}} g_J H_{\perp}[(J \mp M_{J,i})(J \pm M_{J,i} + 1)]^{\frac{1}{2}} = s_{\perp} H_{\perp},\end{aligned} \tag{2.26}$$

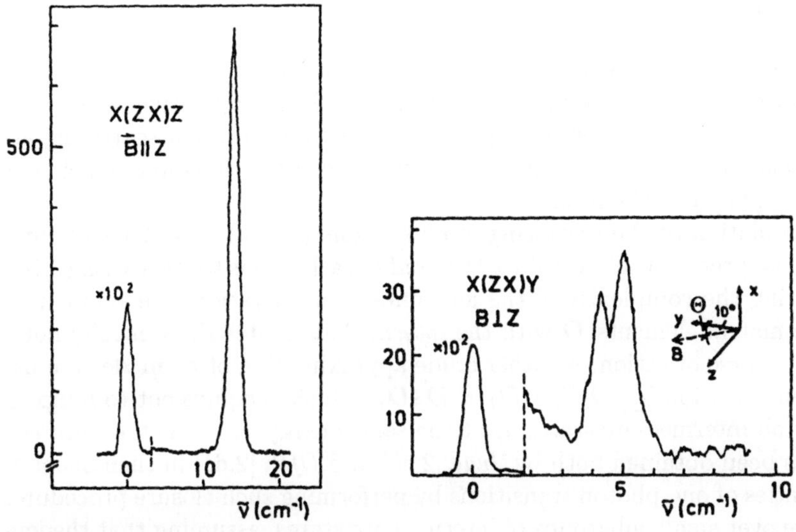

Fig. 2.4. Ground state Zeeman effect of $Ce^{3+}(4f^1)$ in CeF_3, (D^4_{3d}) site: C_2, see Sect. 2.3.1, paramagnetic resonance by Raman scattering in counts per second, $B =$ 12 T, $T = 2$ K. (*Left*): $B \| Z$; (*right*): $B \perp Z$. The inclination of the $\boldsymbol{B}$ direction against one of the local y axes of the three g-tensors ($\Theta = 0, \pm 120^0$) in the basal plane is indicated. Actually three transitions with different g_{eff} (at 5.2, 4.85, and 4.2 cm^{-1}) are observed, $x \| Z$. From [2.142]

where $s_{\|}, s_{\perp}$ are the magnetic splitting factors. Consequently, the sizes of the splitting will be different in the two orientations.

The occurrence of a linear Zeeman splitting is determined by symmetry. For a field component parallel to the axis, a symmetric splitting of some levels will be observed if at least two of the reps of the abelian subgroup induced by the external field (Table 2.A.1) form a Kramers degenerate pair. This is always the case for $p_{\text{subgroup}} \geq 3$, ($n$ even) and for all p_{subgroup}, if n is odd. Nondegenerate levels with n even and $\mu = 0, \frac{p}{2}$ may shift to higher order in the field.

If the field is applied perpendicular to the axis, a linear effect is permitted for all p, provided n is odd. Inspection of (2.26) tells us however that a finite $s_{\perp}$ will only result if the M values in the crystal field state $|\gamma, J, M_{J,i}\rangle$ differ by ± 1, i.e. the condition $-M' = +M \pm 1$ or $n' \cdot p = -2\mu \mp 1$ must be fulfilled, n': integer [2.57]. If n is odd, this is the case for all p when $\mu = \pm\frac{1}{2}$, in addition, for $p = 4, 6$ if $\mu = \pm\frac{p-1}{2}$. For n even, the condition can be fulfilled for $p = 3$, if $\mu = \pm 1$ [2.59]. As an example, the Zeeman effect of the ground state of Ce^{3+}, $(4f^1, {}^2F_{\frac{5}{2}})$ in trigonal CeF_3 is plotted in Fig. 2.4, as observed by electronic Raman scattering [2.135], [2.142].

2.2.4 Intensities

The calculation of intensities of electronic Raman processes and the comparison with experimental results have been a field of very active research in the past. Since two-photon absorption and electronic Raman scattering are closely related processes, pertinent theoretical investigations in one field can be easily applied to the other.

A calculation of the scattering cross sections (s. c. s.) (2.4) for electronic transitions (preferentially within a $4f^n$-configuration) has to start with (2.10). Considering the complexity of the ions under consideration, the summation over the matrix elements $\boldsymbol{D}$ with the intermediate states $|l\rangle$ is usually not a practical course of action. A rather crude approximation often made is to use the closure relation $\sum_l \boldsymbol{D}_S|l\rangle\langle l|\boldsymbol{D}_I = \boldsymbol{D}_S\boldsymbol{D}_I$, which is equivalent to lumping together all intermediate states $|l\rangle$ at the same energy. An essential improvement has been obtained both by *Judd* [2.60] and *Ofelt* [2.61] in their theories of intensities of one-photon transitions by performing such closure procedures piecewise over small subgroups of intermediate states, assuming that the levels of each excited configuration extend over an energy range small compared to the energy of this configuration above the ground state. Excited configurations to be considered are $4f^{n-1}n'd$, $4f^{n-1}n'g$, and the core excitation $4f^{n+1}n''d^9$; $n' \geq 5, n'' = 3, 4$. This technique has been applied by *Axe* [2.62] for the general analysis of two-photon processes and by *Mortensen* and *Koningstein* [2.63] to electronic Raman and Rayleigh scattering, i.e. to intraconfigurational transitions. Effects of configurational mixing by including g-orbitals have been taken into account in [2.64]. The analysis of two-photon processes has been considerably extended by *Downer* et al. [2.65], [2.66]. These authors observed deviations of transition strengths from the predictions of (2.10) in Gd^{3+}:LaF_3, and from the ΔL and ΔJ selection rules. Both findings were interpreted by introducing third- and fourth-order contributions involving spin–orbit and/or crystal-field interactions [2.67] among the intermediate states $4f^6 5d$. This analysis has been generalized for all lanthanide compounds and for the fourth-order combination of spin–orbit and crystal–field interactions in [2.68]. Another important third-order process is the contribution of the ligand polarization [2.69], reflecting the dynamic response of the ligands to the radiation field (dynamic coupling mechanism). A detailed introduction to the field can be found in [2.80].

In the following the method of second quantization is used to present and interpret some formulas on the intensities of two-photon processes in lanthanides. This method, which is in general use now, provides elegant and powerful means for computing matrix elements in atomic theory and has been developed largely by *Judd* et al. [2.60], [2.70], [2.71]. It has been applied by *Downer* et al. [2.65, 2.66, 2.72] and by *Becker* [2.73], see also [2.80] and [2.74]. For this purpose one introduces the operators $\boldsymbol{f}^{\dagger}_{\gamma}, \boldsymbol{f}_{\gamma'}, \boldsymbol{d}_{\gamma''}, \boldsymbol{g}_{\gamma'''}$ etc., which either create or destroy electrons in the f, d, and g configurations, characterized by the quantum numbers $\gamma \equiv (nlm_l m_s)$, $\gamma' \equiv (n'l'm_l'm_s')$, etc. These operators obey the usual fermion anticommutation relations. For a specific configuration (n, l) the various $\boldsymbol{a}^{\dagger}_{\gamma}, \boldsymbol{a}_{\gamma}$ form the components of a double tensor of rank s in spin space and l in orbital angular momentum space. The techniques of tensor coupling operations can be applied as usual. In the notation $(\boldsymbol{a}^{\dagger}\boldsymbol{a})^{(pt)}$, p is the rank of the coupled double tensor in spin space, t that in orbital angular momentum space. For example, $(\boldsymbol{a}^{\dagger}\boldsymbol{a})^{(10)}$ is proportional to $\boldsymbol{S}$ and $(\boldsymbol{a}^{\dagger}\boldsymbol{a})^{(01)}$ is proportional to $\boldsymbol{L}$. Note, that the "scalar" product of two tensors $\boldsymbol{T}$, $\boldsymbol{U}$ can be written as a coupled tensor of rank zero, $(q = -k, -k+1, \ldots, k)$:

$$\boldsymbol{T}^{(k)} \cdot \boldsymbol{U}^{(k)} = \sum_q (-1)^q T_q^{(k)} U_{-q}^{(k)} = (-1)^k (2k+1)^{\frac{1}{2}} (\boldsymbol{T}^{(k)} \boldsymbol{U}^{(k)})^{(k=0)} . \quad (2.27)$$

In what follows the Raman amplitude $(\alpha_{xy})_{fi}$ for scattering by transitions between the two states $|i\rangle$ and $|f\rangle$ will be given in tensor operator form, i.e.

$$\begin{aligned} (\alpha_{\mathrm{S,L}})_{fi} &= (\alpha_{xy})_{fi} \\ &= -\sum_{\mathrm{l}} \left\{ \frac{(\boldsymbol{D}_x)_{f\mathrm{l}}\ (\boldsymbol{D}_y)_{\mathrm{l}i}}{E_{\mathrm{l}} - \hbar\omega_{\mathrm{L},y}} + \frac{(\boldsymbol{D}_y)_{f\mathrm{l}}\ (\boldsymbol{D}_x)_{\mathrm{l}i}}{E_{\mathrm{l}} + \hbar\omega_{\mathrm{S},x}} \right\} . \end{aligned} \quad (2.28)$$

The electric dipole operator for a transition between shells (nl) and $(n'l')$ has the form:

$$\begin{aligned} \boldsymbol{D} = &\ (-1)^l (2)^{\frac{1}{2}} \left[\frac{(2l+1)(2l'+1)}{3} \right]^{\frac{1}{2}} \\ &\times \begin{pmatrix} l & 1 & l' \\ 0 & 0 & 0 \end{pmatrix} \langle nl|r|n'l'\rangle \left[(\boldsymbol{a}^{\dagger}\boldsymbol{b})^{(01)} - (\boldsymbol{b}^{\dagger}\boldsymbol{a})^{(01)} \right] . \end{aligned} \quad (2.29)$$

After some tensor recoupling and operator commutation manipulations, and using the identity

$$(\boldsymbol{a}^{\dagger}\boldsymbol{a})^{(0t)} = -\left[\left(\frac{1}{2}\right)(2t+1) \right]^{\frac{1}{2}} \boldsymbol{U}^{(t)} , \quad (2.30)$$

where $\boldsymbol{U}^{(t)}$ is the unit tensor of rank t in orbital angular momentum space, and with the spherical unit polarization vectors

$(\boldsymbol{e}_{\mathrm{L}}, \boldsymbol{e}_{\mathrm{S}})$ for the laser and Stokes beam

$$(\boldsymbol{e}_{\mathrm{L}}\boldsymbol{e}_{\mathrm{S}})^{(t)} = (-1)^t (\boldsymbol{e}_{\mathrm{S}}\boldsymbol{e}_{\mathrm{L}})^{(t)} , \quad (2.31)$$

one arrives, using $l = 3$, at

$$\begin{aligned} ((\alpha_{x,y})_{fi} = (\alpha_{\mathrm{S,L}})_{fi} = &\sum_{4f^{N-1}n'l'} 7(2l'+1) \begin{pmatrix} 3 & 1 & l' \\ 0 & 0 & 0 \end{pmatrix}^2 \langle 4f|r|n'l'\rangle^2 \\ &\times \sum_t (2t+1)^{\frac{1}{2}} \begin{Bmatrix} 1 & 3 & l' \\ 3 & 1 & t \end{Bmatrix} \\ &\times \left[\frac{1}{E_{n'l'} - \hbar\omega_{\mathrm{L}}} + \frac{(-1)^t}{E_{n'l'} + \hbar\omega_{\mathrm{S}}} \right] (\boldsymbol{e}_{\mathrm{S}}\boldsymbol{e}_{\mathrm{L}})^{(t)} \cdot \boldsymbol{U}^{(t)} . \end{aligned} \quad (2.32)$$

Here the $6j$-symbol [2.40] appears in the curly brackets which originates from the tensor operator recoupling procedures. In the sum over t in (2.32) actually two terms occur: ($t = 0, 1, 2$, but the term with $t = 0$ contributes only to Rayleigh scattering). In the approximation $\hbar\omega_{\mathrm{L}} \approx \hbar\omega_{\mathrm{S}} \ll E_{n'l'}$ the term corresponding to rank $t = 1$ yields:

$$(\alpha_{\mathrm{S,L}}^{(t=1)})_{fi} = 7(3)^{\frac{1}{2}} \sum_{4f^{N-1}n'l'} (2l'+1) \begin{pmatrix} 3 & 1 & l' \\ 0 & 0 & 0 \end{pmatrix}^2 \langle 4f|r|n'l'\rangle^2 \times \begin{Bmatrix} 1 & 3 & l' \\ 3 & 1 & 1 \end{Bmatrix} \frac{2\hbar\omega}{E_{n'l'}^2} (\boldsymbol{e}_{\mathrm{S}}\boldsymbol{e}_{\mathrm{L}})^{(1)} \cdot \boldsymbol{U}^{(1)}. \tag{2.33}$$

For rank $t = 2$ we obtain:

$$(\alpha_{\mathrm{S,L}}^{(t=2)})_{fi} = 7(5)^{\frac{1}{2}} \sum_{4f^{N-1}n'l'} (2l'+1) \begin{pmatrix} 3 & 1 & l' \\ 0 & 0 & 0 \end{pmatrix}^2 \times \langle 4f|r|n'l'\rangle^2 \begin{Bmatrix} 1 & 3 & l' \\ 3 & 1 & 2 \end{Bmatrix} \frac{2}{E_{n'l'}} (\boldsymbol{e}_{\mathrm{S}}\boldsymbol{e}_{\mathrm{L}})^{(2)} \cdot \boldsymbol{U}^{(2)}. \tag{2.34}$$

Evidently because of (2.31), the sign in (2.33) changes for interchanged polarizations, while in (2.34) it does not. Equation (2.33) therefore represents the antisymmetric contribution to the scattering amplitude, while (2.34) represents the symmetric contribution, [cf. (2.14a)].

For practical calculations the matrix elements between the initial and the final states have to be evaluated with the correct wavefunctions. For the lanthanides with n $4f$-electrons the intermediate-coupling wavefunctions, which are linear combinations of Russell–Saunders functions of different $4f^n$ configurations, have been calculated from fits to crystal field levels and/or (preferentially) to magnetic splitting factors and can be found in the literature, (see [2.29, 2.30, 2.42]). The angular dependences of these wavefunctions can be calculated with good precision. Formulas to evaluate the matrix elements of the tensor operators with the help of the Wigner–Eckart theorem can be found in [2.29], [2.42], and [2.75]. The radial parts of the wavefunctions are known with lower reliability and, accordingly, the matrix elements $\langle 4\mathrm{f}|r|n'l'\rangle$ are usually treated as fit parameters. It is therefore convenient to combine the radial matrix elements with the numerical factors in (2.32, 2.33, 2.34) into the factors $F^{t=1}$ and $F^{t=2}$ originally introduced by *Sonnich, Mortensen* and *Koningstein* [2.63]. These factors, which can be determined from fits to experimental results, carry the information on the intermediate states, their energies and the radial matrix elements.

$$F^{(t)} = (-1)^{(t)} \sum_{4f^{N-1}n'l'} 7(2l'+1) \begin{pmatrix} 3 & 1 & l' \\ 0 & 0 & 0 \end{pmatrix}^2 \langle 4f|r|n'l'\rangle^2 \times (2t+1)^{\frac{1}{2}} \begin{Bmatrix} 1 & 3 & l' \\ 3 & 1 & t \end{Bmatrix} \left[\frac{1}{E_{n'l'} - \hbar\omega_{\mathrm{L}}} + \frac{(-1)^t}{E_{n'l'} + \hbar\omega_{\mathrm{S}}} \right]. \tag{2.35}$$

The electronic Raman scattering amplitude is finally obtained as:

$$((\alpha_{x,y})_{fi} \rightarrow)(\alpha_{\mathrm{S,L}})_{fi} = \sum_t (-1)^t \, F^{(t)} \cdot (\boldsymbol{e}_\mathrm{S}\boldsymbol{e}_\mathrm{L})^{(t)} \cdot \boldsymbol{U}^{(t)}, \tag{2.36}$$

with the irreducible representation of the Raman scattering tensor in the simple form: $\alpha_q^{(t)} = F^{(t)} \cdot U_q^{(t)}$. The ratio $F^{(1)}/F^{(2)}$, where the radial integral cancels out, is a measure for the relative intensities of transitions and also their asymmetry; if the $F^{(1)}$ and $F^{(2)}$ contain both symmetric and antisymmetric contributions.

Table 2.1. Comparison of observed and predicted asymmetry ratios $I_{xz,zy}/I_{zx,zy}$ for some electronic Raman transitions in the $(4f^{11}, {}^4I_{\frac{15}{2}})$ configuration of Er^{3+} in $ErPO_4$. From [2.32], [2.73]

Transition energy	$33\,\mathrm{cm}^{-1}$	$53\,\mathrm{cm}^{-1}$	$145\,\mathrm{cm}^{-1}$
Observed asymmetry	5.3	0.2	0.6
Predicted asymmetry ($F^{(1)}/F^{(2)} = 0.25$)	3.5	0.04	1.9
Predicted asymmetry ($F^{(1)}/F^{(2)} = 0.03$)	5.2	0.5	1.1

This theory has been compared with experimental results in $ErPO_4$, $TmPO_4$, and $HoPO_4$ [2.32], [2.73]. These phosphates have the tetragonal zircon structure (D_{4h}^{19}), with 4 RE ions on equivalent sites of D_{2d} symmetry.

Under the assumption that the dominant intermediate states belong to the $4f^{n-1}5d$ configuration at $\approx 10^5\,\mathrm{cm}^{-1}$, the ratio

$$F^{(1)}/F^{(2)} = \tau \approx 1.3 \frac{\hbar\omega_\mathrm{L}}{E_{5d}} \tag{2.37}$$

is obtained, where the numerical factor in (2.37) arises from the $3j$- and $6j$-symbols in (2.35). With $\hbar\omega_\mathrm{L} \approx 2 \times 10^4\,\mathrm{cm}^{-1}$, a value $\tau \approx 0.25$ is expected. In Tables 2.1 and 2.2 this prediction is compared with experimental results on asymmetries and intensities of $ErPO_4$. From the observed asymmetry (I_{xz}/I_{zx}) an experimental value of τ can be derived by using (2.37) and:

$$(I_{xz}/I_{zx}) = \frac{(-m_1F^{(1)} + m_2F^{(2)})^2}{(+m_1F^{(1)} + m_2F^{(2)})^2}, \tag{2.38}$$

where m_t are the matrix elements of the unit tensors $\boldsymbol{U}^{(t)}$. In Tables 2.1 and 2.2 calculated values for $F^{(1)}/F^{(2)} = 0.03$ are also given. The latter ratio agrees much better with the experimental data, especially for the transitions at $33\,\mathrm{cm}^{-1}$.

In Table 2.3 the relative multiplet–multiplet scattering strengths of Tm^{3+}, Er^{3+}, and Ho^{3+} are given, i.e. the sum of the experimental intensities for all

Table 2.2. Comparison of observed and predicted intensities (properly normalized to the xy, yx components of the $33\,\mathrm{cm}^{-1}$ transition) for some electronic Raman transitions in the $(4f^{11}, {}^4I_{\frac{15}{2}})$ configuration of $ErPO_4$; - : not observed, the energies of these levels have been calculated from a crystal field fit. From [2.32], [2.73]

Transition (cm^{-1})	Polarization	Observed intensity	Predicted intensity ($F^{(1)}/F^{(2)} = 0.25$)	Predicted intensity ($F^{(1)}/F^{(2)} = 0.03$)
33	xx, yy	-	0.6	0.6
	xy, yx	15.2	15.2	15.2
	xz, yz	3.0	46.6	5.0
	zx, zy	0.6	13.1	0.9
53	xx, yy	-	0.04	0.04
	zz	-	0.2	0.2
	xy, yx	0.9	14.6	0.2
	xz, yz	0.9	1.8	4.5
	zx, zy	6.1	42.9	9.4
105	xx, yy	-	2.0	2.0
	zz	1.5	7.8	7.8
	xy, yx	-	1.7	0.02
	xz, yz	-	0.6	0.5
	zx, zy	-	0.4	0.5
145	xx, yy	-	0.2	0.2
	xy, yx	1.8	8.4	8.4
	xz, yz	0.6	4.9	3.8
	zx, zy	0.9	2.5	3.5

Table 2.3. Comparison of the observed relative multiplet-to-multiplet Raman scattering intensities for $Tm^{3+}, Er^{3+}, Ho^{3+}$ with theoretical values. From [2.73]

Ion / Transition	Tm^{3+} ${}^3H_6 \rightarrow {}^3H_6$	Tm^{3+} ${}^3H_6 \rightarrow {}^3F_4$	Er^{3+} ${}^4I_{\frac{15}{2}} \rightarrow {}^4I_{\frac{15}{2}}$	Ho^{3+} ${}^5I_8 \rightarrow {}^5I_8$
calc. intens.	100	67	18	51
observ. intens.	100	12	39	4

the transitions observed between individual crystal field states for a specific multiplet-to-multiplet transition, in all polarizations, are given [2.73]. This total intensity is proportional to the reduced matrix elements of $\boldsymbol{U}^{(t)}$ [2.64]:

$$I_{\text{total}} = (F^{(1)})^2 \langle 4f^n\gamma' SL'J' \| U^{(1)} \| 4f^n \gamma SLJ \rangle^2 + (F^{(2)})^2 \langle 4f^n\gamma' SL'J' \| U^{(2)} \| 4f^n \gamma SLJ \rangle^2 \qquad (2.39)$$

Values $F^{(1)}/F^{(2)} = 0$ have been chosen for Tm^{3+} and Er^{3+} and $F^{(1)}/F^{(2)} = -0.22$ for Ho^{3+}, as indicated by the experiments. Also, the energy denominators and the matrix elements are assumed to be identical for all ions. To calibrate the experimental scattering intensities of the different ions relative to one another, all intensities have been normalized to an E_g vibration near $1030\,\mathrm{cm}^{-1}$, (ν_3 of the PO_4 group).

In conclusion, neither the intensities of electronic Raman transitions nor their asymmetries are well described by this theory. While the overall agreement for Tm^{3+} and Er^{3+} might be called fairly good, there are severe discrepancies in the case of Ho^{3+}. The application of the Judd–Ofelt theory to one-photon transitions in lanthanide compounds usually gives quite acceptable results with the exception of the hypersensitive transitions [2.29], whose strengths are particular sensitive to the host. An extension to two-photon (e.g. Raman) transitions appears to be problematic, at least under the usual assumption that the $4f^{n-1}5d$ configuration is mainly responsible for the transition amplitude. This assumption leads to $F^{(1)}/F^{(2)} = 0.25$. At first sight, one would expect a more stringent test of the theory in the case of two-photon transitions, since the electric dipole transitions to and from the intermediate states are parity allowed and second-order perturbation theory should apply. However, the available information concerning the unscreened $5d$- and $5g$ configurations is scarce, giving a weak basis for reliable calculations.

A much better fit to the experimental results for Tm^{3+} and Er^{3+} is obtained using a value of $F^{(1)}/F^{(2)}$ close to zero. This result can be reproduced theoretically if the contribution of the excited $4f^{n-1}5g$ configuration is taken into account. It can be shown [2.73] that the contributions to the scattering amplitude of the d and g configurations have opposite signs and can interfere destructively with each other. Similar conclusions have been established, e.g., in the case of two-photon absorption experiments [2.66], [2.72].

The problem with a sizeable contribution of the g configurations is that in the free ion the energies of these configurations are much too high and the matrix element with the f configuration too small (due to the vanishing overlap of the two configurations) for the required cancelation. The free ion energies and dipole matrix elements clearly indicate that the $4f^{n-1}5d$ configuration is the preponderant intermediate state for all two-photon transitions in trivalent lanthanide ions. Clearly, the free ion approach is insufficient to explain the two-photon transition intensities. The intermediate states in these transitions appear to be no longer localized near the lanthanide nucleus, but instead overlap quite significantly with the neighboring ligands, forming molecular orbitals with energies close to the ionization limit, which varies for each compound. This is a demonstration of the nephelauxetic (i.e., cloud expanding) effect, which describes this expansion of the wavefunctions as they overlap in the crystal with those of the free ion [2.42]. Obviously, from intensity studies of two-photon transitions in lanthanides, important information on the nature of the intermediate states can be derived.

A direct test of the correlation of the shape of the $4f^{n-1}5d$ wavefunction, and correspondingly the $4f^n \rightarrow 4f^{n-1}5d$ oscillator strength, with the intra–$4f^n$ electronic Raman scattering cross sections has been performed for Ce^{3+} in $LuPO_4$ [2.76]. Both the $4f^n \rightarrow 4f^{n-1}5d$ transition rates and the

Table 2.4. Observed and calculated electronic Raman differential scattering cross sections ($\frac{d\sigma}{d\Omega} \times 10^{30}$, [cm^2/steradian]) for Ce^{3+} in $LuPO_4$ for the two multiplet components of $4f^1$. From [2.76]

Transitions	Observed	Calculated Judd–Ofelt	Calculated 5d-wavefunctions	Calculated weighted 5d-wavefunctions.
$^2F_{\frac{5}{2}} \rightarrow {}^2F_{\frac{5}{2}}$	7.7	76.8	105	10.6
$^2F_{\frac{5}{2}} \rightarrow {}^2F_{\frac{7}{2}}$	7.1	9.0	35.5	7.2

Raman cross sections are considerably smaller than expected from calculations based on free-ion estimates of the radial wavefunctions. Ce^{3+} is a very adequate candidate for such investigations because of its simple $4f^1$ configuration, easily amenable to calculations, and the fact that the $4f^n \rightarrow 4f^{n-1}5d$ transitions are located between 30 000 and 52 000 cm^{-1}, where quantitative absorption experiments are feasible. In Table 2.9 observed and calculated Raman scattering cross sections have been compiled for this system. The transitions have been summed over the crystal field levels of each multiplet component and averaged over all polarizations. The calculated values have been obtained by applying first the Judd–Ofelt theory, i.e. giving an average value of 40 000 cm^{-1} to the $5d$ energies and making use of closure over the $4f^{n-1}5d$ configuration. The next improvement was an evaluation of the sum over intermediate states using the angular parts of the $4f^1$ and $5d^1$ wavefunctions obtained from crystal field fits, and scaling the result with $\langle 4f|r|5d\rangle$ and the appropriate tensor matrix element. Finally, a calculation has been made in which each term contributing to (α) in (2.28) is weighted by a factor which takes into account the ratio of the measured and the calculated oscillator strength for that particular $5d^1$ state. Accordingly, the reduction of the radial integral due to the nephelauxetic effect, i.e. the expansion of the $5d^1$ orbital, is taken into account. This last step finally produced a satisfying agreement with the experimental data.

Axe's theory [2.62] was found inadequate to interpret two-photon absorption intensities for the $4f^7$ systems Eu^{2+} and Gd^{3+} in CaF_2, SrF_2 or in $LaF_3, LaCl_3, La(OH)_3$, respectively, due to the exceedingly small size of the leading matrix element. This has been corrected in third (and fourth) order of perturbation theory by considering both spin–orbit and crystal–field intershell and intrashell interactions [2.72]. In a new study, correlation contributions were introduced into the calculations to third order [2.67]; they moderately improved the agreement with the experiments, especially for Ce^{3+} in $LuPO_4$.

Very recently the electronic Raman transition intensities of some lanthanide ions in elpasolite lattices ($Cs_2Na(RE)Cl_6$) have been studied [2.77], (see also Sect. 2.4.2). In these crystals the lanthanide ions $(RE)^{3+}$ are situated at pseudo-octahedral sites in the $(RE)Cl_6{}^{3-}$ anions, so that threefold vibrational and up to fourfold electronic degeneracies occur, providing relatively simple vibrational and electronic spectra. For Ce^{3+} the intensity calculations based on the Judd–Ofelt theory did not give satisfactory results. Comparing the observed and the theoretical relative intensities, it becomes again apparent that the magnitude of the parameter F_1 does not play a significant role in accounting for the observed intensities. For Pr^{3+} and Eu^{3+} in $Cs_2Na(RE)Cl_6$, the mean ratio $F_1/F_2 = 0.22$, close to the value of 0.25 quoted in [2.73], has been found (see above), again indicating that here the $4f^{n-1}5d^1$ configuration is the only one contributing. In some cases, strong electron–phonon coupling was found to have serious consequences on the electronic Raman intensities [2.77]. This is not surprising, since in cases of near degeneracies between electronic and phonon transitions of the same type of symmetry a hybridization of the transitions may occur (see Sect. 2.3.3).

The scattering intensities measured for $TmPO_4$ were compared with a direct second order calculation according to Equ. (2.28) in [2.78], considering the detailed energy level structure of the intermediate excited configuration $4f^{11}5d^1$ and the contributions to the scattering intensities from various perturbations within the $4f^{11}5d^1$ configuration (cross terms due to the $4f^{11}$ core spin–orbit interaction, the $5d^1$ crystal field and the $5d^1$ spin–orbit coupling). Inclusion of these terms improves the agreement between theory and experiment, but significant differences still remain.

Another recent test of the Judd–Ofelt theory in second order [2.62] involves the study of the polarization behavior of electronic Raman intensities in ground multiplet transitions of Pr^{3+} and Nd^{3+} in the RE-vanadates (zircon structure) [2.79], where the ratio F_1/F_2 is extracted from a large number of data points associated with the intensities of electronic Raman scattering at various polarization angles for light linearly polarized with respect to the crystal axis. An earlier method only took into account two data points for parallel and perpendicular polarization [2.80].

The data analysis is again based on the application of second quantization techniques and recoupling manipulations of tensorial operators [2.80]. The scattering amplitude is expressed in spherical coordinates as

$$(\alpha_{\mathrm{S,L}}) = \sum_{t=0}^{2} \sum_{q=-t}^{t} (\lambda_q^t)^* \alpha_q^{(t)}, \tag{2.40}$$

where the $\alpha_q^{(t)}$ are the spherical tensors of (2.99), (see Appendix) and the λ_q^t transform according to the $\alpha_q^{(t)} \Rightarrow \alpha_{ij}$ of (2.99), if the α_{ij} are replaced by $(x_i x_j)$, $x_i = (x, y, z) = (\sin\theta \cdot \cos\varphi, \sin\theta \cdot \sin\varphi, \cos\theta)$. Here the θ, φ are the polarization angles of the experiment[10]. Inserting the relation for the irreducible representations of the Raman tensor, (2.40) can be recast into

[10] $\theta_i, \varphi_i, \theta_j, \varphi_j$ describe the polarization state, where θ represents the polar angle of the polarization vector measured from the z axis, φ the azimuth angle measured from the x axis in the xy plane.

$$(\alpha_{\mathrm{S,L}})^{(2)} = \sum_{t=0}^{2} \sum_{q=-t}^{t} (\lambda_q^t)^* F^{(t)} U_q^{(t)}, \tag{2.41}$$

which is the most general expression for the polarization dependence of the electronic Raman amplitude in second order perturbation theory [2.80].

Some representative experimental results [2.79] are displayed in Fig. 2.5. The fitted values for $\tau = F_1/F_2$ were found to be 1.0 and 0.48, for $PrVO_4$ and $NdVO_4$, respectively. These values have been compared with theoretical ones derived using both second- and third-order perturbation theory, with the inclusion of spin–orbit interaction in the latter. The second-order theory, where only the contributions of the d configuration were taken into account, was found to be adequate, the third-order spin–orbit and other contributions are approximately one order of magnitude smaller than the second-order ones for both compounds. Obviously the τ values are at variance with the previous results for Er^{3+}:YPO_4 and Tm^{3+}:YPO_4 [2.32].

Clearly, the problem of intensities in electronic Raman scattering, especially in $4f^n$ systems, is not yet satisfactorily solved. Results for different compounds are too divergent to be considered as the final answer. The problem is also on the agenda for the n-type high-T_c superconductors of perovskite-like structure $(RE)_{2-x}Ce_xCuO_4$, (RE = Pr, Nd, Sm; $x \approx 0.15$) and their nonsuperconducting parent compounds ($x = 0$), where detailed Raman experiments have been performed [2.163], [2.165], (Sect. 2.4.1).

2.2.5 Intra-configurational Raman Resonances

Resonance Raman spectroscopy, where the energy of the incident and/or the scattered light coincides with an allowed transition from the ground state, is an important source of information in many fields of solid state physics and in biophysics. It requires, of course, that strong interband transitions lie in the range accessible to continuously tunable lasers. Such is the case for many semiconductors. In transition metal and lanthanide compounds this condition is fulfilled only in exceptional cases. In Ce^{3+} compounds the inter configurational transition $4f^1 \rightarrow 5d^1$ starts near 30 000 cm^{-1} (see above), with even higher threshold energies in the other lanthanides. It has been observed however that also intra-configurational ($4f^n \rightarrow 4f^n$) electric dipole transitions, which are parity forbidden in lower order, may lead to a resonance enhancement [2.81, 2.82, 2.83, 2.84].

This phenomenon was studied in detail in single crystals of $ErPO_4$ [2.81], where the transitions between the multiplet components ${}^4I_{\frac{15}{2}} \rightarrow {}^4F_{\frac{7}{2}}$ of the $4f^{11}$ configuration occur near 480 nm. Incident laser excitation frequencies were selected to be in near coincidence with transitions from the ground state to the lower two crystal–field levels of the ${}^4F_{\frac{7}{2}}$ multiplet component. The polarizations of the incident and scattered light were selected such that, through the D_{2d} crystal field electric dipole selection rules, only resonance

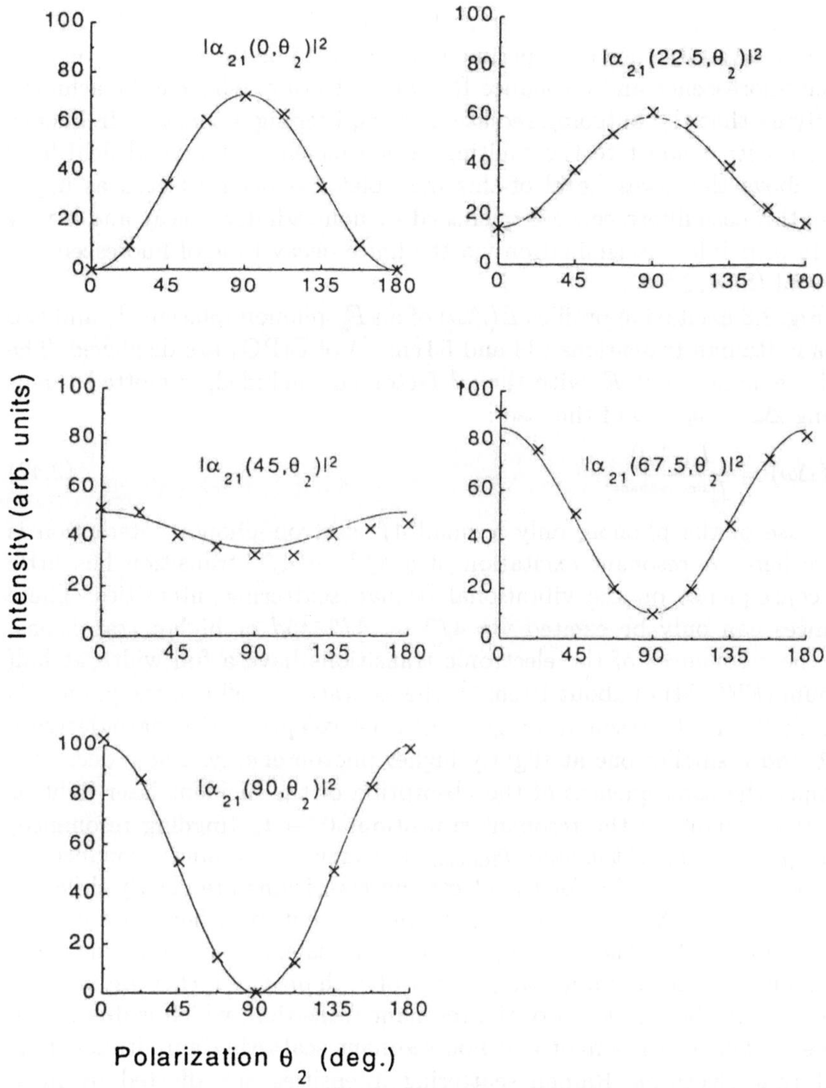

Fig. 2.5. Polarization dependence of the intensity of the electronic Raman transition at 84 cm^{-1} of Pr^{3+} in $PrVO_4$, $\lambda_L = 514.5$ nm at low-temperature. The notation α_{21} refers to the scattered and incident radiation, respectively. Θ_1, Θ_2 are the polar angles of the polarization vectors with respect to the z-axis in the crystal coordinate system. The azimuthal angles φ_1 and φ_2 are kept fixed with respect to the x-axis in the xy-plane: $\varphi_1 = -45^0, \varphi_2 = 45^0$. Θ_1 is incremented in successive plots by $22.5^0; 0^0 \leq \Theta_1 \leq 90^0$. Solid lines are fits using the theory sketched in the text. From [2.79]

involving the crystal field level at 20 492.7 $cm^{-1}(=\bar{\nu}_r)$, the higher of the two, was allowed. In such an experiment, care has to be taken to discriminate between fluorescence and resonance Raman scattering. This can be achieved by verifying that the outgoing radiation shifts, keeping a constant frequency difference with respect to the exciting laser frequency. A crystal field level clearly above the lowest level of this multiplet has been selected as upper level; in this case fluorescence is quenched by non-radiative decay and is very weak. In principle, discrimination via the finite decay time of fluorescence is also useful (Sect. 2.2.8).

In Fig. 2.6 excitation profiles $E(\Delta\omega)$ of an E_g-phonon (303 cm^{-1}) and two electronic Raman transitions (33 and 53 cm^{-1}) of $ErPO_4$ are displayed. The intensity enhancement E, with the ω^4 factor not included, is plotted versus detuning $\Delta\omega = \omega_r - \omega$ of the laser:

$$E(\Delta\omega) = \frac{I_s(\Delta\omega)}{I_s^{\text{non-resonant}}} \tag{2.42}$$

In the case of the phonon, only a small $4f$ electron–phonon interaction is effective here. A resonant excitation of a $4f^n \rightarrow 4f^n$ transition has little or no consequence on the vibrational Raman scattering intensities: These resonances can only be excited via $4f^n \rightarrow 4f^{n-1}5d$ or higher transitions. While the resonances of the electronic transitions have a full width at half maximum (FWHM) of about 10 cm^{-1}, the excitation profile of the phonon is essentially flat in the resonance region with the exception of a sharp decrease at $\approx \bar{\nu}_r$ and a smaller one at slightly higher photon energy. These decreases are simply the consequences of the absorption of the incident laser light or the scattered light by the resonant transitions $0 \rightarrow \bar{\nu}_r$ (ingoing resonance) and/or $\bar{\nu}_{\text{Raman}} \rightarrow \bar{\nu}_r$, if the state $\bar{\nu}_{\text{Raman}}$ is occupied (outgoing resonance).

The excitation profiles for the electronic transitions are clearly different from the phonon behavior: A resonance enhancement by either a factor 47 or 140 is observed [2.81], the double peak structure is again the consequence of resonant absorption. An interesting feature of both profiles is their asymmetry with respect to the frequency of the resonant transition, which is due to the interference between resonant and non-resonant scattering amplitudes. It is evident that electronic Raman scattering intensities are affected by intra configurational resonances for excitation frequencies as far as 50 cm^{-1} off resonance.

The modeling of the excitation profiles with standard theory is straightforward and starts from (2.28) [2.81]: $(I_{xy})_{fi} \propto |(\alpha_{xy})_{fi}|^2$. The sum over the intermediate states now comprises crystal field states from excited $4f^n$ multiplet components. Equation (2.28) is separated into two parts, one corresponding to resonant transition, the other representing the sum of non-resonant terms:

$$(I_{s,xy})_{fi} \propto \left| -\sum_{l \neq r} \left\{ \frac{(D_x)_{fl}\ (D_y)_{li}}{E_l - \hbar\omega_L} + \frac{(D_y)_{fl}\ (D_x)_{li}}{E_l + \hbar\omega_S} \right\} + \frac{(D_x)_{fr}\ (D_y)_{ri}}{E_r - \hbar\omega_L - i\Gamma_r} \right|^2 . \tag{2.43}$$

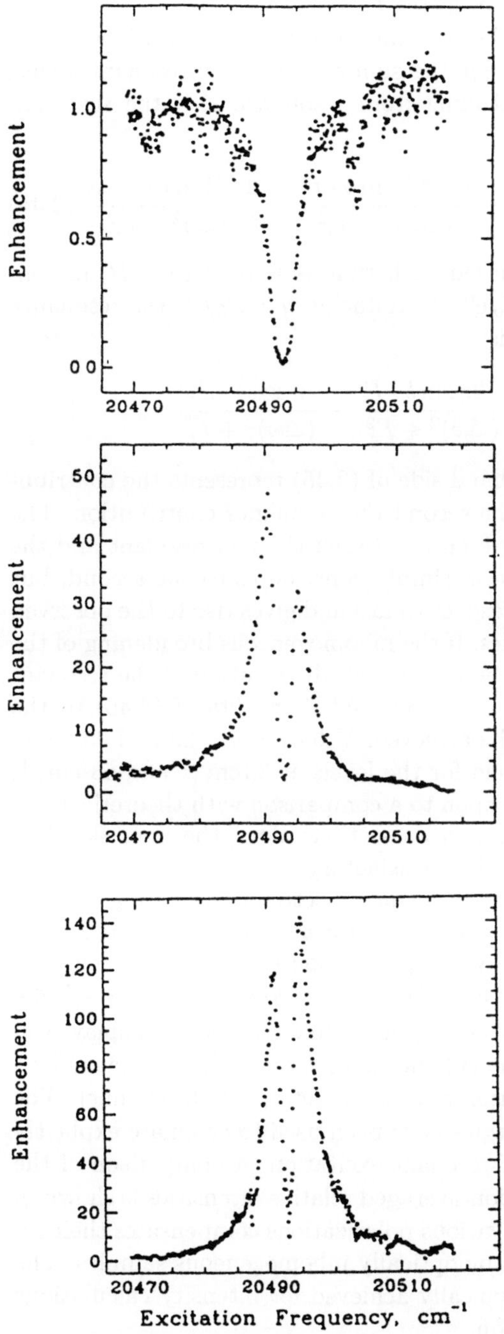

Fig. 2.6. Raman excitation profiles in $ErPO_4$, T $\approx$ 10 K, $y(xz)x$. *Top*: Excitation profile for the E_g-phonon at 303 cm^{-1}. *Center*: Excitation profile for the Δ = 33 cm^{-1} electronic transition. *Bottom*: Excitation profile for the Δ = 53 cm^{-1} electronic transition. Excitation profiles for phonons and for electronic transitions differ markedly. From [2.81]

The phenomenological damping parameter Γ_{r} has been added to account for the finite homogeneous width of the resonant crystal field excitations. Introducing the shorthand notation A for the sum over the non-resonant terms, $B = (\boldsymbol{D}_x)_{fr}\ (\boldsymbol{D}_y)_{ri}$ and the detuning from resonance $\Delta\omega$, the scattered intensity I_{s} is given by:

$$(I_{s,xy})_{fi} \propto A^2 + \frac{B^2}{(\Delta\omega)^2 + \Gamma_{\mathrm{r}}^2} + \frac{-2AB\sin\delta \cdot \Gamma_{\mathrm{r}}}{(\Delta\omega)^2 + \Gamma_{\mathrm{r}}^2} + \frac{2AB\cos\delta \cdot \Delta\omega}{(\Delta\omega)^2 + \Gamma_{\mathrm{r}}^2}. \quad (2.44)$$

Here the phase of A has been picked such that A is real; then B, in general, will remain complex, $B = B_0\,\mathrm{e}^{i\delta}$. Introducing $\eta = B_0/A$, the resonance enhancement $E(\Delta\omega) = I/A^2$ reads:

$$E(\Delta\omega) = 1 + \frac{\eta^2}{(\Delta\omega)^2 + \Gamma_{\mathrm{r}}^2} + \frac{-2\eta\sin\delta \cdot \Gamma_{\mathrm{r}}}{(\Delta\omega)^2 + \Gamma_{\mathrm{r}}^2} + \frac{2\eta\cos\delta \cdot \Delta\omega}{(\Delta\omega)^2 + \Gamma_{\mathrm{r}}^2}. \quad (2.45)$$

The first term ("1") on the right hand side of (2.45) represents the contribution of the non-resonant terms, the second the resonance contribution. The third and fourth result from interference between the non-resonant and the resonant amplitudes. The third term simply contributes to the second, but the fourth is antisymmetric with respect to $\Delta\omega$ and gives rise to the observed asymmetry of the excitation profiles. If the inhomogeneous broadening of the resonant transition is assumed to be a Gaussian distribution of the resonant energies in the crystal ($\Delta' \approx 2\,\mathrm{cm}^{-1}$), very satisfying fits of (2.45) to the experimental data of Fig. 2.6 can be achieved. Values of $\eta = 22$ and 35.5 and $\delta = \frac{\pi}{3}$ and $\frac{2\pi}{3}$ have been determined for the levels at $33\,\mathrm{cm}^{-1}$ and $53\ \mathrm{cm}^{-1}$, respectively. These values are also open to a comparison with theoretical calculations for the absolute scattering cross sections along the lines sketched above. The results obtained are rather satisfactory.

In $(Lu_{0.80}, Ce_{0.20})PO_4$ the near-resonance electronic scattering via the parity-allowed electric dipole $4f^n \longrightarrow 4f^{n-1}5d$ transitions in Ce^{3+} compounds has been studied in detail [2.31] by comparing spectra of crystal field levels of ${}^2F_{\frac{5}{2}} \longrightarrow {}^2F_{\frac{5}{2}}$ and ${}^2F_{\frac{5}{2}} \longrightarrow {}^2F_{\frac{7}{2}}$ taken at two wavelengths, $\lambda_1 = 514.5$ nm, $\lambda_2 = 355$ nm (tripled Nd^{3+}:YAG output). Some results are displayed in Fig. 2.7. For the observed lines, the enhancement ratios indicate that indeed the $5d^1$ configuration plays a major role as an intermediate channel. The known $5d^1$ wavefunctions and energies have been used to calculate explicitly the expected intensities from near-resonant excitation. A comparison of the measured and calculated polarization-averaged relative intensities is shown in Table 2.5. This averaging over the various polarizations compensates their redistribution due to leakage caused by optically inhomogeneous samples. The agreement lies within the range typically achieved for intensity calculations using the Judd–Ofelt approximation.

The time dependence of resonant electronic scattering will be discussed in Sect. 2.8.

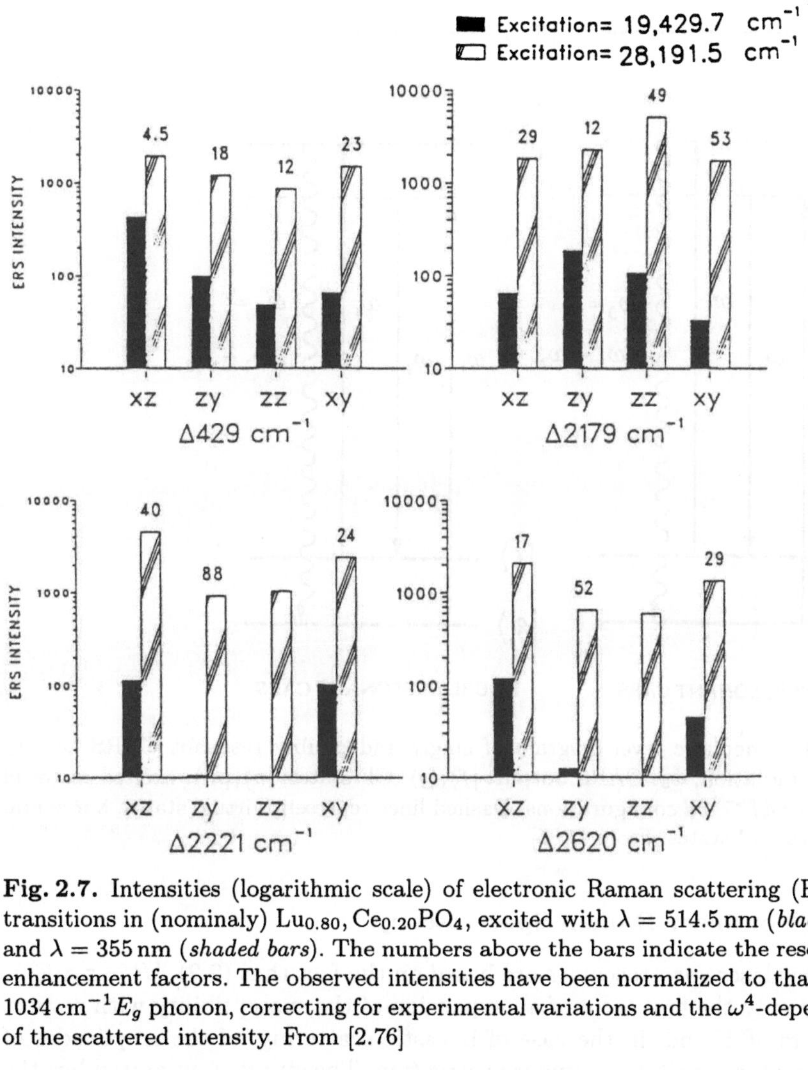

Fig. 2.7. Intensities (logarithmic scale) of electronic Raman scattering (ERS) of transitions in (nominaly) $Lu_{0.80}, Ce_{0.20}PO_4$, excited with $\lambda = 514.5$ nm (*black bars*) and $\lambda = 355$ nm (*shaded bars*). The numbers above the bars indicate the resonance-enhancement factors. The observed intensities have been normalized to that of the $1034\,cm^{-1}$ E_g phonon, correcting for experimental variations and the ω^4-dependence of the scattered intensity. From [2.76]

Table 2.5. Observed and calculated polarization-averaged electronic Raman scattering intensities in $(Lu_{0.80}, Ce_{0.20})PO_4$ (nominal concentration) for near-resonant excitation ($\lambda_{exc.} = 355$ nm). From [2.84]

Transitions (cm^{-1})	Observed Intensities	Calculated Intensities
240	0	1543
429	2738	8008
2179	5591	4514
2221	4462	2913
2620	2315	2447
2676	0	193

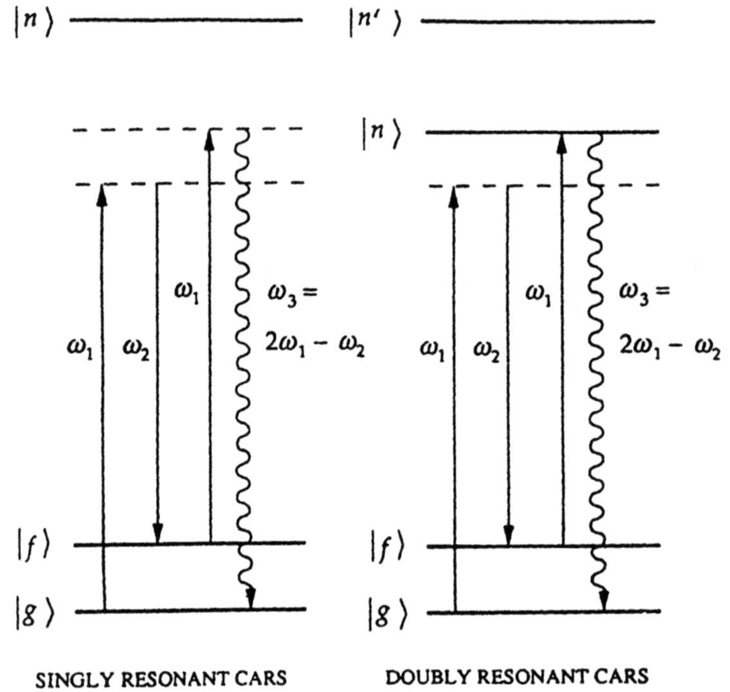

Fig. 2.8. Schematic level diagram of singly and doubly resonant CARS. ω_1, ω_2: incident radiation, ω_3: CARS output. $|f\rangle, |g\rangle$: CF states; $|n\rangle, |n'\rangle$: excited states of the $4f^n$ or $4f^{n-1}5d$ configurations. Dashed lines represent virtual states, horizontal solid lines real states. From [2.85]

2.2.6 Nonlinear Raman Spectroscopy

Nonlinear laser spectroscopies are based on the fact that (2.2), $[P = \varepsilon_0 \chi E]$ is actually only the first term of an expansion of the susceptibility with respect to powers of E and, in the case of inelastic scattering, of the amplitudes of the elementary excitations under observation. The study of, in particular, the third order susceptibilities $\chi^{(3)}$, where in the frequency domain

$$P^{(3)}(k_4, \omega_4) = \chi^{(3)}(k_4, \omega_4) \vdots E_1(k_1, \omega_1) E_2(k_2, \omega_2) E_3(k_3, \omega_3), \tag{2.46}$$

$$k_4 = k_1 + k_2 + k_3, \quad \omega_4 = \omega_1 + \omega_2 + \omega_3, \tag{2.47}$$

allows a wealth of information to be extracted from the experiment.

An often successfully applied spectroscopic technique is the three-wave mixing or Coherent Anti-Stokes Raman Spectroscopy (CARS) (CARS). In this case two strong incident laser beams at ω_1 and ω_2 interact with the material system to produce coherent output at $\omega_p = 2\omega_1 - \omega_2$ in the direction of phase matching $k_p = 2k_1 - k_2$ [2.22]. In the transparent region, $\chi^{(3)}$ is generally a slowly varying function, except when an intermediate frequency is close to a two-photon resonance of the material. Such resonances can occur in practice once or twice in a CARS process. In Fig. 2.8 schematic level diagrams for singly and doubly resonant CARS are shown. It is evident that no energy is transferred to the material from the light field in a CARS process, hence the susceptibilities are real and the observed resonances are dispersive.

The CARS signal is proportional to the square of $\chi^{(3)}$. It contains the sum of a nonresonant part involving virtual transitions and a resonant (R), frequency dependent part carrying the spectroscopic information: $\chi^{(3)} = \chi^{(3),\mathrm{NR}} + \chi^{(3),\mathrm{R}}$. In most cases $|\chi^{(3),\mathrm{R}}| < |\chi^{(3),\mathrm{NR}}|$.

The resonant third-order susceptibility which describes a singly resonant CARS experiment is [2.22, 2.74, 2.85, 2.86, 2.87]:

$$\chi^{(3),1\mathrm{R}}_{ijkl}(-\omega_3;\omega_1,\omega_1,-\omega_2) = \chi^{(3),NR}_{ijkl} + \frac{ML}{24}\sum_f \tag{2.48}$$

$$\times \frac{\tilde{\alpha}_{ij,f}(\omega_3,-\omega_1)\alpha_{kl,f}(\omega_1,-\omega_2) + \tilde{\alpha}_{ik,f}(\omega_3,-\omega_1)\alpha_{jl,f}(\omega_1,-\omega_2)}{\hbar(\Omega_{fg} - (\omega_1-\omega_2))},$$

$$\tilde{\alpha}_{ij,f}(\omega_3,-\omega_1) = -e^2\sum_n \frac{\mu^{(i)}_{gn}\mu^{(j)}_{nf}}{\hbar(\Omega_{ng}-\omega_3)} + \frac{\mu^{(j)}_{gn}\mu^{(i)}_{nf}}{\hbar(\Omega^*_{ng}+\omega_1)}, \tag{2.49}$$

$$\alpha_{ij,f}(\omega_1,\omega_2) = -e^2\sum_n \frac{\mu^{(i)}_{fn}\mu^{(j)}_{ng}}{\hbar(\Omega_{ng}+\omega_2)} + \frac{\mu^{(j)}_{fn}\mu^{(i)}_{ng}}{\hbar(\Omega_{ng}-\omega_1)}. \tag{2.50}$$

Here i,j,k,l refer to the polarization directions of the ingoing and the three outgoing beams $\omega_{1,2,3}$, respectively. $\alpha_{ij}, \tilde{\alpha}_{ij}$ are polarizability tensor elements, where $\mu^{(j)}_{gn}$ represents the jth component of the transition dipole matrix element $\langle g|r_j|\mathrm{n}\rangle$. $\Omega_{ng} = \omega_{ng} - \mathrm{i}\Gamma_{ng}$ where ω_{ng} and $\mathrm{i}\Gamma_{ng}$ are the associated resonance frequencies and widths of the level n, respectively, M is the density of resonant scattering centers, L is a suitable local-field correction factor. As in Sect. 2.2.5, $\Delta_f = (\omega_{fg} - (\omega_1-\omega_2))/\Gamma_{fg}$, $C = \chi^{\mathrm{R}}_f(\Delta_f = 0)/\chi^{\mathrm{NR}}$, the anti-Stokes (scattered) intensity I_S normalized to the off-resonance intensity is written:

$$I_\mathrm{S}/I^\mathrm{NR}_\mathrm{S} = \left|\frac{\chi^{(3),\mathrm{NR}} + \chi^{(3),\mathrm{R}}}{\chi^{(3),\mathrm{NR}}}\right|^2 = 1 + 2C\frac{\Delta}{\Delta^2+1} + C^2\frac{1}{\Delta^2+1}. \tag{2.51}$$

Usually $C < 1$ (often $\ll 1$) and the detectability of the anti-Stokes signal depends critically on the strength of the non-resonant background, which can be suppressed in special cases by a suitable choice of polarizations.

The signal can be enhanced considerably, when a second two-photon transition can be tuned into resonance with intermediate electronic states $|\mathrm{n}\rangle$ (Fig. 2.8b). In this case of a doubly resonant CARS experiment the third order susceptibility $\chi^{(3),2\mathrm{R}}_{ijkl}(-\omega_3;\omega_1,\omega_1,-\omega_2)$ is formally identical to (2.48) [2.74], [2.85], however in $\tilde{\alpha}_{ij,f}(\omega_3,-\omega_1)$ (2.49) only the resonant terms, with the denominator $(\Omega_{ng} - \omega_3) - \mathrm{i}\Gamma_{ng}$, are considered, while the non-resonant term $\alpha_{ij,f}$ (ω_1, ω_2) is unchanged.

The selection rules, which have to be obeyed by the fourth rank tensors $\chi^{(3)}_{ijkl}$ in four-wave mixing spectroscopy follow from (2.49) and (2.50). It is evident that each of the polarizabilities $\tilde{\alpha}_{ij,f}(\omega_3, -\omega_1)$ and $\alpha_{ij,f}(\omega_1, \omega_2)$ must be Raman active. Hence only those tensor components $\chi^{(3)}_{ijkl}$ will occur in CARS, for which the $ijkl$ are products of Raman active second order components (see Table 2.A.1) and the product itself transforms again as a product of two polar vector components [see (2.46)]. These products have been compiled in the literature, [2.22], [2.88]. It should be noted that the $\chi^{(3)}_{ijkl}$ reflect the symmetry of the unit cell because of the coherence in the scattering phenomenon. There are other symmetry relations: Kleinman's conjecture [2.22], on the basis of the postulate of the existence of a scalar energy density, suggests that $\chi^{(3)}_{ijkl}$ remains unchanged when the Cartesian indices are permuted. This relation is better obeyed if the frequencies involved are far from resonances.

In Fig. 2.9 singly and doubly resonant CARS spectra have been plotted for several RE compounds. The spectra of PrF_3 (left) were excited by a frequency doubled Nd^{3+}:YAG laser and a tunable dye laser ($720\,\mathrm{nm} \leq \lambda \leq 750\,\mathrm{nm}$) [2.74]. A tripled Nd^{3+}:YAG laser and a dye laser tuned near 385 nm were used in the case of Ce^{3+}:$LuPO_4$ (right), thus achieving the (second) resonance with the lowest $5d^1$ band. The increase of a resonance enhancement by almost one order of magnitude is evident in the latter, doubly resonant case. The resonance effects can be further enlarged by reducing the linewidths Γ_{ij} in (2.48), which determine the size of the energy denominators under exact resonance. Singly resonant CARS has been observed in pure CeF_3 [2.89]. Resonances involving the ground state of the $^2F_{\frac{5}{2}}$ multiplet component and all crystal field levels of the $^2F_{\frac{7}{2}}$ multiplet component have been detected at $T = 3.8\,\mathrm{K}$. The resonance enhancement of the third-order susceptibility due to the electronic transitions was found to be a factor of 4.8. The Ce^{3+} electronic Raman cross section for $^2F_{\frac{7}{2}}$ was derived accurately by modelling the virtual intermediate states as a single degenerate state with a center of gravity $45\,000\,\mathrm{cm}^{-1}$ above the ground state.

CARS intensities have been calculated more generally following the lines given in Sect. 2.2.4 [2.85]. Based on the same approximations, the agreement obtained is similar to that found for other intensity calculations of electronic Raman transitions.

Other degenerate four-wave mixing experiments have been performed in Nd^{3+}-doped crystals and glasses. Again the $4f \rightarrow 5d$ transitions are responsible for the effect, the value of the $\langle 4f|r|5d\rangle$ radial integral being essential for the size of the calculated result [2.90].

Some other techniques of multiresonant four-wave mixing induced by two-photon absorption have been applied by *Cone* et al. to $Tb(OH)_3$, $LiTbF_4$, and TbF_3 crystals [2.91, 2.92, 2.93]. In the same context, the multiwave mixing (stimulated-photon-echo) experiments in Pr^{3+}:LaF_3 by *Moshary* et al. [2.94]

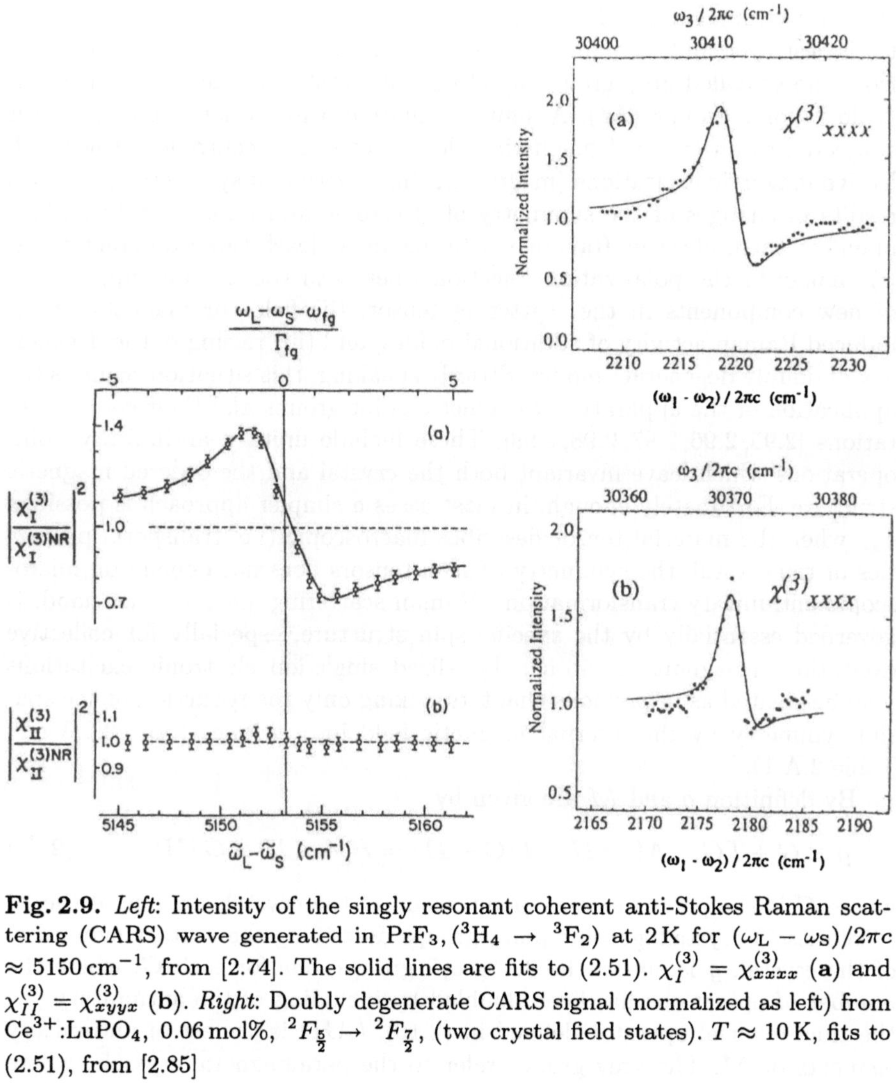

Fig. 2.9. *Left*: Intensity of the singly resonant coherent anti-Stokes Raman scattering (CARS) wave generated in PrF_3, $(^3H_4 \rightarrow {}^3F_2)$ at 2 K for $(\omega_L - \omega_S)/2\pi c \approx 5150\,\mathrm{cm}^{-1}$, from [2.74]. The solid lines are fits to (2.51). $\chi_I^{(3)} = \chi_{xxxx}^{(3)}$ (**a**) and $\chi_{II}^{(3)} = \chi_{xyyx}^{(3)}$ (**b**). *Right*: Doubly degenerate CARS signal (normalized as left) from Ce^{3+}:$LuPO_4$, 0.06 mol%, $^2F_{\frac{5}{2}} \rightarrow {}^2F_{\frac{7}{2}}$, (two crystal field states). $T \approx 10\,\mathrm{K}$, fits to (2.51), from [2.85]

and again the fundamental two-photon absorption work of *Downer* et al. [2.65], [2.66], [2.72] should be mentioned.

2.2.7 Raman Scattering in Magnetically Ordered Crystals

Light scattering in magnetically ordered materials deserves some special attention for several reasons. The symmetry properties of a magnetically ordered crystal, where the time averaged magnetic moment density is nonzero, are basically different from those of a nonmagnetic solid. As a crystal is cooled below the transition temperature and magnetic order sets in [ferromagnetic (fe), antiferromagnetic (af), ferrimagnetic (fi), etc.], the time reversal symmetry is destroyed and the point group of the paramagnetic (pm) crystal changes from the so-called grey group ($\boldsymbol{g}$) of the pm crystal to one of the magnetic "color" point groups ($\boldsymbol{M}$). A similar situation arises when a pm crystal is subjected to an external magnetic field. Besides the occurrence of new collective magnetic excitations (magnons), this lowering of symmetry ($\boldsymbol{g} \rightarrow \boldsymbol{M}$) results in changes of the symmetry of electronic and phonon states, which manifest themselves in Raman scattering in at least three different ways: (i) change in the polarization selection rules with the possible appearance of new components in the scattering tensor, (ii) field- or magnetic-order-induced Raman activity of additional modes, and (iii) raising of the degeneracy of doubly degenerate modes. Strictly speaking, this situation requires the application of the apparatus of magnetic point groups and their corepresentations [2.95, 2.96, 2.97, 2.98, 2.99]. These include unitary–antiunitary point operations which leave invariant both the crystal and the ordered magnetic structure. Fortunately enough, in most cases a simpler approach is possible, e.g. when the material tensor describes macroscopic (i.e. transport) properties of the crystal, the symmetry of such tensors does not depend on microscopic antiunitary transformations. Raman scattering, on the other hand, is governed essentially by the specific spin structure, especially for collective excitations (magnons, excitons). Localized single-ion electronic excitations may be treated as in previous chapters, taking only the reduction of the spatial symmetry by the internal magnetic field into account (Sect. 2.2.3 and Table 2.A.1).

By definition $\boldsymbol{g}$ and $\boldsymbol{M}$ are given by

$$\boldsymbol{g} \equiv \boldsymbol{G} + T\boldsymbol{G}, \quad \boldsymbol{M} \equiv \boldsymbol{H} + T(\boldsymbol{G} - \boldsymbol{H}) = \boldsymbol{H} + A\boldsymbol{H} \equiv \boldsymbol{G}[\boldsymbol{H}], \tag{2.52}$$

where $\boldsymbol{G}$ is the parent unitary point group of the crystal, i.e. one of the 32 ("colorless") crystallographic point groups describing the unitary symmetry of the crystal. $\boldsymbol{g}$ is one of the 32 "grey" groups, one for each $\boldsymbol{G}$, where the number of operations has been doubled by including of the antiunitary T, the time reversal operator defined in (2.16). ($A\boldsymbol{H}$) is the set of antiunitary elements of $\boldsymbol{M}$. The grey groups refer to the paramagnetic crystals, where the time average of the magnetic moment is zero. $\boldsymbol{H}$ is one of the invariant halving unitary subgroups of $\boldsymbol{G}$ and of the 58 "black and white" groups $\boldsymbol{M}$, i.e. $\boldsymbol{H}$ is one of the subgroups of $\boldsymbol{G}$ with half the number of group elements. For a given $\boldsymbol{G}$, the pertinent point groups $\boldsymbol{H}$ (generally more than one) have been tabulated [2.50, 2.95, 2.97]. They have been included for convenience in the first column of Table 2.A.1. In $\boldsymbol{M}$, the operator T only occurs in combination with rotation and rotation–reflection operations. The $\boldsymbol{M}$ groups (signature: $\boldsymbol{G}[\boldsymbol{H}]$) are the domain of fm, af, and fi structures. Altogether, 132 magnetic point groups can be constructed (32 $\boldsymbol{G}$, 32 $\boldsymbol{g}$, 58 $\boldsymbol{M}$). In combination with translational operations the 1651 Shubnikov magnetic space groups are obtained.

For the analysis of magnon excitations or excitons by Raman or neutron scattering, especially of the degeneracies at the boundaries of the Brillouin zone, the use of magnetic groups is required, i.e. the usual selection rules (Sect. 2.2.3) apply, if the coreps of $\boldsymbol{g}$ or $\boldsymbol{M}$ enter the relations (2.23). The reps of $\boldsymbol{M}$ and of $\boldsymbol{g}$ are called irreducible corepresentations (coreps), they are based on the reps of $\boldsymbol{H}$, not of $\boldsymbol{G}$. Accordingly, the number of coreps in a given $\boldsymbol{M}$ is determined by the number of classes of group elements in $\boldsymbol{H}$. If $\Gamma^{(i)}$ is a given rep of $\boldsymbol{H}$, the corep of $\boldsymbol{M}$ derived from $\Gamma^{(i)}$ is labeled $\mathrm{D}\Gamma^{(i)}$. Depending on $\Gamma^{(i)}$, the corep $\mathrm{D}\Gamma^{(i)}$ can be one of three types:

(a) $\Gamma^{(i)} \rightarrow \mathrm{D}\Gamma^{(i)}$ with the same dimensionality,
(b) $\Gamma^{(i)} \rightarrow \mathrm{D}\Gamma^{(i)}$ with twice the dimensionality,
(c) $\Gamma^{(i)}$ and $\Gamma^{(j)}$, which are non-equivalent reps of $\boldsymbol{H}$, stick together to produce a degenerate corep $\mathrm{D}\Gamma^{(k)}$ of $\boldsymbol{M}$.

Besides the trivial case of groups $\boldsymbol{G}$, where only the symmetry reduction due to $\boldsymbol{B}$ as tabulated in Table 2.A.1 has to be considered, it has been shown for grey groups $\boldsymbol{g}$ (whose coreps are tabulated in [2.50] and [2.97]) that the Raman tensor $\alpha_{\rho\sigma}$ persists in the usual form known for the groups $\boldsymbol{G}$ [2.95], as is intuitively obvious. If in the case of black and white groups, A (2.52) can be chosen to be $T \cdot I$, where I is the space inversion, the Raman tensors will behave exactly as in the corresponding grey group, where $A \equiv T$ [2.95]. When $TI \notin T\,\boldsymbol{H}$, it is still possible to write $A = T\boldsymbol{S}$, where $\boldsymbol{S}$ is some point group operation $\notin \boldsymbol{H}$ and the Raman tensors can be obtained by a suitable transformation, depending on the type (a–c) of the corep and using the real orthogonal matrix corresponding to $\boldsymbol{S}$. Details can be found in [2.95]. In most groups $\boldsymbol{M}$, the trivial case (a) applies, and selection rules can be derived using the reps of $\boldsymbol{H}$. In 11 groups $\boldsymbol{M}$ coreps of type (b) or (c) occur. In that case Refs. [2.95] and [2.97] should be consulted.[11] The raising of degeneracies in a magnetic crystal can be determined via the compatibility relations between $\boldsymbol{G}$ and $\boldsymbol{H}$ [2.97].

As an example, we discuss the correlation between the reps of D_{4h} and the coreps of the magnetic group $D_{4h}[C_{4h}]$ which applies to $TmVO_4$ with a Jahn–Teller phase transition at $T_D = 2.14\,\mathrm{K}$. This phase transition can be suppressed by an external magnetic field $\boldsymbol{B} \geq 0.54\,\mathrm{T} \| C_4$ (see Sect. 2.2.2) which produces a fe saturation:

D_{4h}	A_{1g}, A_{2g}	B_{1g}, B_{2g}	A_{1u}, A_{2u}	B_{1u}, B_{2u}	E_g	E_u
$D_{4h}[C_{4h}]$	$D(A_g)$	$D(B_g)$	$D(A_u)$	$D(B_u)$	$D(E_g), D(E_g^*)$	$D(E_u), D(E_u^*)$

Obviously, the symmetry reduction ($\boldsymbol{g} \rightarrow \boldsymbol{M}$) is identical to the reduction expected by the application of $\boldsymbol{B} \parallel C_4$, (see Table 2.A.1). Another less trivial example is $\boldsymbol{G}[\boldsymbol{H}] = D_{6h}[D_{3d}]$ where $\boldsymbol{H}$ is not the field-induced subgroup C_{6h} (Table 2.A.1):

[11] These are:
$C_4[C_2]$, $S_4[C_2]$, $C_{4h}[C_{2h}]$, $D_4[D_2]$, $C_{4v}[C_{2v}]$, $D_{2d}[D_2]$, $D_{2d}[C_{2v}]$, $D_{4h}[D_{2h}]$, $C_{3h}[C_3]$, $C_6[C_3]$, $C_{6h}[C_{3i}]$.

D_{6h}	A_{1g}, A_{2g}	B_{1g}, B_{2g}	E_{1g}, E_{2g}	A_{1u}, A_{2u}	B_{1u}, B_{2u}	E_{1u}, E_{2u}
$D_{6h}[D_{3d}]$	$D(A_{1g}), D(A_{2g})$	$D(A_{1g}), D(A_{2g})$	$D(E_g)$	$D(A_{1u}), D(A_{2u})$	$D(A_{1u}), D(A_{2u})$	$D(E_u)$

The silent (in D_{6h}) transitions B_{1g}, B_{2g} may become Raman active, with antisymmetric scattering in B_{2g} (and in E_{2g}), the silent B_{2u} and E_{2u} transitions may become infrared active in the ordered state.

A more phenomenological method to describe magnetic structures and their static properties has been introduced by Bertaut (see e.g. [2.100]). It is widely used in neutron diffraction but less adequate for treating dynamical properties.

Beside the enormous amount of Raman work on magnons performed in the last 20 years [2.20], which will not be covered in this review, there exist several remarkable results of electronic scattering from localized excitations in magnetically ordered systems:

$FeCl_2 \cdot 2H_2O$, (FC2) and $FeCl_2 \cdot 2D_2O$, (FC2D) are *metamagnets*. Below the Néel temperature of FC2, $T_N = 21.5$ K, (FC2D: $T_N = 23$ K) in increasing magnetic fields, a uniaxial af structure is followed by a uniaxial fi phase. The transition into the fe phase occurs if a magnetic field $H_\alpha^{c1} = 3.9$ T is applied along the easy direction α and the temperature is below 11.2 K. The fi phase has one third of the fe magnetization at saturation. H_α is inclined 32.9° with respect to the c axis of the crystal lying in the a–c plane of the monoclinic crystal with the space group $C2/m$ with a base-centered unit cell [2.101]. In the fi phase the Fe ions form parallel linear fe chains, which are af coupled to each other (intrachain nearest-neighbor exchange constant: $J_0 \geq 1\,\text{cm}^{-1}$, interchain exchange: $J_1 \approx -0.3\,\text{cm}^{-1}$). At $H_\alpha^{c2} = 4.6$ T the saturated pm or field induced fe phase is reached. The transitions at H^{c1} and H^{c2} are of first order with a large thermal hysteresis due to crystal field anisotropy below the tricritical points at $T_1^* = 11.5$ K and $T_2^* = 8.97$ K, respectively, on the phase boundaries. Both in the af and in the fi phase the magnetic point group is $C_{2h}[C_i]$. As stated above, in this case the coreps of the grey groups apply and the Raman tensors are equivalent to those of the pm phase. Selection rules are obtained by considering the A_g symmetry of the excited state [2.102]. From the A_g rep of the unitary halving subgroup C_i the two reps A_g, B_g of M (C_{2h}) are induced for the exciton levels in the af phase.

In Fig. 2.10 the influence of the magnetic order on the lowest electronic Raman transition of the Fe^{2+} ion near $620\,\text{cm}^{-1}$ at 2 K is shown in FC2D. The ground state of the free ion is ^5D. The fivefold orbital degeneracy is completely raised in the approximately rhombic crystal field of intermediate strength. The electronic nature of these transitions was confirmed by their temperature dependence, their absence in the Raman spectra of the two isomorphous compounds $CoCl_2 \cdot 2H_2O$ and $MnCl_2 \cdot 2H_2O$, and their small frequency shift (3%) upon deuteration [2.103]. At 2 K and zero field (af phase), this transition splits into a lower (A_g) and a higher frequency (B_g)component, as predicted by the selection rules. This splitting increases slightly and nonlinearly with increasing field. At the two metamagnetic phase transitions the

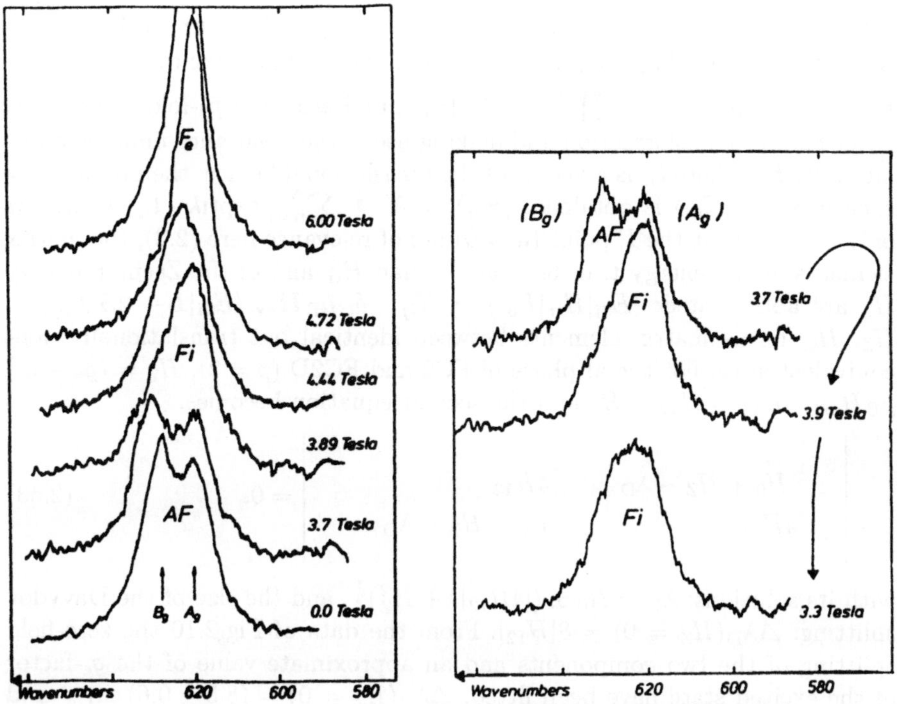

Fig. 2.10. *Left*: Electronic Raman spectrum of $FeCl_2 \cdot 2D_2O$, ^5D of Fe^{2+}, at 2 K. The spectra demonstrate the influence of the metamagnetic phase transitions AF $\rightarrow$ F_i $\rightarrow$ F_e. On the *right* the hysteresis of the F_i phase is demonstrated, from [2.103]

spectra change markedly: Interpreting the spectra in the fi phase as consisting of three overlapping unresolved lines and the spectrum in the fm phase as showing only one line, the behavior of these electronic transitions in the af phase can be attributed to a magnetic Davidov splitting of the corresponding excitations [2.104].

The occurrence of Davidov splitting of electronic levels in magnetically ordered crystals is direct evidence of the excitonic nature of the electronic excitations. The exciton of Frenkel type is transfered between ions on identical sites by an interionic interaction, which detunes the excitations analogous to coupled oscillators in resonance. The Davidov splitting of orbitally non-degenerate levels is possible if the magnetic unit cell contains more than one translationally non-equivalent magnetic ion, i.e., if the rep of the excited electronic state in site symmetry induces more than one rep of the factor group in the magnetic space group. In a two-sublattice antiferromagnet Davydov splitting is due to the resonant energy transfer between ions in different but equivalent sublattices.

For a quantitative analysis of the Davydov splitting [2.105] in FC2D the ground state $|G\rangle$ of a crystal with N unit cells and p magnetic ions in the cell is considered as a product of single-ion states $|\gamma_{ni}\rangle$, (with $1 \leq n \leq N, 1 \leq i \leq p$), so that $|G\rangle = \prod_{n=1,i=1}^{N,p} |\gamma_{ni}\rangle$. The excited states $E_{n'i'}$ (excitation ϵ on ion $n'i'$) are $|E_{n'i'}\rangle = |\epsilon_{n'i'}\rangle \prod_{(n=1,i=1)}^{(n',i' \notin N,p)} |\gamma_{ni}\rangle$, and are $(N \times p)$-fold degenerate. After Fourier transformation and making use of the invariance under a translation by $\boldsymbol{t}_n$, where $\boldsymbol{t}_n$ is a vector of the translational lattice, the propagating exciton state $|E_{ki}\rangle$ is obtained: $|E_{ki}\rangle = N^{-\frac{1}{2}} \sum_{n=1}^{N} \exp(\mathrm{i}\boldsymbol{k} \cdot \boldsymbol{t}_n)\ |E_{ni}\rangle$. As only the states at the Γ point ($\boldsymbol{k} = 0$) are of relevance here (2.8), the matrix elements of the energy transfer Hamiltonian H_{tr} and of the Zeeman energy H_{Z} are abbreviated: $\langle E_{0i}|H_{\mathrm{tr}}|E_{0j}\rangle = H_{ij}$, $\delta_{ij}\mu_{\mathrm{B}}\boldsymbol{H}_{\mathrm{ext}}\langle E_{0i}|\boldsymbol{L} + 2\boldsymbol{S}|E_{0j}\rangle = H_{\mathrm{Z}}$. H_{tr} forms matrix elements between identical but translationally non-equivalent sites. For the af phase of FC2 and FC2D ($p = 2$), $H_{\mathrm{Z}} = (g_\gamma - g_\epsilon)\ \mu_{\mathrm{B}}H_{\mathrm{ext}}$, $H_{11} = H_{22} = H_0$ and the secular equation becomes:

$$\begin{vmatrix} H_0 + H_{\mathrm{Z}} - \lambda_{\mathrm{D}} & 4H_{12} \\ 4H_{21} & H_0 - H_{\mathrm{Z}} - \lambda_{\mathrm{D}} \end{vmatrix} = 0, \tag{2.53}$$

with its solutions: $\lambda_{\mathrm{D}} = H_0 \pm (|4H_{12}|^2 + H_{\mathrm{Z}}^2)^{\frac{1}{2}}$, and the size of the Davydov splitting: $\Delta\lambda_{\mathrm{D}}(H_{\mathrm{Z}} = 0) = 8|H_{12}|$. From the data of Fig.2.10 the zero-field splitting of the two components and an approximate value of the g_ϵ-factor of the excited state have been fitted: $\Delta\lambda_{\mathrm{D}}(H_{\mathrm{Z}} = 0) = (8.5 \pm 0.6)\,\mathrm{cm}^{-1}$ and $g_\epsilon \approx 2.1$.[12] The ground state g_γ factor ($g_\gamma = 2.34$) has been determined from magnetization measurements or far-infrared spectroscopy [2.104]. From this the matrix element of the intersublattice energy transfer $|H_{12}^{\mathrm{exp}}| = (1.1 \pm 0.1)\,\mathrm{cm}^{-1}$ is derived for the af phase.

It appears reasonable to assume that the intersublattice energy transfer is due to exchange interaction between ions in the ground state (spin $\boldsymbol{S}_\gamma$) and the excited state ($\boldsymbol{S}_\epsilon$) according to $H_{\mathrm{tr}} = -\sum_i J'\boldsymbol{S}_\epsilon \boldsymbol{S}_\gamma{}^i$ [2.106], ($|\boldsymbol{S}_\epsilon| = |\boldsymbol{S}_\gamma| = 2$, J': excited state exchange integral). In this case $J' = -(0.28 \pm 0.025)\,\mathrm{cm}^{-1}$ is obtained, which matches the value of $J_1 (\approx -0.3\,\mathrm{cm}^{-1})$. J' is negative because the B_g-component of the Davydov doublet has the higher energy. The same relation ($J' \approx J$) has been found in the case of $RbMnF_4$ [2.106]. It is remarkable that both the af magnetic structure and the exciton propagation are governed by the same interionic interaction.

In the fi phase, where the number of ions in the magnetic unit cell is tripled with respect to the fm phase, resonant energy transfer is only possible between ions too far separated for any exchange interaction to become relevant. Thus the condition mentioned in footnote [2.12] is not valid here and the band in the fi phase of Fig. (2.10) is shaped by the unresolved superposition of the incoherent signals from three magnetically inequivalent ionic sites.

Another quasi-onedimensional ferromagnet which is a prototype system for investigations on spin dynamics in low dimensional spin systems is $CsNiF_3$

[12] This simple theory implies that the lifetime of the exciton state is large compared to the single-ion dwelling time of the exciton.

[2.107]. In this compound $Ni^{2+}(3d^8)$ ions are located at the center of slightly distorted fluorine octahedra, which align along the hexagonal c-direction. The distortion lowers the site symmetry from $\mathrm{O_h}$ to D_{3d}, hence the spin degeneracy of the $(^3A_{2g})$ ground state is partially raised by the combined effect of spin–orbit coupling and trigonal crystal field and is described by a Hamiltonian $\sum_i D\,(S_i^z)^2$. The fe coupling along the c-chains (distance between Fe ions: $\frac{1}{2}c = 2.61$ Å) is of the Heisenberg type $-2\,J\sum_i \boldsymbol{S}_i\,\boldsymbol{S}_{i+1}$; $J = (8.0 \pm 0.035)\,\mathrm{cm}^{-1}, D = (6.2 \pm 0.14)\,\mathrm{cm}^{-1}$. The interchain coupling is weak due to the large intervening Cs^+ ions (Fe–Fe distance: 6.23 Å). The three-dimensional ordering occurs at $T_N = 2.63$ K, the spins are lying in the basal plane (planar anisotropy). Above T_N, no long-range order exists but short-range order is encountered in the chains combined with large magnetic fluctuations up to about 50 K.

In this compound the Raman transitions to the lowest excited levels originating from the $^3T_{2g}$ – ($\approx 7000\,\mathrm{cm}^{-1}$) and $^3T_{1g}^a$ – ($\approx 11500\,\mathrm{cm}^{-1}$) states of the free ion have been studied. The lowest sublevel of $^3T_{1g}$ is a $\Gamma_1^+, (\mathrm{D_{3d}})$ singlet, that of $^3T_{2g}$ a Γ_3 doublet. The ground state is a superposition of unresolved Γ_1 and $\Gamma_3, (D_{3d})$ levels. According to Table 2.A.1 all polarizations are allowed. In Fig. 2.11 some of the Raman spectra and an absorption spectrum are depicted. The zero-phonon lines (magnetic-dipole transitions in absorption) and the asymmetry in the Raman spectrum are evident. The broad vibrational band in absorption is due to odd-phonon-induced electric dipole transitions. Davydov splitting in the af phase is not observed here due to the weakness of the (exchange) interaction between translationally non-equivalent Ni ions.

Above the three-dimensional ordering at T_N, the zero-phonon lines in absorption (c) and in scattering shift to lower energies and their halfwidths increase linearly with increasing T (slopes: 0.2 and $0.6\,\mathrm{cm}^{-1}K^{-1}$). This behavior reflects the magnetic short-range order (magnetic energy) and spin dynamics, as can be shown by applying a magnetic field in the easy plane. This field shifts the lines according to a linear Zeeman effect ($g \approx 2$) and orders the spins, thereby decreasing the linewidths. The effect of short-range order on the line shift can be understood qualitatively by realising that the magnetic energy of the crystal represents a change (increase) in the energy of the ground state of the magnetic ions.

Above T_N, i.e., in the region of spin disorder, the Raman as well as the absorption intensities of many transitions in the temperature region of one-dimensional short-range order can be expressed as linear combinations of static spin correlation functions. Since the Raman transitions are assumed to be single-ion effects, only autocorrelation functions will appear in low order of perturbation theory.

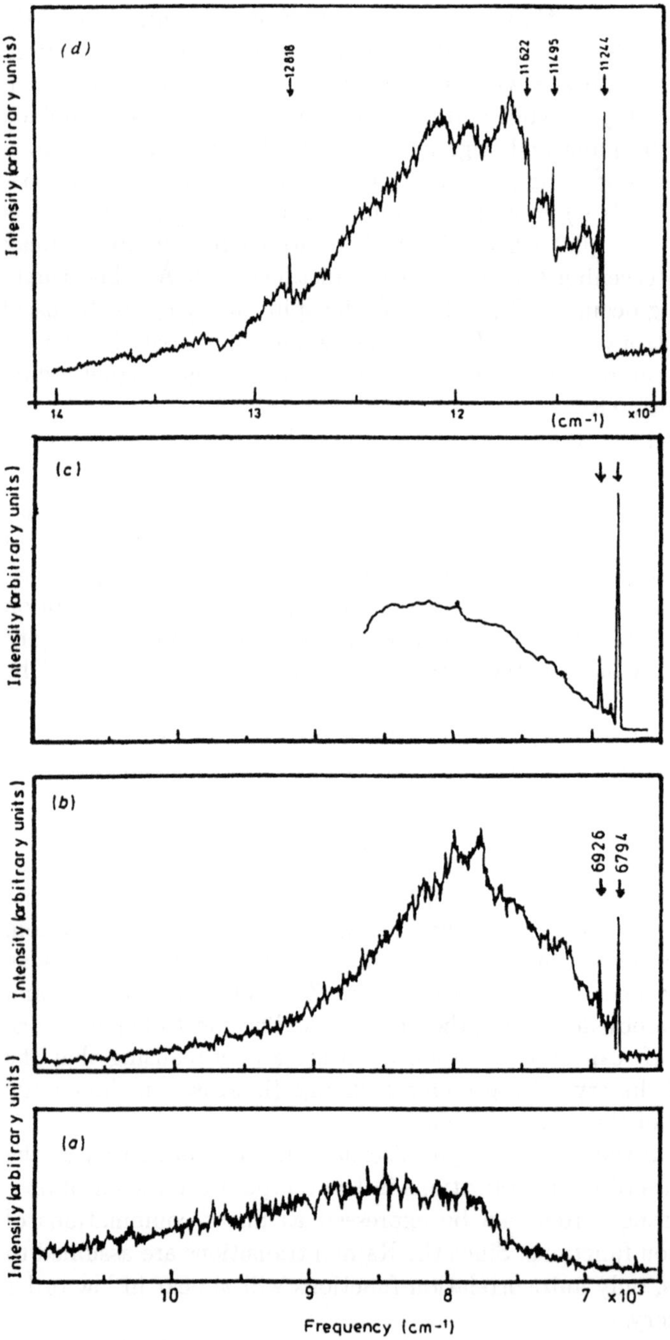

Fig. 2.11. Raman spectrum of $CsNiF_3$ at $T = 2\,K$, showing the electronic ${}^3A_{2g} \rightarrow {}^3T_{2g}$ transition in yz (**a**) and zx (**b**) polarization; (**c**): absorption spectrum. The electronic ${}^3A_{2g} \rightarrow {}^3T^a_{1g}$ transition for (zz) polarization is shown in (**d**), from [2.107]

The calculations start from (2.6) and (2.13). The integrated scattered intensity is proportional to [2.107]:

$$\int_{-\infty}^{\infty} I_{xy}(\omega)\,\mathrm{d}\,\omega = \langle \chi_{xy}^{\dagger} \chi_{xy} \rangle = \sum_{i,i'} \langle i | \chi_{xy}^{\dagger} | f \rangle \langle f | \chi_{xy} | i' \rangle \langle i' | \rho | i \rangle , \tag{2.54}$$

where ρ is the density operator $\exp(-\beta\mathcal{H})$ taken between the $^3A_{2g}$ ground state(s) $|i\rangle$ and $|i'\rangle$, and where the right side is the thermal average of the operator $\Xi = \chi_{xy}^{\dagger}|f\rangle\langle f|\chi_{xy}$. In the lowest order the intensities, depending on temperature and magnetic field in the easy plane, are obtained either as $\propto (1-\langle S^{i^2}\rangle)$ or $\propto \langle S^{i^2}\rangle$. The zz intensity of the line at 11 244 cm^{-1}, displayed in Fig. (2.11 d), is expected to behave according to $a_1(1-\langle S^{z^2}\rangle)$, giving $a_1/3$ for complete spin disorder ($T \to \infty$), and $a_1/2$ both for spin disorder in the easy plane and for spins aligned along y by a magnetic field B_y. In this case no intensity change to be induced by an external field in y-direction is expected in agreement with the observations. The same behavior is predicted and observed for the line at 6794 cm^{-1}.

2.2.8 Time Resolved Scattering

The availability both of pulsed lasers (pulsewidths $\tau \leq 100$ ps) and of fast detection electronics have opened the field to studies of the time evolution of electronic Raman transitions, which offer information e.g. on ion–ion interactions, dephasing mechanisms in the (resonance) scattering process etc. Such processes are being studied in detail using methods of nonlinear optics (photon echos, free induction decay, etc.) [2.105]. The potential of Raman scattering in this respect has been demonstrated only qualitatively up to now [2.4], [2.108], but appears to be promising.

The technique used by *Koningstein* et al. [2.4], as mentioned in the introduction, uses a pulsed laser (pulsewidth 10 ns) and two boxcar integrators. The authors have been able to separate the off-resonance Raman signal, which follows the laser pulse without any time delay, from fluorescence with a decay time ≥ 10 ns. The signals from the photomultiplier tube, with a rise and decay time of the signal of ≈ 2 ns, are fed into the boxcar inputs. The gates of one integrator are triggered in coincidence with the laser pulse, the other with a delay of 25 ns with respect to the pulse. While a fluorescence signal shows up at both boxcar outputs, the Raman signal appears only at the undelayed output.

With the technique just described, one has been able to separate Raman scattered light from fluorescence only a few wavenumbers apart in the case of cubic terbium aluminum garnet ($Tb_3Al_5O_{12}$, or $3Tb_2O_3 \cdot 5Al_2O_3$), (O_h^{10} with RE site symmetry D_2) [2.108]. In Fig. 2.12, part of the laser emission spectrum is plotted with a superposition of Raman and fluorescence radiation. There is a transition from off-resonance to on-resonance scattering and, associated with this transition, a change in the time dependence of the scattering. Off resonance only the Raman scattered light is observed and follows

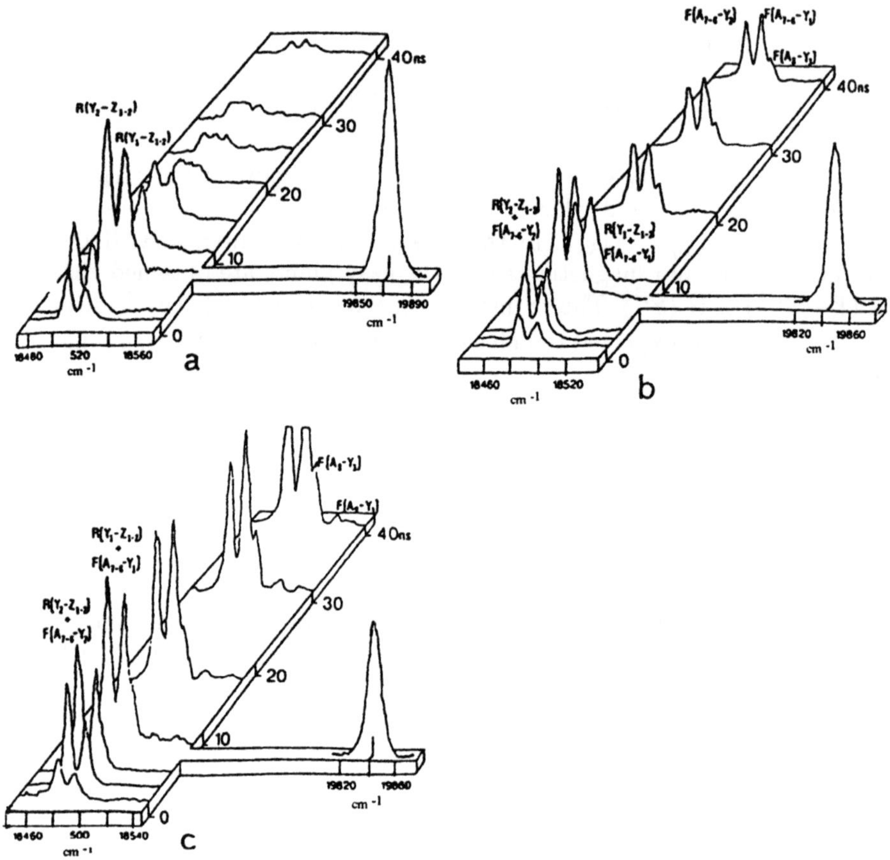

Fig. 2.12. Time-resolved (0–40 ns) Raman spectra (R) and fluorescence spectra (F) of terbium aluminum garnet $Tb_3Al_5O_{12}$ at $T \approx 80\,K$. The spectra have been excited with a dye laser pumped by a XeCl excimer laser. The wavelength of the dye laser is tuned to the $^5D_4$ crystal field levels of Tb^{3+}, $(4f^8)$. Fluorescence from these levels and Raman radiation to the lower crystal field levels of $^7F_5$ at 2135 and 2145 cm^{-1} almost coincide but are separated by their time dependence. (**a**) $\nu_{\text{laser}} = 20\,649\,cm^{-1}$, (off resonance), (**b**) $\nu_{\text{laser}} = 20\,622\,cm^{-1}$, (near resonance) and (**c**) $\nu_{\text{laser}} = 20620\,cm^{-1}$, (on resonance). The 779-cm^{-1} phonon at, e.g. 19870 cm^{-1} in (**a**) is also shown as intensity reference. Its intensity is diminishing near and on resonance due to absorption of the incident radiation, from [2.4] and [2.108]

the laser pulse without delay, near resonance both types of radiation can still be disentangled by their different time decay, on resonance both processes merge into a single emission event accompanied in its way by a slowing down of the pure Raman scattering process. No resonance enhancement of the Raman radiation is observed in these experiments, the inverse of the fluorescence excitation profile, however, closely follows the absorption spectrum.

The theoretical approach to interpret such data relies heavily on concepts developed for resonant light scattering by molecules in solutions or in solids [2.109]. Basically, spontaneous off-resonance Raman scattering arises from an optically driven, electronically phased ensemble (2.44). On approaching resonance, this coherent superposition decays due to coherence loss processes into a fluorescent population (resonance fluorescence, hot luminescence) and the induced polarization of this ensemble, usually described by density matrix techniques, acts as the source for the scattered radiation. When the exciting light frequency is close to the transition frequency between the ground state $|g\rangle$ and the excited state $|i\rangle$, the ions are driven into a nonstationary state, which is a superposition of the states $|g\rangle$ and $|i\rangle$. The spontaneous emission from this average population is Raman scattering.

For a simplified discussion, a three-level system is assumed consisting of the ground state, the excited electronic state in resonance with the laser and the final state $|f\rangle$ of the Raman (fluorescence) transition. Quasielastic coherence loss processes are caused, e.g., by interactions of the electronic system with the fluctuations of the heat bath, where the phase coherence within the ensemble is destroyed but the occupation of the levels is preserved. Inelastic collisions, e.g. due to phonon emission or absorption, and radiative or nonradiative decay will alter the level population towards thermal equilibrium. The fluorescence from this dephased ensemble is resonance fluorescence. It contains the full width of level $|i\rangle$, whereas the Raman emission contains only the laser width convoluted with the width of the final state $|f\rangle$ of the scattering process. Raman scattering and fluorescence exhibit different spectral features because the two processes originate from different sources. By introducing phenomenological decay parameters for the diagonal and off-diagonal elements of the density matrix, the following relation is obtained [2.109]:

$$\frac{1}{T_\mathrm{A}} = \frac{1}{2T_1} + \frac{1}{T_\mathrm{R}} \tag{2.55}$$

Here T_A is the total decay time due to lifetime, inelastic and quasielastic collisions, T_R the width due to dephasing processes alone, and T_1 the population decay time due to radiative and nonradiative processes inherent to the ions in the crystal.

In the weak signal limit, and for a monochromatic source, the photon counting rate $w(\nu_\mathrm{L})$ amounts to [2.109]:

$$w(\nu_\mathrm{L}) \propto \frac{|\langle f|\boldsymbol{D}|i\rangle|^2 \cdot |\langle i|\boldsymbol{D}|f\rangle|^2}{(\nu_\mathrm{L}-\nu_{ig})^2 + \left(\frac{1}{T_\mathrm{A}}\right)^2} \tag{2.56}$$
$$\times \left\{ \pi\delta(\nu_\mathrm{L}-\nu_\mathrm{S}-\nu_{fg}) + \left(\frac{2T_1 - T_\mathrm{A}}{T_\mathrm{A}^2}\right) \frac{1}{(\nu_\mathrm{S}-\nu_{if})^2 + \left(\frac{1}{T_\mathrm{A}}\right)^2} \right\}.$$

The spectrum of (2.56) consists of two types of emission: The δ-shaped Raman emission and the Lorentzian fluorescence centered at the resonance frequency ν_{if} of the ion. The latter only occurs for $T_\mathrm{A} <$ or $\ll 2T_1$. Integration of the line shape function (2.56) with respect to ν_S yields the ratio of the total amount of fluorescence I_F to the total amount of Raman emission I_R:

$$\frac{I_F}{I_\mathrm{R}} = \frac{2T_1 - T_\mathrm{A}}{2T_\mathrm{A}} \approx \frac{2T_1}{T_\mathrm{R}}. \tag{2.57}$$

T_1 is essentially determined by the decay due to one-phonon emission. This decay is slow in the lower crystal field levels of a multiplet component, separated by a large gap from the nearest multiplet below, when only radiative and non-radiative decay due to multi-phonon emission occurs, but fast in the higher levels. It follows from (2.57) that for comparable electronic dephasing times the fraction of resonant Raman scattering in the total emission is smaller for resonance with the lower crystal field levels of a multiplet component than for the higher levels. This is in agreement with the experimental results (Fig. 2.12) [2.108]. If the uppermost crystal field levels of $^5D_4$ come into resonance, the intensity of the fast fluorescence drops below the detection limit, the signal is solely due to electronic Raman scattering. The fact that the Raman excitation spectrum does not follow the absorption spectrum is attributed to the predominance of pure electronic dephasing over the vibrational decay process of the $^5D_4$ crystal field states of Tb^{3+} under study. The electronic dephasing times, which decrease strongly with rising temperature, are much shorter than the pulsewidth (5–10 ns) and will amount to a few picoseconds at 77 K.

2.3 Effects of Localized Electron–Phonon Interaction

2.3.1 Vibronic States

In the previous sections the lattice has been treated as a static host, providing the static crystalline electric field. The excitations of the lattice have at best been lumped into the concept of the heat bath or of specific decay channels provided for the ions by the lattice. In many cases this is not an adequate approach. Effects of interactions between the *d*- or *f*-ions and the acoustic and (or) optic phonons are numerous: they will in some cases affect the electronic energies drastically. Let us name just a few of these effects, which are not of specific interest in the present context: the strong temperature dependence of the electronic line widths (Fig. 2.1), the phonon sidebands appearing in one-photon absorption and emission spectra, the multiphonon relaxation processes of excited states in general, the phonon assistance in non-resonant energy transfer, the drastic effects of vibrational excitations of specific molecular groups, e.g., of the hydration shell in certain crystals like the RE ethylsulfates with a strong reduction of lifetimes of the electronic states. A review of the research topics of present interest in these fields can be found in [2.110].

These effects are generally considered as signatures of the presence of a vibronic interaction. This type of interaction usually applies in cases where the electronic energy levels of the system coupled to $\boldsymbol{Q}$, where $\boldsymbol{Q}$ represents symbolically all the vibrational coordinates of the lattice, are widely separated in energy as compared to the energies of (optical) phonons.

Usually the Born–Oppenheimer approximation is considered as the starting point for the discussion of the physical concepts of vibronic effects [2.27], [2.38], [2.111]. The vibronic eigenfunctions of the coupled system may be written in the approximate form:

$$\Psi(\boldsymbol{r}, \boldsymbol{Q}) = \psi(\boldsymbol{r}; \boldsymbol{Q}) \cdot \chi(\boldsymbol{Q}). \tag{2.58}$$

Here, $\boldsymbol{r}$ denotes the electronic coordinates of the ion, $\psi(\boldsymbol{r}; \boldsymbol{Q})$ is an electronic wavefunction depending parametrically on $\boldsymbol{Q}$ through the Schrödinger equation:

$$\mathcal{H}_{\mathrm{e}}(\boldsymbol{r}; \boldsymbol{Q})\psi(\boldsymbol{r}; \boldsymbol{Q}) = E_{\mathrm{e}}(\boldsymbol{Q})\psi(\boldsymbol{r}; \boldsymbol{Q}) \tag{2.59}$$

where $\mathcal{H}_{\mathrm{e}}(\boldsymbol{r}; \boldsymbol{Q})$ is the electronic Hamiltonian and $E_{\mathrm{e}}(\boldsymbol{Q})$ the eigenvalue for fixed positions $\boldsymbol{Q}$ of the lattice points. The vibrational wavefunction $\chi(\boldsymbol{Q})$ satisfies a wave equation, where the potential energy term $\mathcal{U}(\boldsymbol{Q})$ in the Hamiltonian is

$$\mathcal{U}(\boldsymbol{Q}) = E_{\mathrm{e}}(\boldsymbol{Q}) + \int \psi^*(\boldsymbol{r}; \boldsymbol{Q}) \sum_{\nu} \frac{P_\nu^2}{2M_\nu} \psi(\boldsymbol{r}; \boldsymbol{Q}) \, \mathrm{d}\boldsymbol{r}. \tag{2.60}$$

$\mathcal{U}(\boldsymbol{Q})$ is determined by the electronic energies $E_{\mathrm{e}}(\boldsymbol{Q})$ and the weighted average of the lattice kinetic energy. The expectation value of an electronic operator $\mathcal{O}$ with respect to the eigenstate of the vibronic system is found by using (2.58):

$$\langle \Psi^* | \mathcal{O} | \Psi \rangle = \int \chi^*(\boldsymbol{Q}) \left[\int \psi^*(\boldsymbol{r}; \boldsymbol{Q}) \; \mathcal{O} \; \psi(\boldsymbol{r}; \boldsymbol{Q}) \, \mathrm{d}\, \boldsymbol{r} \right] \chi(\boldsymbol{Q}) \, \mathrm{d}\, \boldsymbol{Q}, \tag{2.61}$$

where the electronic matrix element $\mathcal{O}(\boldsymbol{Q})$ in square brackets again depends parametrically on $(\boldsymbol{Q})$. The expectation value of $\mathcal{O}$ is the average of $\mathcal{O}(\boldsymbol{Q})$ taken with the weight function $|\chi(\boldsymbol{Q})|^2$. Thus, if $|\chi(\boldsymbol{Q})|^2$ is centered at $\boldsymbol{Q}_{\mathrm{e}}$, generally $\langle \Psi^* | \mathcal{O} | \Psi \rangle \neq \mathcal{O}(\boldsymbol{Q})_{\mathrm{e}}$, the value obtained for a static lattice.

If the separation of the relevant electronic energy levels (with energies $E_{\mathrm{e},1}$ and $E_{\mathrm{e},2}$) is comparable with some phonon energies, i.e. $|E_{\mathrm{e},1} - E_{\mathrm{e},2}| \approx \hbar\omega_{\mathrm{phonon}}$, the Born–Oppenheimer approximation breaks down, a situation often encountered in the ground state multiplet components. In such a case, the vibronic eigenfunctions of the system will, to a good approximation, acquire the form [2.112]:

$$\Psi(\boldsymbol{r}; \boldsymbol{Q}) = \psi_1(\boldsymbol{r}; \boldsymbol{Q})\chi_1(\boldsymbol{Q}) + \psi_2(\boldsymbol{r}; \boldsymbol{Q})\chi_2(\boldsymbol{Q}), \tag{2.62}$$

or a suitable sum over such products, which in general cannot be transformed into a product of the type displayed in (2.58).

Matters simplify considerably if the interaction between electronic states and local lattice distorsions by phonons can be assumed to be weak (the vibronic character of the electronic states will be taken into account, as given by (2.58)) and energy renormalizations due to this interaction are small as compared to all electronic and phonon excitation energies involved. Hence perturbation methods can be applied in this case. We shall first consider this case for a single ion. If $Q_{\Gamma\alpha}$ represents a symmetry-adapted combination of displacements of the ligands surrounding the metal ion and belonging to the α component of the rep Γ, then by expanding the crystal field Hamiltonian $\mathcal{H}_{\mathrm{CF}} = \sum_{\Gamma\alpha} B_{\Gamma\alpha} \mathrm{O}_{\Gamma\alpha}$ to first order in the lattice displacements, and lumping together the electronic coordinates into an appropriate electronic operator $\mathrm{O}_{\Gamma\alpha}$, the interaction Hamiltonian may be written:

$$\mathcal{H}_{\mathrm{s}} = \sum_{\Gamma\alpha} b_{\Gamma\alpha} Q_{\Gamma\alpha} \mathrm{O}_{\Gamma\alpha}. \tag{2.63}$$

This is the Jahn–Teller (JT) interaction, bilinear in the electronic and vibrational operators. The coupling constant $b_{\Gamma\alpha}$ is conceptually interpreted as the first derivative of the crystal field paramater $B_{\Gamma\alpha}$ with respect to $Q_{\Gamma\alpha}$:$b_{\Gamma\alpha} \equiv \frac{\partial B_{\Gamma\alpha}}{\partial Q_{\Gamma\alpha}}$, the dynamic crystal field.

To first order in Q the electronic energies must be corrected by the diagonal elements of (2.63) whereas the new states become

$$|\psi\rangle \Rightarrow |\psi\rangle + \sum_{\psi' \neq \psi} |\psi'\rangle \frac{\mathcal{H}^{l}_{\psi,\psi'}}{\hbar\,(\omega_{\psi} - \omega_{\psi'})} \cdot Q_{\mathrm{l}}, \tag{2.64}$$

the new energies of the vibrational states are derived from (2.60). $\mathcal{H}^{l}_{\psi,\psi'}$ are the off-diagonal matrix elements of (2.63) where a specific lattice excitation Q_{l} is considered.

Vibronic coupling also conveys more complexity to the selection rules for electronic scattering [2.113], (see also [2.114]). Equation (2.23) remains valid if

$$\Gamma_{\mathrm{R}} \equiv (\Gamma_i^{(v)} \otimes \Gamma_f^{(v)} \otimes \gamma_i \otimes \gamma_f), \tag{2.65}$$

where $\Gamma_{i,f}^{(v)}$ are the reps of the vibrational parts of the product (2.58), while $\gamma_{i,f}$ are the reps of the electronic functions. We take Γ_{S} to indicate, as in (2.23), the set of reps which comprise all components of the scattering tensor. Usually in vibrational scattering, $\gamma_i = \gamma_f$ is one-dimensional, hence (2.65) reduces to (2.23). In the case of degeneracy, however, the direct product $\gamma_i \otimes \gamma_f$ contains other reps beside Γ_1 and the selection rules have to be derived from (2.65) [2.113]. We have to distinguish between processes even upon time reversal (symmetric Raman scattering) and time-odd ones (antisymmetric scattering); also between systems with odd or even numbers of electrons (Kramers degeneracy or the lack of it).

For a discussion of the new situation, where Placzek's polarizability theory [2.3] does not apply, we study a matrix element of an operator $\mathcal{S}$, (rep $\Gamma_{\mathcal{S}}$), with the product states $|\mu_j\rangle$ and $|\bar{\mu}_i\rangle$, where $|\bar{\mu}_i\rangle$ is the time reversed (Kramers conjugate) state $|\mu_i\rangle$. This matrix element can be written [2.38]:

$$\langle\bar{\mu}_i|\mathcal{S}|\mu_j\rangle = \frac{1}{2}\left(\langle\bar{\mu}_i|\mathcal{S}|\mu_j\rangle + \varepsilon_{\mathcal{S}}\varepsilon_{\boldsymbol{T}}\langle\bar{\mu}_j|\mathcal{S}|\mu_i\rangle\right). \tag{2.66}$$

$\varepsilon_{\mathcal{S}} = +1$ if $\mathcal{S}$ is even upon time reversal and -1 if $\mathcal{S}$ is odd, while $\varepsilon_{\boldsymbol{T}}$ is the square of the time-reversal operator $\boldsymbol{T}$. $\varepsilon_{\boldsymbol{T}} = +1$ for even numbered electron systems and $\varepsilon_{\boldsymbol{T}} = -1$ for odd numbered systems, (2.17). Now the bras $\langle\bar{\mu}_i|$ span the same set of reps as the kets $|\mu_j\rangle$, in identically the same form. Thus, it can be shown [2.38] that depending on the sign of $\varepsilon_{\mathcal{S}}\varepsilon_{\boldsymbol{T}}$, the matrix element $\langle\bar{\mu}_i|\mathcal{S}|\mu_j\rangle$ belongs to the representations $[\gamma_i \otimes \gamma_j]_{\mathrm{S}} \otimes \Gamma_{\mathcal{S}}$ (for $\varepsilon_{\mathcal{S}}\varepsilon_{\boldsymbol{T}} > 0$) or to $\{\gamma_i \otimes \gamma_j\}_{\mathrm{A}} \otimes \Gamma_{\mathcal{S}}$ (for $\varepsilon_{\mathcal{S}}\varepsilon_{\boldsymbol{T}} < 0$), where $[\gamma_i \otimes \gamma_j]_{\mathrm{S}}$ is the symmetric and $\{\gamma_i \otimes \gamma_j\}_{\mathrm{A}}$ is the antisymmetric product representation.[13] For $\langle\bar{\mu}_i|\mathcal{S}|\mu_j\rangle \neq 0$, the totally symmetric rep Γ_1 must be contained either in $[\gamma_i \otimes \gamma_j]_{\mathrm{S}} \otimes \Gamma_{\mathcal{S}}$ or in $\{\gamma_i \otimes \gamma_j\}_{\mathrm{A}} \otimes \Gamma_{\mathcal{S}}$, depending on the sign of $\varepsilon_{\mathcal{S}}\varepsilon_{\boldsymbol{T}}$.

For an odd number of electrons the double group representations apply for the γ_i and in general $\{\gamma \otimes \gamma\}_{\mathrm{A}} \equiv \gamma_1$, the identity representation. In systems like Ce^{3+}, $(4f^1)$, and for $\mathcal{S}$ representing the electric dipole transition operator $\boldsymbol{r}$ or $(\alpha^{\mathrm{s}}_{\rho,\sigma})$ (symmetric scattering), we find $\varepsilon_{\mathcal{S}} = +1$, $\varepsilon_{\boldsymbol{T}} = -1$, $(\varepsilon_{\mathcal{S}}\varepsilon_{\boldsymbol{T}} < 0)$ for one-photon absorption or emission and also for symmetric Raman scattering. These processes are allowed by symmetry, whenever $\Gamma^{(')}_{\mathcal{S}} \otimes \Gamma^{(v)}_f$ contains $\{\gamma_i \otimes \gamma_j\}_{\mathrm{A}} \equiv \gamma_1$. Antisymmetric scattering can be observed, if $\Gamma^{('')}_{\mathcal{S}} \otimes \Gamma^{(v)}_f$ contains $[\gamma_i \otimes \gamma_j]_{\mathrm{S}}$. $\Gamma^{(')}_{\mathcal{S}}$ (and $\Gamma^{('')}_{\mathcal{S}}$) are those reps, to which the symmetric ($\Gamma^{(')}_{\mathcal{S}}$) (or antisymmetric, ($\Gamma^{('')}_{\mathcal{S}}$)) part of the polarizability tensor or the electric (or the magnetic) transition dipole moment belongs. Thus the selection rules for infrared absorption and symmetric scattering are unaltered with respect to those of the non-degenerate case. However, antisymmetric scattering by phonons is expected to occur (off-resonance!) for those reps of vibrational states (v), $\Gamma^{(v)}$, which obey:

$$\Gamma^{(v)}_f \in [\gamma_i \otimes \gamma_j]_{\mathrm{S}} \otimes \Gamma^{('')}_{\mathcal{S}}. \tag{2.67}$$

Antisymmetric electronic scattering again follows the usual selection rules derived from (2.67) for $\Gamma^{(v)}_f \equiv \Gamma_1$. In Sect. 2.3.3.2 (strong coupling) an

---

[13] The two representations $\mathcal{D}$ of order p of $|\mu_j\rangle$ and $|\bar{\mu}_j\rangle$ span the reducible representation $(\mathcal{D})^2 = \mathcal{D} \otimes \mathcal{D}$ and can be broken into a symmetric subset that contains $\frac{1}{2}p(p+1)$ terms and an antisymmetric subset that contains $\frac{1}{2}p(p-1)$ terms. The characters $\chi(R)$ of the symmetric and the antisymmetric parts of $(\mathcal{D})^2$ are (R is any element of the point group) [2.38]:

$$\chi^{\mathcal{D}^2}(R) = (\chi^{\mathcal{D}}(R))^2; \qquad \chi^{\mathcal{D}^2}_{\mathrm{S}}(R) = \frac{1}{2}(\chi^{\mathcal{D}}(R))^2 + \frac{1}{2}(\chi^{\mathcal{D}}(R^2));$$

$$\chi^{\mathcal{D}^2}_{\mathrm{A}}(R) = \frac{1}{2}(\chi^{\mathcal{D}}(R))^2 - \frac{1}{2}(\chi^{\mathcal{D}}(R^2)).$$

example for this asymmetric phonon scattering will be discussed; see also Refs. [2.113], [2.115].

The situation in systems with an even number of electrons and a degenerate ground state may also be rather complex. Symmetric phonon scattering and phonon absorption are allowed for representations of the vibrational states such that

$$\Gamma^{(v)} \in [\gamma_i \otimes \gamma_j]_{\mathrm{S}} \otimes \Gamma_S^{(')}, \tag{2.68}$$

while antisymmetric scattering requires

$$\Gamma^{(v)} \in \{\gamma_i \otimes \gamma_j\}_{\mathrm{A}} \otimes \Gamma_S^{('')}. \tag{2.69}$$

These selection rules do not, of course, give any hint about the transition strength of the additionally allowed transitions.

The physical reason for this increased complexity in the case of degenerate electronic ground states is obvious: This degeneracy will permit additional electronic transitions via the intermediate excited states. For example, in the RE fluorides REF_3 (RE $= La^{3+}, Ce^{3+}, Pr^{3+}, Nd^{3+}$) of tysonite structure, the space group is approximately D_{6h}^3, with La^{3+} site symmetry: D_{3h}, (rigorously actually C_2 in crystals twinned by merohedry with a space group $D_{3d}^4(P\bar{3}c1)$) for the domains[14]) [2.116], [2.117]. In the following discussion we use, for simplicity, approximate D_{6h} symmetry both for the phonons and the (delocalized) electronic components of the vibronic states.

For N odd, the electronic ground state may transform either as $\bar{\Gamma}_7, \bar{\Gamma}_8$, or as $\bar{\Gamma}_9$ of D_{6h} (Table 2.A.1), the symmetric squares of $\bar{\Gamma}_7$ and $\bar{\Gamma}_8$ contain $\Gamma_2^+(\mathrm{A_{2g}})$ and $\Gamma_5^+(\mathrm{E_{1g}})$, or $\Gamma_2^+, \Gamma_3^+(\mathrm{B_{1g}}), \Gamma_4^+(\mathrm{B_{2g}})$ of $\bar{\Gamma}_9$ [2.50]. The antisymmetric part of the polarizability tensor Γ_S'' comprises $A_{2g}(M_z), E_{1g}(M_{x,y})$. According to (2.67) and for $\bar{\Gamma}_7$ or $\bar{\Gamma}_8$ as ground state, antisymmetric scattering can be expected in A_{2g} (xy) polarization for A_{1g} and E_{2g} phonons, and in E_{1g} polarization (xz, yz) for A_{1g}, A_{2g}, and E_{2g} phonons. Thus the selection rules predict an antisymmetric intensity spillover of A_{1g}, A_{2g}, and E_{2g} phonons in either A_{2g} or E_{1g} polarization. For a $\bar{\Gamma}_9$ ground state, antisymmetry is expected in A_{2g} polarization for A_{1g}, B_{1g}, and B_{2g} phonons, in E_{1g} polarization for E_{1g} and E_{2g} phonons. While an intensity spillover in an otherwise forbidden polarization might be difficult to detect, the antisymmetry of the E_{1g} phonons due to electron–phonon interaction should be clearly evident (see Sect. 2.3.3.2).

Let us again consider D_{6h} as an example for systems with an even number of $4f$ electrons: The degenerate electronic ground state may belong to one of the two reps: $E_{1g}, E_{2g}; (\Gamma_5, \Gamma_6)$. In any case the symmetric products span A_{1g}, E_{2g}, the antisymmetric product gives A_{2g} [2.113]. Due to the occurrence of E_{2g} in $[\gamma_i \otimes \gamma_j]_{\mathrm{S}}$, new transitions may gain intensity in infrared absorption or Raman spectra: $E_{2g} \otimes A_{2u} = E_{2u}$, i.e. silent E_{2u} phonon transitions should

[14] The generating operations for the twins are those symmetry elements of the space group D_{6h}^3 which are not contained in D_{3d}^4

appear in the A_{2u} spectrum (E‖z), and $E_{2g} \otimes E_{1u} = B_{1u} + B_{2u} + E_{1u}$, the E_{1u} spectrum will be enriched by excitations of the otherwise silent B_{1u} and B_{2u} symmetry types. In the Raman spectrum besides the usual A_{1g}, E_{1g}, and E_{2g} spectra there should be a spillover from E_{2g} in A_{1g}, from B_{1g}, B_{2g} in E_{1g} and from A_{1g}, A_{2g} in E_{2g}. Antisymmetric phonon scattering is to be expected for $A_{2g} \otimes (A_{2g} + E_{1g}) = A_{1g}$ (in A_{2g} polarization) and E_{1g} (in E_{1g} polarization).

These anomalies in the phonon spectra, which are produced by the coupling effects with the degenerate electronic ground state, will gradually disappear with growing temperature when the excited electronic states become thermally populated.

2.3.2 Jahn–Teller Effects

Raman spectroscopy has been one of the first experimental tools used to detect Jahn–Teller (JT) distortions in solid compounds with $3d$ or $4f$ ions [2.118, 2.119, 2.120]. From the wealth of experimental data only a few sets, in which the electronic levels are directly affected by the JT interaction, will be selected.

From an experimental point of view, the static and dynamic JT effects (JTEs) have to be considered. The *static* effect, which is in fact an exceptional case, arises when the (orbital, non-Kramers) degeneracy of a vibronic state (occurring at some $\boldsymbol{Q}_s$ in a symmetric configuration) is lifted linearly in $(\boldsymbol{Q} - \boldsymbol{Q}_s)$. For sufficiently strong JT coupling the energy separation between the split components may be large enough to revitalize for low lying states the Born–Oppenheimer approximation and (2.58) will be valid. The size of the vibronic splitting (neglecting spin–orbit coupling) is determined through the compensation of the energy, gained by the lowering of the vibronic ground state, through the lattice deformation which is proportional to $|\boldsymbol{Q} - \boldsymbol{Q}_s|^2$. The symmetric configuration is therefore unstable with respect to this lattice distortion [2.111].

The (two) *dynamic* JTEs occur for coupling energies which are lower than in the static case but comparable with the kinetic energy of the lattice ions. These cases cannot be treated within the framework of (2.59). Vibronic wavefunctions of type (2.62) apply in this case and the potential energy $V(\boldsymbol{r}, \boldsymbol{Q})$ in the Hamiltonian does no longer depend parametrically on $\boldsymbol{Q}$; $V(\boldsymbol{r}, \boldsymbol{Q})$ displays the full symmetry of the symmetric configuration of the lattice site where the JT ion is situated. As a consequence, the new vibronic eigenstates have to be classified according to the point symmetry group of the symmetric configuration and display the degeneracies of the reps of this group. Contrary to the static case, in the dynamic effect the JT coupling does *not* lift the degeneracies of these vibronic states, forming components belonging to the same reps of the group as the original symmetric configuration [2.111].

The first dynamic JTE involves a thermally activated reorientation of a complex which, at low temperatures, exhibits in its ground state a static

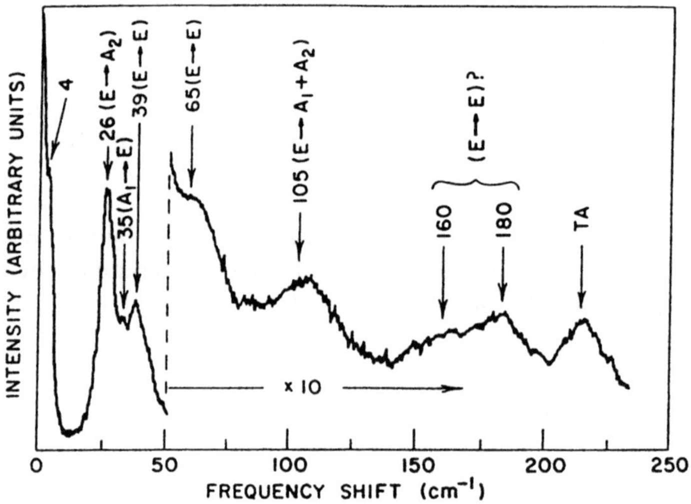

Fig. 2.13. Hindered-rotational Raman spectrum (E_g) of cubic CaO:Cu^{2+}, T = 4.2 K, spectral resolution: 1.5 cm^{-1}. The observed transitions have been attributed to hindered-rotational levels $\frac{1}{2} \leq |j| \leq \frac{11}{2}$. From [2.121]

JTE. In the second kind of dynamic JTE a single vibronic eigenstate (2.62) is involved and the JT energy is small as compared to the phonon energy. In this case even at the lowest temperature a stabilization of a distortion is not possible (i.e. a 'slow' experiment will not detect the distortion) because the zero-point motion will carry the system from one local energy minimum to another. Tunneling between the local energy minima must also be taken into account [2.111], [2.118].

Local JT distortions have been detected for example in single crystals of CaO (rocksalt structure), doped with small concentrations of Cu^{2+}, ($3d^9$, ^2E_g ground state at a Ca site) [2.121]. The E_g spectrum below 250 cm^{-1} at 4.2 K is shown in Fig. 2.13. No spectra are observable in F_{2g} or A_{1g} scattering geometries. Four sharp lines are observed at 4, 26, 35, and 38.5 cm^{-1}, while three broader peaks appear at 65, 105, and $\approx$ 180 cm^{-1}. At 77 K, the three transitions between 26 and 38.5 cm^{-1} appear as a single peak, since the spectrum becomes broader. The energy levels have been successfully attributed to a hindered rotation (rotational quantum number j) of the distorted complex (Cu–O_6 molecular cluster) between equivalent minima of the effective nuclear potential. These minima correspond to tetragonal distortions along the cube axes of the neighboring oxygens for the Cu^{2+} ion in the CaO lattice. The observed energy levels have been compared with levels calculated from a coupled electronic–vibrational Hamiltonian for a doubly degenerate electronic ground state interacting with a doubly degenerate (molecular) vibration including effects of anharmonicity [2.122]. On the other hand, the coupling to a continuum of lattice vibrational modes is simplified by using the fact that, for each vibrational branch and each wave vector $\boldsymbol{q}$ degenerate pairs of symmetrized linear combinations of the normal modes from the star of the q-vector can be generated, which transform as E_g at the impurity site. A cluster mode is found to have a frequency which is a weighted average of these lattice modes involved and an averaged coupling V to the electronic states and to have a projection onto each individual lattice mode.

The parameters that characterize the hindered-rotational states are the rotational energy splitting $E_j = \alpha(j^2 + \frac{1}{4})$ and the height 2β of the potential barrier. They were best fitted with $\alpha = 5.2\,\mathrm{cm}^{-1}$ and $2\beta = 45\,\mathrm{cm}^{-1}$ for CaO:Cu^{2+} [2.121]. The JT stabilization energy E_{JT} was found to be $900\,\mathrm{cm}^{-1}$. The model also supports the observed polarizations of the transitions.

The application of uniaxial strain [2.121] reduces the site symmetry: from $\mathrm{O_h}$ at the Ca site in CaO to D_{4h} for stress along $\langle 001\rangle$, to D_{2h} for stress along $\langle 110\rangle$ while the heights of the potential barriers of hindered rotation vary from case to case. Degeneracies of the vibronic levels are partially lifted and their energies are nearly linearly shifted (up to $\approx 0.2\,\mathrm{cm}^{-1}\,\mathrm{MPa}^{-1}$). These data were again used to determine parameters of the model applied to CaO:Cu^{2+} The strain coupling constant $|V_E| = 2.6 \cdot 10^4\,\mathrm{cm}^{-1}$ was however found to be too small (by a factor of about 2.5) to be consistent with the other Raman data and the fits to the vibronic levels, obtained with the cluster model. A similar inconsistency was encountered for E_{JT}. The most likely explanation, given in [2.121], is that the coupling of the JT ion is not restricted to the inner coordination shell of the lattice (the strongly polarizable oxygen ions) but includes the second-neighbor calcium ions. This conclusion is supported by intensity measurements of the impurity-induced one-phonon Raman spectrum of CaO:Cu^{2+}, following approximately the density of states. It is found that substantial coupling to the second-neighbor Ca^{2+} ions and to the oxygen polarization, as well as to the oxygen cores, are required to explain the observed spectra.

Similar experiments have been performed on Ni^{3+}, $3d^7$ configuration in Al_2O_3 [2.123], where the Ni^{3+} substitutes at an Al^{3+} site. Again the complex reorients or tunnels with a frequency corresponding to 60 cm^{-1}. Raman scattering transitions between the vibronic levels can be observed in this compound at impurity concentrations in the ppm range. The large cross section for these transitions results from the spatial reorientation of the electronic-ground-state orbitals, accompanied by large changes in the optical polarizability, as the distortion of the complex reorients. The experimental results, also those obtained from uniaxial stress experiments, are closely related to the results for CaO:Cu^{2+}. The cluster-model parameters differ for Ni^{3+}:Al_2O_3 from the previous ones due to the hardness of the host lattice: $\alpha = 43\,\mathrm{cm}^{-1}$, $2\beta = 120\,\mathrm{cm}^{-1}$, $V_{\mathrm{E}} = 3.6 \cdot 10^4\,\mathrm{cm}^{-1}$, $E_{\mathrm{JT}} = 1100\,\mathrm{cm}^{-1}$. In the Ni compound a discrepancy between the fit values from the cluster model and the vibronic levels was, perhaps fortuitously, not encountered.

A large number of Raman experiments has concentrated on the cooperative Jahn–Teller effect (CJTE) [2.118], [2.124]. This effect occurs in concentrated systems for small displacements of ligand ions, which change the crystal field acting on the JT active ions and lowers their electronic energy by orienting the non-spherical charge distributions (electronic multipole moments). These ligand displacements will interact with each other, e.g., via elastic strain in the crystal, thus providing an effective long-range interaction between the JT ions. At a sufficiently high concentration, the entire crystal can become unstable with respect to these displacements under the cooperative influence of the elastic interactions. A structural phase transition results, leading to a parallel alignment of all the electronic multipoles and the lattice distortions (ferrodistortive) or to more complicated antiferrodistortive arrangements. Depending on the range of the interaction between the JT ions, the phase transition of the order–disorder type can be often easily treated theoretically by mean-field approximations. Accoustic phonons with wavevector $\boldsymbol{q} \neq 0$ and optical phonons, including those with $\boldsymbol{q} = 0$, will also produce local distortions at the site of the JT ion, the latter with a limited interaction range. For coupling with phonons the CJTE can be considered as a special case of the ion–ion interaction mechanism due to the exchange of virtual phonons [2.118] ([2.125]; see also next section).

The crystal field must allow an appropriate level degeneracy of the ground state, or cause splittings which are smaller than the cooperative interaction energies. Either magnetic ordering effects or an external magnetic field will, if strong enough, compete with the JT ordering and suppress the latter when they are accompanied by a different arrangement of ordered electronic multipoles.

The Hamiltonian for a system with JT ions at sites l can be written in terms of phonon creation and annihilation operators [2.118], [2.119]:

$$\begin{aligned}\mathcal{H} = {} & \mathcal{H}_{\mathrm{c}} + \sum_{j,q} \hbar\omega_j(\boldsymbol{q}) \left[c_j^+(\boldsymbol{q}) c_j(\boldsymbol{q}) + \frac{1}{2} \right] \\ & + \sum_{j,q,m} b_j^m(\boldsymbol{q}) \left[c_j^+(\boldsymbol{q}) + c_j(-\boldsymbol{q}) \right] \boldsymbol{O}^m(l, \boldsymbol{q}). \end{aligned} \tag{2.70}$$

Here $\mathcal{H}_{\mathrm{c}}$ is the effective crystal field Hamiltonian (see Sects. 2.4.1, 2), $\mathcal{H}_{\mathrm{c}} = \sum_{lm} B^m \boldsymbol{O}^m(l)$, B^m are the empirically determined crystal field parameters (for a definition see footnote 21). The second term is the uncoupled phonon system, the third the JT coupling term (see (2.63)). Using the Fourier transform of the electronic multipole operators: $\boldsymbol{O}^m(\boldsymbol{q}) = N^{-\frac{1}{2}} \sum_l \boldsymbol{O}^m(l) \times \exp(\mathrm{i}\boldsymbol{q} \cdot \boldsymbol{r}(l))$ and the transformation to displaced oscillators: $\gamma_j(\boldsymbol{q}) = \left[c_j(\boldsymbol{q}) + \sum_m \frac{b_j^m(\boldsymbol{q})}{\hbar\omega_j(\boldsymbol{q})} \boldsymbol{O}^m(\boldsymbol{q}) \right]$, the Hamiltonian (2.70) can be formally decoupled to

$$\mathcal{H} = \mathcal{H}_c + \sum_{jq} \hbar\omega_j(\boldsymbol{q}) \left[\gamma_j^+(\boldsymbol{q})\gamma_j(\boldsymbol{q}) + \frac{1}{2} \right] + \sum_{l,l'} J^{m,m'}(l,l') \boldsymbol{O}^m(l) \boldsymbol{O}^{m'}(l'), \tag{2.71}$$

where the phonon operators γ no longer commute with $\mathcal{H}_c$. The γ_j operators define the excitations related to the local distortion, whereas the c_j operators define excitations which correspond to those of the undistorted phase. The phonon frequencies are not changed by the interaction in this approximation.

The effective interaction between different electronic multipoles reads:

$$J^{m,m'}(l,l') = (1/N) \sum_{j,q} b_j^m(\boldsymbol{q}) b_j^{m'}(\boldsymbol{q}) / \hbar\omega_j(\boldsymbol{q}) \cdot \exp[\mathrm{i}\boldsymbol{q} \cdot (\boldsymbol{r}(l) - \boldsymbol{r}(l'))]. \tag{2.72}$$

For $l \neq l'$ this is the ion–ion interaction driving the phase transition, the term $l = l'$ is the JT energy of a single unit cell and is usually referred to as a self-energy and may be considered as a dynamic contribution to the crystal field.

In simple cases, where only one electronic multipole operator $\boldsymbol{O}(l)$ determines the alignment of the multipoles, it is advantageous to consider the frequency dependent susceptibility $g(\omega)$ in a field $h(l)$ with an interaction energy $\sum_l h(l)\boldsymbol{O}(l)$ for the non-interacting ions under $\mathcal{H}_c$:

$$g(\omega) = \sum_{n,m} \frac{|O_{n,m}|^2 (\mu_n - \mu_m)}{(\omega + \mathrm{i}\varepsilon + \omega_n - \omega_m)}. \tag{2.73}$$

μ_n and μ_m are the thermal occupation probabilities of crystal field states $|n\rangle$ and $|m\rangle$. The (static) order parameter susceptibility of the interacting multipoles is:

$$G(\boldsymbol{q}, \omega = 0) = \frac{g(0)}{1 - J(\boldsymbol{q}) g(0)}. \tag{2.74}$$

Note that (2.74) diverges at the phase transition temperature.

If only the electronic ground state is occupied, and it happens to be an orbitally degenerate doublet, the $\boldsymbol{O}(l)$ may be simplified by the use of pseudospins $\sigma(l) = \frac{1}{2}$ and all operators are linear functions of $\sigma^x(l), \sigma^y(l), \sigma^z(l)$. $\boldsymbol{O}(l)$ may be chosen as $\sigma^z(l)$, while a residual crystal field splitting Δ may be introduced as $\Delta \cdot \sigma^x(l)$. The interaction part in (2.71) simplifies to the Hamiltonian of the Ising model in a transverse field:

$$\mathcal{H}_{\text{int}} = -\frac{1}{2} \sum_{l \neq l'} J(l,l') \sigma^z(l) \sigma^z(l') \left(+ \sum_l \Delta \cdot \sigma^x(l) \right). \tag{2.75}$$

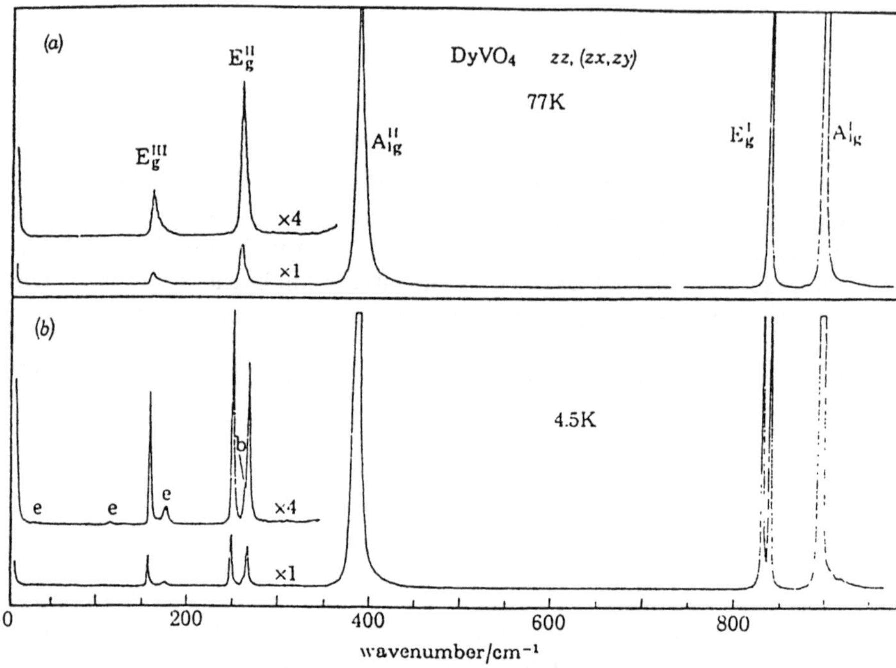

Fig. 2.14. Phonon Raman spectra of $DyVO_4$ at 77 K (**a**), and 4.5 K (**b**); polarizations zz and (zx, zy) are superimposed. Lines marked e are of electronic origin, b is due to a leakage of a B_{2g} phonon. The splitting of the E_g phonons in the orthorhombic phase is clearly shown. From ref. [2.119]

This Hamiltonian is fundamental in magnetism and in ferroelectrics. For $\Delta = 0$, a second-order phase transition results. Within the molecular field approximation, $\mathcal{H}_{\text{int}}$ simplifies to

$$\mathcal{H}_{\text{MF}} = -\lambda \langle \sigma^z \rangle \sum_{l} \sigma^z(l), \tag{2.76}$$

$\lambda = \sum_{l \neq l'} J(l, l') = J(0)$. For self-consistency

$$\langle \sigma^z \rangle = \tanh(\lambda \cdot \langle \sigma^z \rangle / kT), \tag{2.77}$$

leading to a Brillouin function for a temperature dependence of the pseudospin eigenvalue $\langle \sigma^z \rangle$ of the vibronic system, which falls from unity at $T = 0$ to zero at the transition temperature $kT_{\text{D}} = \lambda$.

This simple theory is applicable to $DyVO_4$ ($T_{\text{D}} = 14$ K) and $DyAsO_4$ ($T_{\text{D}} = 12.2$ K) [2.119] which crystallize in the tetragonal zircon structure (D_{4h}^{19}, $Z = 2$ in the physical unit cell, RE site: D_{2d}; below T_{D} : D_{2h}^{28}, C_{2v}, (orthorhombic)). In both compounds there exist two low lying Kramers doublets Γ_6, Γ_7 of Dy^{3+}, which are nearly degenerate (separation in $DyVO_4 \approx 9\,\text{cm}^{-1}$). This pseudo- degeneracy is lifted below T_{D}. The splitting increases

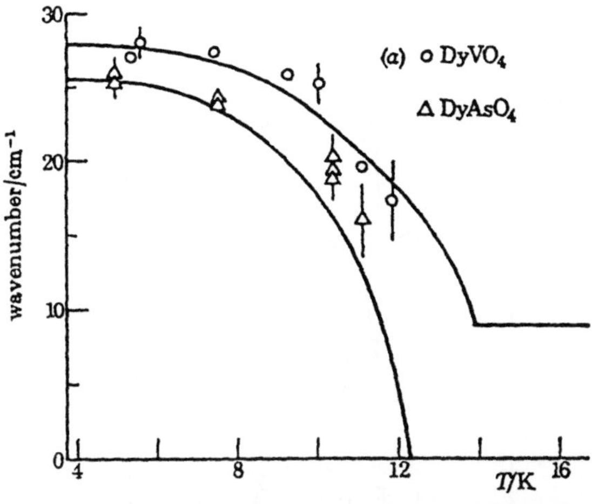

Fig. 2.15. Temperature dependence of the Raman frequencies of the lowest electronic modes in $DyVO_4$ and $DyAsO_4$ below T_D. The solid lines are calculated from molecular field theory, following (2.77). From [2.119]

at 4 K towards 27.5 cm^{-1} in $DyVO_4$, and 25.2 cm^{-1} in $DyAsO_4$. Both systems display ferro-quadrupolar alignment of the RE ions, indicating that $\boldsymbol{q} = 0$ lattice modes are involved, producing B_2 distortions on the RE sites which correspond to lattice modes of B_{1g} symmetry. In Fig. 2.14 some Raman spectra of $DyVO_4$ are depicted while in Fig. 2.15 the softening of the electronic mode, associated with the JT splitting of the lower electronic states, is plotted and compared with molecular field theory. In both compounds E_g phonon modes split below T_D (see e.g. Fig. 2.14), again displaying a Brillouin function dependence (2.77) with temperature.

In $TbVO_4$ the scheme of the low lying electronic levels is more complicated. A doublet (E) at $\approx 8\,cm^{-1}$ is sandwiched between two singlets (A_1 (lowest), and B_1) separated by 22.9 cm^{-1} at $T = 36$ K. Here a very large electronic splitting of $\approx 50\,cm^{-1}$ is observed below $T_D = 34$ K which shifts apart also the singlets. The theoretical interpretation uses matrix operators in a four-dimensional space [2.119]. The most general solutions for the coupled electron–phonon modes are obtained by solving approximately (e.g. within random phase approximation)) the equations of motion for the electronic operators. Here the structural distortion is isomorphic to a $\boldsymbol{q} = 0$ mode of B_{2g} symmetry in D_{4h}. This soft mode is active in light scattering both above and below T_D, the magnitude of the B_{2g} distortion of the tetragonal phase may be considered as the order parameter, equivalent to the net alignment of electronic quadrupole moments. *Harley* et al. [2.126] have studied, applying Brillouin spectroscopy, the fluctuation dynamics of the electronic ground state in $TbVO_4$ near T_D, using the molecular iodine filter technique [2.127] to attenuate the elastic component of the scattered light by $\approx 10^7$, see Fig. 2.16.

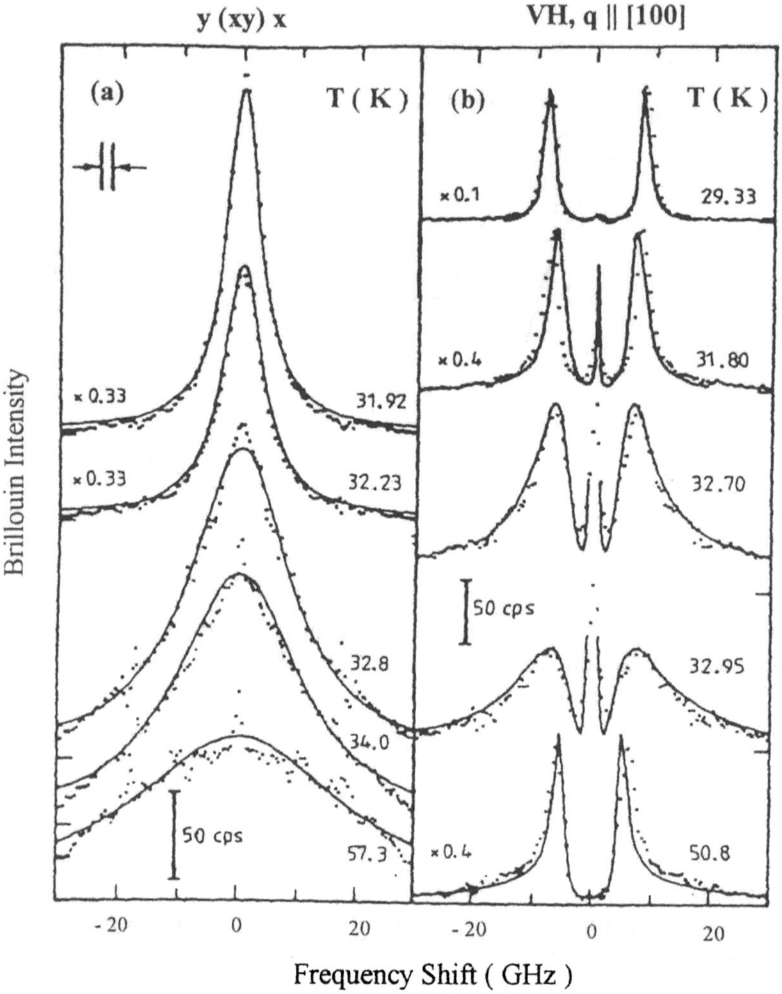

Fig. 2.16. Brillouin scattering at low frequencies in $TbVO_4$ by electronic fluctuations in the weak coupling case (left, $\boldsymbol{q} \parallel (110)$) and in the case of strongly coupled electron-acoustic phonon (TA) modes (right, $\boldsymbol{q} \parallel (100)$), light incident at $\approx 45°$ to z, scattered at $\approx -45°$ to z around $T_D = 32.60 \pm 0.30\,K$. At $T \gg T_D (50.8K)$ and $T < T_D (29.33\,K)$ the TA-phonon of asymmetric shape due to coupling is the only feature observed. The solid lines are theoretical fits. From [2.126]

The electronic levels and the lattice mode are only weakly coupled when the wave vector $\boldsymbol{q}$ lies along the (110) direction, whereas the coupling is maximal for propagation in the (100) direction. On the left side of Fig. 2.16 the spectra for the uncoupled case display the dynamic susceptibility associated with the purely electronic degrees of freedom. The linewidth increases strongly with increasing T. On the right-hand side, for the strong coupling

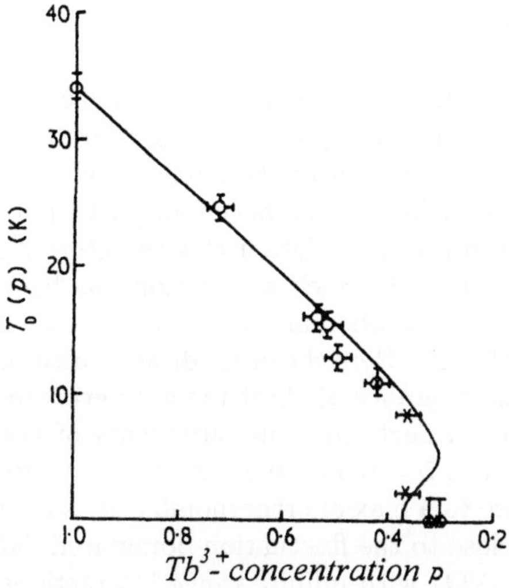

Fig. 2.17. Concentration (p) dependence of the JT-phase transition temperatures T_D in $Tb_pGd_{1-p}VO_4$ as observed from the splitting of E_g phonons in the Raman spectra (circles) and from optical birefringence (crosses). The solid line is calculated from molecular field theory. From [2.128]

case the broad asymmetric lines suggest strong interference effects between acoustic phonon and electronic modes, which are presumably due to magnetic fluctuations in the ground state. Despite their complex shapes, the spectra can be quantitatively interpreted along the lines discussed above [2.126]. The coupling constant between the electronic excitation and the optical field has the opposite sign of the coupling constant with the TA phonon. This is expected for magnetic dipole transitions.

In general, the CJTEs form a class of phase transitions for which the mechanism driving the transition is fully understood on a microscopic level. This is not valid for the large majority of other structural phase transitions.

Besides the coupling to optical phonons at $\boldsymbol{q} = 0$ taken into account up to now, the JT ions can also couple to a macroscopic elastic strain and to acoustic phonon modes of $\boldsymbol{q} \neq 0$. Strain coupling will be important whenever the elastic constants display any strong anomaly as a function of temperature. In many cases a definitive discrimination between coupling of electronic states with acoustic or optic phonons or both is difficult to perform [2.118].

The role of dilution of the JT active ions has been studied in a series of mixed crystals $Tb_cGd_{1-c}VO_4$ as a function of the concentration c by observing $T_D(c)$ by Raman spectroscopy.[15] Such experiments also provide information on the range of the JT interaction. The result is shown in Fig. 2.17 [2.128], together with the results of mean-field calculations made with the ansatz $\lambda_c = c \cdot \lambda(c = 1)$ (virtual crystal approximation, VCA). Upon dilution, T_D decreases, and reaches 0 K near $c = 0.365$. The peculiar behavior near this critical concentration is due to the unique sequence of low-lying electronic

[15] The ground state of the Gd^{3+} ion $(4f^7)$ is an orbital singlet.

states (singlet–doublet–singlet, see above). VCA is thus a good approximation for long range interactions.

On the other hand, when the interactions are short range, molecular-field theory is inadequate. Dilution produces large perturbations over small regions, isolated clusters of magnetic sites are formed; they are cut off from the full network of magnetic sites and the isolated clusters cannot participate in cooperative ordering. Here the concepts of percolation theory enter: T_{D} will decrease to zero near the percolation limit. Such a situation has been studied in $Dy_cY_{1-c}VO_4$ [2.128], where optical phonon modes are known to contribute substantially to JT coupling [2.118]. Here the order disappears at a critical concentration $c \approx 0.4$. The preponderance of short-range interaction in the concentrated compound became evident from measurements of the critical exponent β of the order parameter, $\beta = 0.34 \pm 0.02$ [2.129], given by the distortion of the tetragonal structure to a biaxial orthorhombic structure below T_{D}. The value of β comes very close to the fluctuation-dominated, 3d Ising behavior ($\beta = 0.31$), while for $\mathrm{TbVO_4}$ with infinite range interactions $\beta = 0.50 \pm 0.02$, the mean-field value, was found.

Clearly the application of external uniaxial stress in specific directions, or of a magnetic field, will assist or hinder JT distortions depending on the type of multipole alignments induced by the external morphic effects (such as those produced by external fields or mechanical strain) as compared to the CJTE ordering. The situation is particularly clear for $\mathrm{TmVO_4}$ in an external magnetic field, where the Hamiltonian in molecular field approximation is (strain-coupling omitted, see (2.76)):

$$\mathcal{H}_{\mathrm{mag}} = -\lambda\langle\sigma^z\rangle \sum_l \sigma^z(l) + \frac{1}{2} g\mu_{\mathrm{B}} H \sum_l \sigma^x(l). \tag{2.78}$$

The energies corresponding to this Hamiltonian are $W = \pm\lambda \tanh(W/kT)$, if $W > \frac{1}{2}g\mu_{\mathrm{B}}H$ and $W = \pm\frac{1}{2}g\mu_{\mathrm{B}}H$ otherwise. The ground state splitting of $\mathrm{TmVO_4}$ as a function of the magnetic field, as observed in the absorption spectrum by a transition to an excited crystal field level, is shown in Fig. 2.18 [2.130].

2.3.3 Resonant 4*f*-Electron–Phonon Interaction

In the previous section the local or cooperative static distortions of crystal lattices, the most prominent source of electron–phonon interaction in magnetic insulators, have been discussed. Dynamic effects, e.g., the renormalization of excited phonon and electronic states due to this interaction, are almost as conspicuous, and again Raman spectroscopy, which detects both types of excitations with comparable sensitivities, is the most versatile method to investigate them. This area of research can be considered as a subsection of the work on dynamical interactions between RE ions and their environment, e.g. [2.118], [2.119].

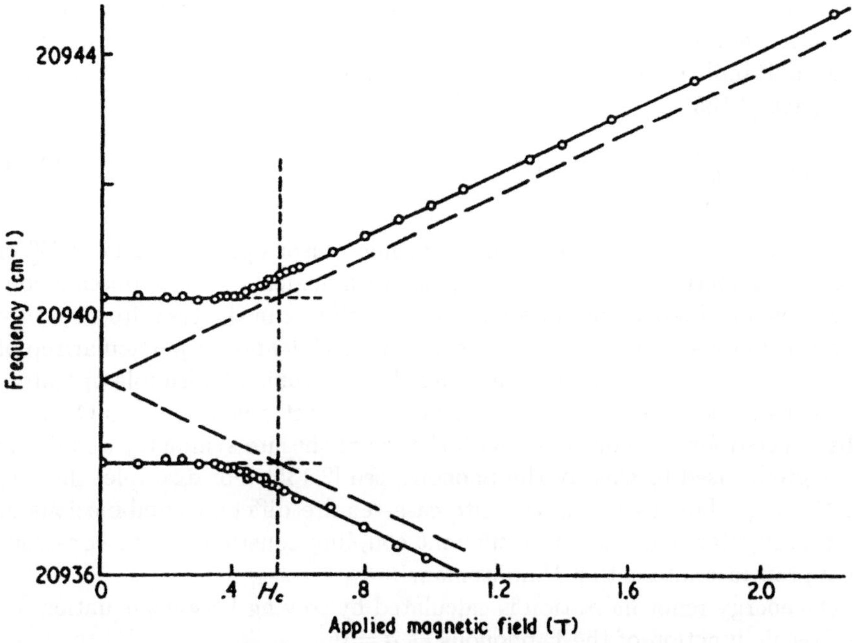

Fig. 2.18. Splitting of the electronic ground-state in $TmVO_4$, T = 1.4 K, as a function of an external magnetic field applied along [001], as measured by optical absorption to an excited crystal field level ($^3H_6 \rightarrow {}^1G_4$), lowest level. Broken lines: molecular field theory, solid lines: demagnetization effects and dipole fields considered. From [2.130]

From an experimental point of view the observed phenomena have to be separated into two classes, the case of weak interaction, where the coupled excitations are distinguishable according to their prevailing character, either as electronic or as phonon type excitations. In the case of strong coupling, the two types are inextricably mixed. Electronic transitions in the weak coupling case now have to be attributed to vibrons, obeying selection rules of the type discussed in Sect. 2.3.1.

In high-T_c superconductors (RE-cuprates, $REBa_2Cu_3O_x$, $6 \leq x \leq 7$) both coupling types have been observed. Details will be discussed in Sect. 2.4.1.2.

The theoretical treatment of renormalization effects starts either from the frequency dependent susceptibilities (2.73), generalizing these by considering the mixing of the electronic and phonon modes in the region where the dispersion curves cross. The renormalized excitation frequencies are given by the poles of the susceptibility functions, usually calculated in random-phase approximation. The method has been discussed in [2.132], see also [2.119]. Another approach uses Green's function techniques applied to the concept of magnetoelastic interactions [2.133], [2.134] and calculates the self-energies of the phonons in lowest (second) order perturbation theory. The self-energies of the two degenerate components of an E-phonon depend differently on the magnetic field B, thus giving rise, e.g., to the magnetic phonon splitting: Here the Hamiltonian (2.70) is used, where the interaction part is written in real space [2.135]:

$$\mathcal{H}_{\mathrm{ME}} = \sum_{l,\Gamma} b_\Gamma(l)\varphi_\Gamma(l)\mathrm{O}_\Gamma(l) \tag{2.79}$$

φ_Γ is a phonon operator, O_Γ is an electronic tensor operator (sum of O_q^k of suitable symmetry at site l), b_Γ is a phenomenological electron–phonon coupling constant, l labels the RE sites per unit cell (six in the tysonite structure of approximate structure D_{6h}^3, see Sect. 2.3.1). Γ denotes a particular rep of the site symmetry group. As usual, only the electronic quadrupole operators in (2.79) will be considered for simplicity. The electronic operators O_Γ have to be selected for and correlated with the rep of the site symmetry and of the factor group used to classify the phonons, see [2.135]. For example, the E_{1g} and the E_{2g} phonons in the tysonite case require different combinations of the quadrupolar operators and different coupling constants b_Γ to construct the appropriate interaction Hamiltonian.

The energy renormalization is calculated by solving Dyson's equation for the Green's function of the E-phonons at $\boldsymbol{q} = 0$:

$$\mathcal{D}^{-1}(\omega) = \mathcal{D}_0^{-1}(\omega) + 2\omega_0 \mathcal{S}(\omega), \tag{2.80}$$

where ω_0 is the uncoupled phonon frequency and

$$\mathcal{D}_0^{-1}(\omega) = \begin{pmatrix} (\omega^2 - \omega_0^2) & 0 \\ 0 & (\omega^2 - \omega_0^2) \end{pmatrix} \tag{2.81}$$

is the unperturbed Green's function.

Finally $\mathcal{S}(\omega)$ is a 2×2 matrix representing the frequency dependent self-energy due to the electron–phonon interaction:

$$\mathcal{S}(\omega) = \begin{pmatrix} \Sigma_{\Gamma,\Gamma} & \Sigma_{\Gamma,\Gamma'} \\ \Sigma_{\Gamma',\Gamma} & \Sigma_{\Gamma,\Gamma} \end{pmatrix}; \quad \Sigma_{\Gamma,\Gamma'} = \Sigma^*_{\Gamma',\Gamma}. \tag{2.82}$$

Γ, Γ' are the reps of the electronic states in site symmetry correlated with the symmetry types of the phonons in factor group symmetry.

The $\Sigma_{\Gamma,\Gamma'}(\omega)$ are abbreviations for the functions:

$$\Sigma_{\Gamma,\Gamma'}(\omega) = 2b_\Gamma b_{\Gamma'} \sum_{n,m,l} \frac{(\mu_n - \mu_m)\langle m|\mathrm{O}_\Gamma(l)|n\rangle\langle n|\mathrm{O}_{\Gamma'}(l)|m\rangle}{\omega - (\omega_m - \omega_n)}. \tag{2.83}$$

Again, $|m\rangle, |\mathrm{n}\rangle$ label the crystal field levels of the ground state multiplet components at energies ω_m, ω_n, having thermal occupation numbers μ_m, μ_n. The resonant character of the self-energies is obvious, linewidth effects, which are of importance in the case of resonance, have not been considered in (2.83). The self-energy has the form of a quadrupolar susceptibility, the response of the $4f$-shell charge distribution to the lattice deformation by the phonon. The interaction mechanism through emission and reabsorption of virtual phonons, typical for a second-order perturbation theory, is evident. As a consequence of the resonant behavior of the interaction, the effects of the phonon self-energies can also be discussed qualitatively in terms of the familiar concept of anticrossings.

After summation over the electronic states which can interact with optical phonons and over all equivalent RE sites in the unit cell, one arrives at the following expressions for the diagonal ($B = 0$) and the off-diagonal ($B > 0$) parts of (2.83), which have been adapted to the special case of Ce^{3+}, ($4f^1$), $^2F_{\frac{5}{2}}$ (3 levels with transition energies $\hbar\omega_{i,j}$), $\Sigma_{\Gamma,\Gamma} = \Sigma_{\Gamma',\Gamma'}, \Sigma_{\Gamma,\Gamma'} = \Sigma_{\Gamma',\Gamma}$, $kT \ll \hbar\omega_{1,2}$:

$$\Sigma_{\Gamma,\Gamma}(\omega) = b_\Gamma^2 \times \left[\frac{\mu_2 - \mu_1}{\omega^2 - \omega_{1,2}^2} \cdot \omega_{1,2}\, Z_{1,2}^{\Gamma,\Gamma} + \frac{\mu_3 - \mu_1}{\omega^2 - \omega_{1,3}^2} \cdot \omega_{1,3}\, Z_{1,3}^{\Gamma,\Gamma} + \frac{\mu_3 - \mu_2}{\omega^2 - \omega_{2,3}^2} \cdot \omega_{2,3}\, Z_{2,3}^{\Gamma,\Gamma} \right]; \tag{2.84}$$

$$\Sigma_{\Gamma,\Gamma'}(\omega) = \mathrm{i} b_\Gamma^2 \cdot \omega \left[\frac{\mu_{1-} - \mu_{1+}}{\omega^2 - \omega_{1,2}^2} \cdot Z_{1,2}^{\Gamma,\Gamma'} + \frac{\mu_{1-} - \mu_{1+}}{\omega^2 - \omega_{1,3}^2} \cdot Z_{1,3}^{\Gamma,\Gamma'} \right]. \tag{2.85}$$

The $Z_{N,N'}^{\Gamma,\Gamma'}$ are sums of products of quadrupolar matrix elements [see (2.83)] which can be calculated with the crystal field states. In (2.84) the Zeeman effect has been neglected with respect to the crystal field splitting, but not in (2.85) in the difference of occupation numbers. This difference varies $\sim \tanh\{\hbar(\omega_{1-} - \omega_{1+})/k_B T\}, \omega_{1-} - \omega_{1+}$ is the ground state Zeeman splitting.

The diagonal elements (2.84) of the self-energy do not depend on the magnetic field and only slowly depend on temperature. They describe the energy shifts of the phonon states at $B = 0$ with respect to the uncoupled case. The off-diagonal elements (2.85) are the additional contributions to the phonon self-energies due to the paramagnetic saturation of the crystal in the field, i.e. the phonon response to a field-induced ordering of the electronic quadrupoles.

Only approximate solutions of the eigenvalue problem of (2.80) can be given. The frequency shift $\Delta\omega$ due to magneto-elastic interactions is:

$$\Delta\omega = \omega - \omega_0 = \Sigma_{\Gamma,\Gamma}(\omega, T) \pm |\Sigma_{\Gamma,\Gamma'}(\omega, T, B)|, \tag{2.86}$$

if $\Delta\omega \ll \omega, \omega_0$. A more elaborate solution has been given in [2.133]. The size of the magnetic phonon splitting $|\Sigma_{\Gamma,\Gamma'}(\omega, T, B)|$ saturates at $B \to \infty$.

Magnetic-field-dependent anticrossing effects between optical phonons and Zeeman components of crystal field states can be expected in crystals with a large magnetic splitting factor of the ground state doublet. The chances are that the magnetic field sweeps the electronic levels across one or more phonon levels. The theoretical interpretation of anticrossing effects has to start from the same model of magnetoelastic self-energies as before. However near the hypothetical crossing points of the uncoupled modes some of the approximations made above will break down. These are: (a) The Zeeman splitting of the excited electronic level has to be taken into account, especially in the resonance denominators. (b) The frequencies in the expressions (2.84) and (2.85) for the self-energies have to be considered as variables in the eigenvalue problem, not as fixed numbers. The number of roots increases accordingly and will describe correctly, for example, the complex behavior of a Zeeman doublet crossing a degenerate phonon pair, where each Zeeman component interacts solely with one phonon component of the correct symmetry (polarization). (c) The finite widths of the coupled modes must be considered by means of adjustable parameters representing their lifetimes.

Anticrossing phenomena are often treated by a phenomenological coupling model [2.136]: The Green's functions $G_{i,j}$ derived from the two uncoupled modes at ω_1 and ω_2 and at the new eigenfrequencies ω_+, ω_-, are eigenfunctions of:

$$\begin{pmatrix} \omega^2 - \omega_1^2 + \mathrm{i}\omega\Gamma_1 & \Delta^2 \\ \Delta^2 & \omega^2 - \omega_2^2 + \mathrm{i}\omega\Gamma_2 \end{pmatrix} \cdot \begin{pmatrix} G_{1,1} & G_{1,2} \\ G_{2,1} & G_{2,2} \end{pmatrix} = \begin{pmatrix} 1 & 0 \\ 0 & 1 \end{pmatrix}; \tag{2.87}$$

$\omega_\pm$ are the roots of the secular equation [real part of (2.87)]. According to the fluctuation–dissipation theorem the scattered intensity $I(\omega)$ of the two coupled oscillators with P_1, P_2 as scattering amplitudes of the noninteracting modes is given by:

$$I(\omega) \sim \mathrm{Im}\chi(\omega) = \mathrm{Im}[P_1^2 G_{1,1} + 2P_1P_2G_{1,2} + P_2^2 G_{2,2}], \tag{2.88}$$

This model is equivalent to the previous microscopic model as long as the use of a phenomenological coupling constant Δ^2, independent of frequency and magnetic field, is an acceptable approximation.

Obviously, the $4f$-electron–phonon interaction should also manifest itself in the *electronic* excitation spectra by a renormalization of electronic energies,, but in many cases these effects are masked by static crystal field effects, e.g., in diluted compounds. In a noteworthy case, however, phonon mediated Davydov splittings between excited crystal field levels in the lowest multiplet component of PrF_3 have been detected [2.137]. Thus, the excitations of the $4f$-system cannot be treated in the single-ion approximation but rather as a Frenkel-type exciton with the dynamic crystal field causing the interionic interaction. In addition, induced collective magnetic moments of the system of Pr^{3+} ions in PrF_3 have been detected, in spite of the fact that the magnetic moments of the single Pr^{3+} ions have been quenched completely in PrF_3 by the crystal field of low symmetry [2.138].

The theoretical treatment of coupled electronic excitations [2.137] uses the equation of motion technique for the Green's functions [2.139]. The JT Hamiltonian (2.70), with the Zeeman interaction added, is transformed into the projection operator representation $\mathcal{P}^{\alpha,m}_{M,N} = |\alpha m M\rangle \langle \alpha m N|$, where α runs over the equivalent sites in the unit cell, m enumerates all unit cells in the crystal and $|M\rangle, |N\rangle$ are crystal field states. The excitation energies of the system are derived from the poles of the following Green's functions in time or frequency domain: $\langle\langle \mathcal{B}_{q,p}, \mathcal{B}^+_{q',p'} \rangle\rangle$, $\langle\langle \mathcal{B}_{q,p}, \mathcal{P}^{\alpha,m}_{R,S} \rangle\rangle$, $\langle\langle \mathcal{P}^{\beta,n}_{M,N}, \mathcal{B}^+_{q',p'} \rangle\rangle$, $\langle\langle \mathcal{P}^{\beta,n}_{M,N}, \mathcal{P}^{\alpha,m}_{R,S} \rangle\rangle$; $\mathcal{B}_{q,p} = (c^+_{q,p} + c_{q,p})$, the sum of phonon operators at wavevector q and branch p. The equation of motion of the operators $\mathcal{O}_1, \mathcal{O}_2$, $\omega\langle\langle \mathcal{O}_1, \mathcal{O}_2 \rangle\rangle_\omega = \langle [\mathcal{O}_1, \mathcal{O}_2]_\pm \rangle + \frac{1}{\hbar}\langle\langle [\mathcal{O}_1, \mathcal{H}]_-, \mathcal{O}_2 \rangle\rangle_\omega$ is solved in the random-phase approximation. Here the single brackets indicate the expectation value, the square brackets indicate the commutator and the index ω the frequency domain.[16] This results in a system of linear equations for the Green's functions, written as a matrix Dyson equation, which can be further simplified (block-diagonalized) by introducing phase-adjusted exciton functions, i.e. linear combinations of electronic projection operators which transform irreducibly under the operations of the point group of the q vector of the exciton function under consideration. An example can be found in [2.137]. The solution of the eigenvalue problem gives simultaneously the energies and eigenfunctions of coupled phonon and excitonic $4f$-states. The phonon relations are identical with the previous results [e.g. (2.86)]. The eigenfunctions may be used to calculate oscillator strengths or scattering cross sections for the coupled excitations.

2.3.3.1 Weak Coupling. Renormalization effects of low-lying electronic and optical phonon states, as well as anticrossing effects, can be best observed in crystals with a low number of phonon branches with low damping in the energy region of interest, i.e. with a unit cell comprising a small number of atoms, among which the magnetic ions should cover a large percentage of the occupied sites. Hydrated crystals are not a good choice in this respect, because this percentage is reduced and the vibrational modes of the water molecules, especially the external ones, introduce additional and effective decay channnels of the excited electronic states [2.29].

Three phenomena have been found which display self-energy effects of excited electronic or vibrational levels in a magnetic crystal: Shifts of energy levels with respect to the case of zero interaction which can be approached asymptotically in diluted crystals, phonon–electron anticrossing effects and, most conspicuously, the splitting of degenerate vibrational levels in a magnetic field (external or internal, i.e. due to magnetic order).

In crystals of uniaxial or higher symmetry degenerate optical phonon states of E or F-type symmetry may split under the application of a strong external magnetic field along the direction of high symmetry. A similar phenomenon is found in low-symmetry, biaxial crystals if quasi-degeneracies exist, i.e. if the separation between two nearby optical phonon modes at $q = 0$

[16] Remember: dimension of $\langle\langle \mathcal{O}_1, \mathcal{O}_2 \rangle\rangle_\omega$ = dimension of $(\mathcal{O}_1 \times \mathcal{O}_2)\times$ time

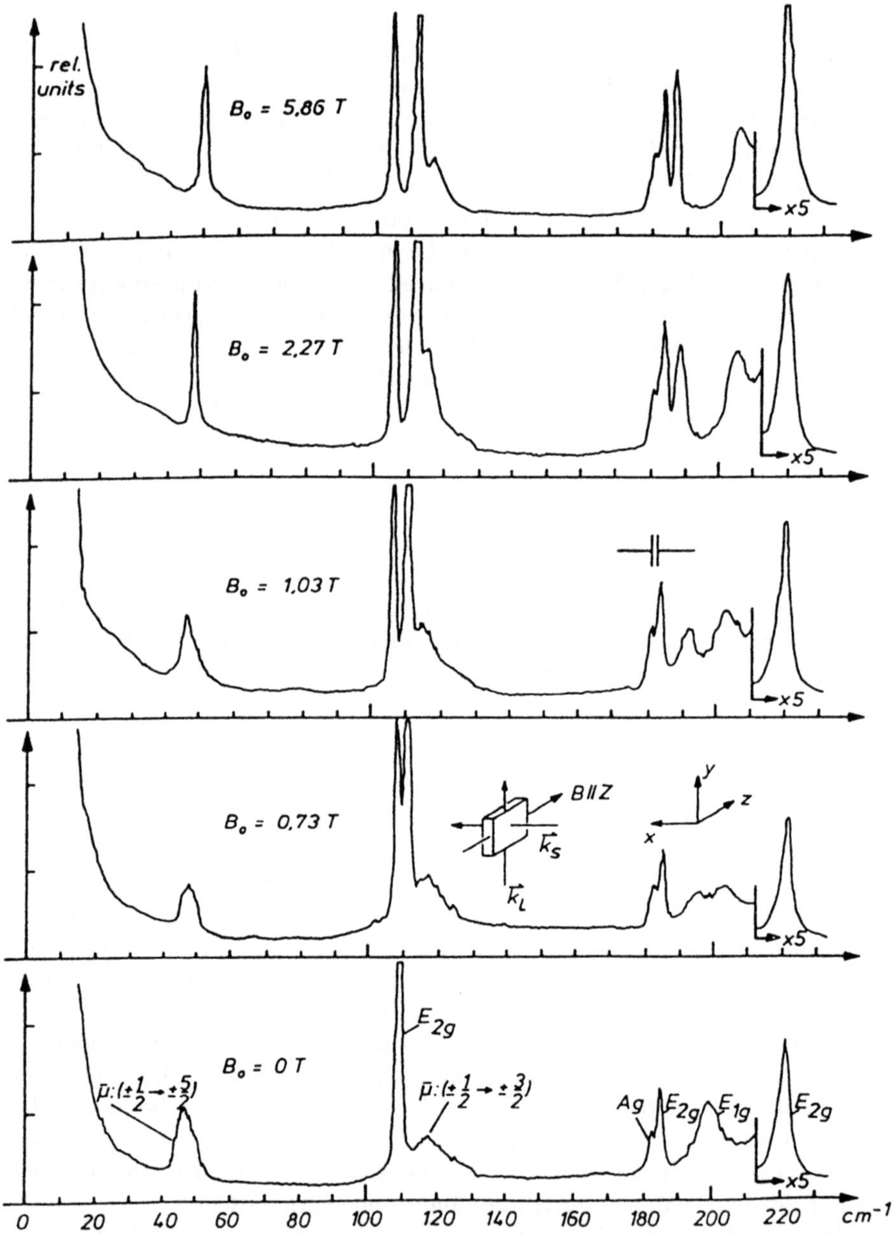

Fig. 2.19. Raman spectra of hexagonal $CeCl_3$, $T = 2\,K$; $y(xy + xz)x$ at different magnetic fields $B \| z$. The electronic transitions within the $4f^1(^2F_{\frac{5}{2}})$ configuration are indicated by the crystal quantum numbers $\bar{\mu}$ of the excited states (see Table 2.A.1), the phonon transitions by the symmetry types. From [2.140]

is smaller than the interaction energy. These effects are easily detectable signatures for the presence of electron–phonon interaction. This splitting can be understood qualitatively from symmetry considerations alone [2.98], [2.99]: It is a consequence of the interaction, caused by removing the time-reversal symmetry, but not by lowering the crystallographic symmetry of the system. As the crystal is magnetically saturated at low temperatures in the external field, the point group of the system changes from a grey group to one of the magnetic point groups (Sect. 2.2.7).

The structural prerequisites mentioned are well met in the RE halides ($RECl_3, REF_3$), at least for the light REs, where the chlorides display the hexagonal space group $C_{6h}^2, P6_3/m$, site symmetry C_{3h} and the fluorides the tysonite structure (Sect. 2.3.1). In Fig. 2.19 we have plotted Raman spectra of paramagnetic single crystals of $CeCl_3$ in a magnetic field along the hexagonal axis. Electronic and phonon transitions are clearly separated. Both the E_{2g} phonon at $109\,cm^{-1}$ and the E_{1g} phonon at $197\,cm^{-1}$ display a splitting $\Delta E_{phon.}(B)$ into two components which saturates at high fields. The saturation splitting amounts to $\pm 3.5\,cm^{-1}, (E_{2g})$ and $(+8.0, -10.0)\,cm^{-1}$, (E_{2g}) [2.140], [2.141]. The splitting follows the paramagnetic saturation [$\sim \tanh \Delta W_{el}(B)/k_B T$] of the two-level electronic ground state (energy difference ΔW_{el} in field B) as becomes also obvious from an example in CeF_3, Fig. 2.20 [2.135]. The effect at saturating fields $\Delta E(B \to \infty)$ is not correlated to the size of the linear Zeeman effect $\Delta W_Z(B)$ of the ground state [2.142].

An appreciable reduction of the field dependent part of the line width of the split phonon components, especially of the component at lower energies ($\sim 1 - \tanh^2[\Delta W_{el}(B)/kT]$ is also observed. Physically, the reduction of this part of the line width is due to the freezing of the fluctuating magnetic fields at a RE site, i.e. due to a suppression of magnetic noise, which originates from the interionic interactions [2.133]. This linewidth reduction is paralleled by a field-induced increase of the magnetic relaxation times in these systems. The effect is smaller for the upper phonon component, here the effect of suppression of fluctuations is counterbalanced by additional decay channels via the system of interacting magnetic moments.

Figure 2.20 also demonstrates the considerable dependence of the phonon self-energies on the concentration c of the Ce^{3+} ion. Both the saturation splitting and the zero-field shift vary linearly with c, the relative sizes of the two self-energies are not directly related to each other. The temperature dependence of the zero-field shift is determined by the thermal occupation numbers of the crystal–field levels (see 2.84) and its size can be described by the same coupling constants as the magnetic phonon splitting. This has been demonstrated experimentally [2.135] by comparing diamagnetic LaF_3 with $Ce_cLa_{1-c}F_3$ and taking the lanthanide contraction and the lattice anharmonicity into account, and is a consequence of (2.84) and (2.85). Magnetic phonon splitting also occurs in magnetically ordered crystals. Fig. 2.21 shows Raman spectra of $Tb(OH)_3$ [2.143]. Above $T_c = 3.72\,K$ (Fig. 2.21a) the split-

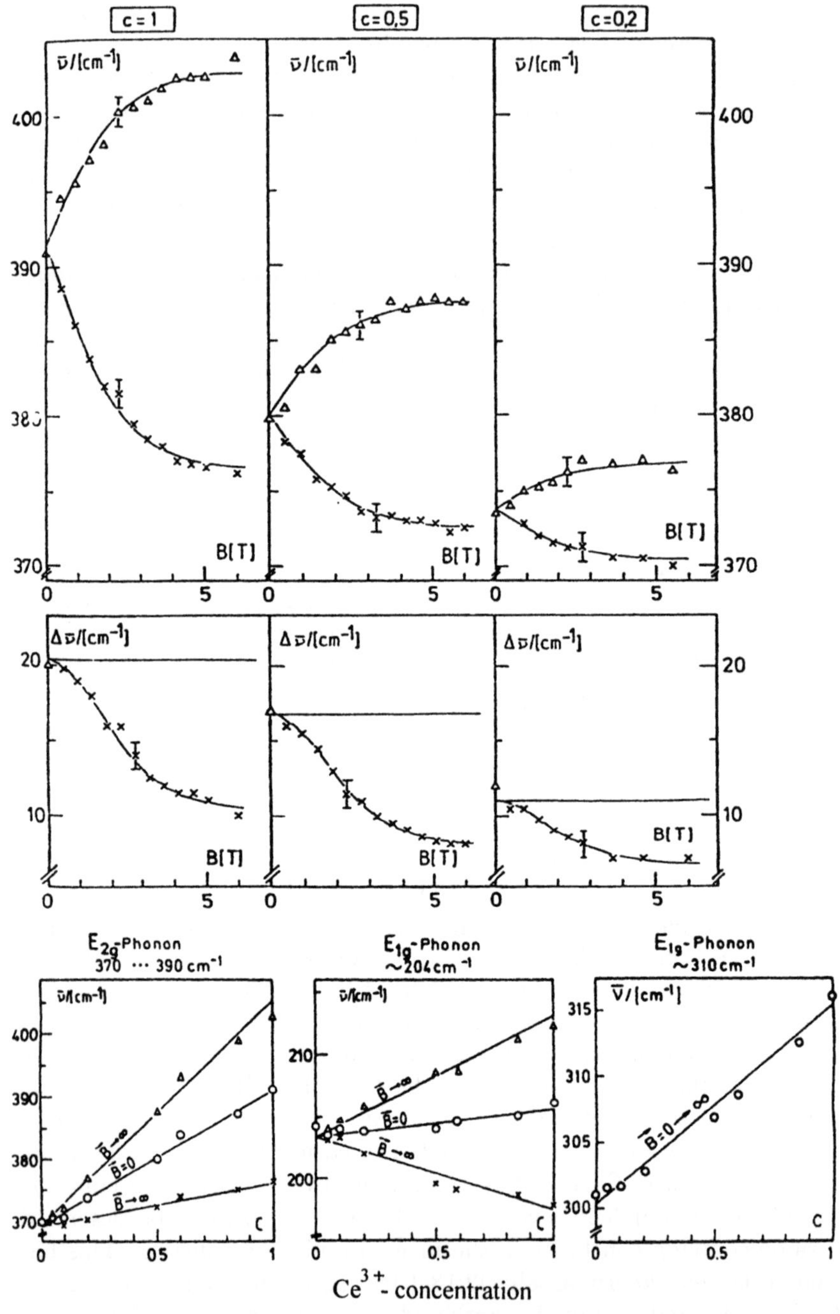

Fig. 2.20. (*Above*): Magnetic splitting ($\bar{\nu}$) of the E_{2g} phonon near 390 cm^{-1} (upper row) and magnetic linewidth reduction $\Delta\bar{\nu}$ (FWHM) of the lower branch of the split pair for different concentrations c in $Ce_cLa_{1-c}F_3$ at $T = 2.1$ K. (Below): Zero-field shift of optical phonons and magnetic phonon splitting (extrapolated to $B \to \infty$ according to a tanh behavior), as a function of c. The E_{1g} phonon near 310 cm^{-1} does not split under an external magnetic field. From [2.135]

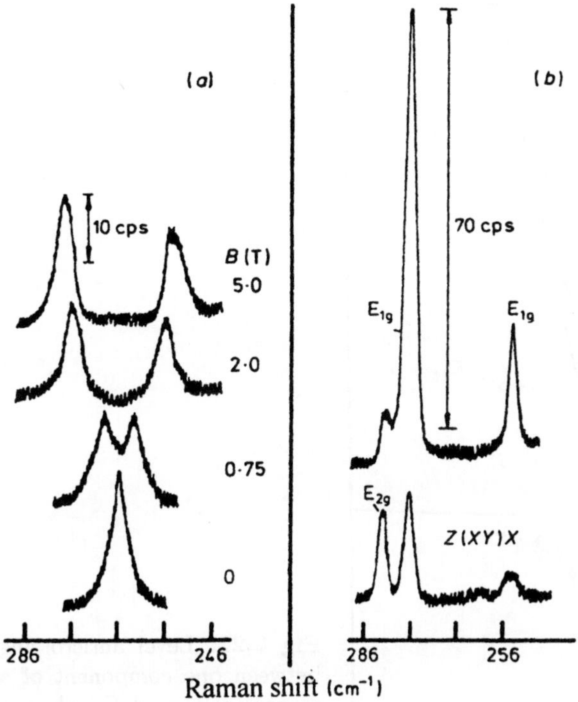

Fig. 2.21. Magnetic splitting of an E_{1g} phonon in $Tb(OH)_3$ (**a**) above $T_c = 3.72\,K$ in an external field $B \| z, z(xz)x$ and (**b**) spontaneous splitting below T_c in the ferromagnetic state ($T = 1.8\,K$, $B_{ext.} = 0$), $z(xz)x$ and $z(xy)x$. The weak E_{2g} phonon at $282\,cm^{-1}$ does not split. From [2.143]

ting of an E_{1g} phonon near $266\,cm^{-1}$ is induced by an external magnetic field, below T_c (Fig. 2.21b) a spontaneous splitting is observed with different scattering cross sections of the two components.

It is required by symmetry that the Raman radiation scattered by the split phonon components in Faraday configuration of the magnetic field be circularly polarized with orthogonal polarizations of the two phonon components [2.98], [2.99]. This is also obvious from the (2×2) Dyson matrix for the Green's functions of the E phonons with imaginary off-diagonal elements (2.82) [2.133], [2.135]. This has been confirmed experimentally [2.144], however double refraction and Faraday rotation in the crystal, together with the finite aperture of the cone of scattered light and an inevitable small misalignment of the crystal, may mask this effect.

In cubic crystals triply degenerate phonons occur. Magnetic phonon splitting by applying a magnetic field in any direction of the crystal has been predicted theoretically for these states [2.98], [2.99] but, due to symmetry, not for doubly degenerate states in cubic crystals. The energy of one phonon component of this triplet remains unshifted and the component is polarized in the

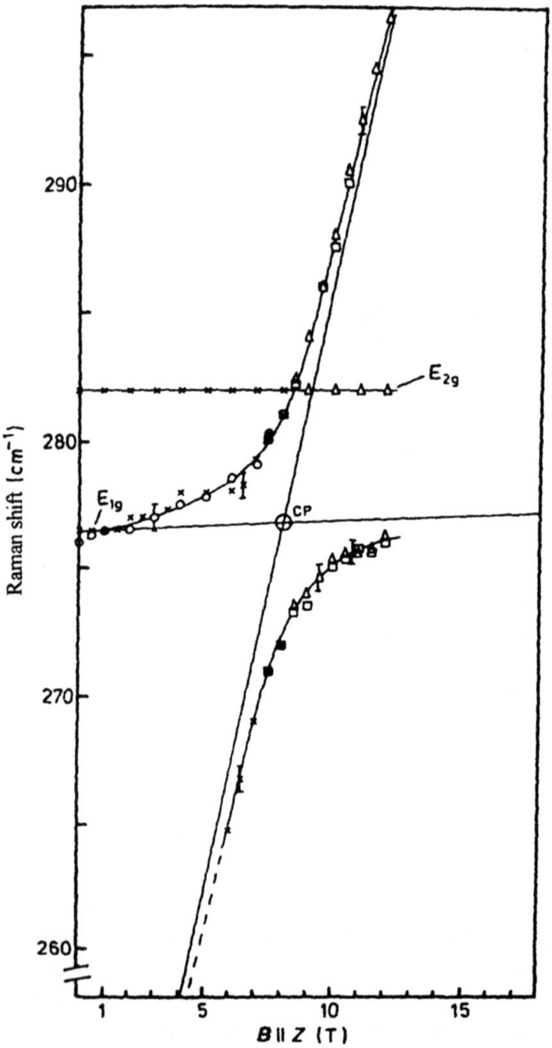

Fig. 2.22. Level anticrossing between one component of a magnetically split E_{1g} phonon and an electronic state (μ = 1) of $^7F_6$ in Tb(OH)$_3$, starting at 236 cm^{-1} (B = 0 T) with an effective g factor of 9.4. The different symbols refer to measurements in a superconducting system ($B \leq 8$ T) and in a Bitter coil ($B \leq$ 12 T). CP: hypothetical crossing point. Full curves: best fits. From [2.143]

direction of the applied field, two other components will shift symmetrically, becoming right and left circularly polarized in the ploane perpendicular to the field. This effect has been detected in dysprosium-aluminum-garnet (DAG, $Dy_3Al_2(AlO_4)_3$; O_h^{10}, (Ia3d)) [2.145]. Saturation splittings of 0.4 cm^{-1} and 0.75 cm^{-1} have been observed for F_{2g} phonons at 260 and 239 cm^{-1}.

Striking examples of Raman-detected anticrossing phenomena of a 4f-transition with optical phonons in Tb(OH)$_3$ (isomorphous with $CeCl_3$ [2.143]) and $LiTbF_4$ (scheelite structure, C_{6h}^2, number of formula units per unit cell $Z = 2$, site symmetry C_{3h} [2.146]) are shown in Figs. 2.22–24. The crystal field state, derived from $^7F_6$ with the largest magnetic splitting factor, is lowest in both compounds. Figure 2.22 demonstrates that only one component of

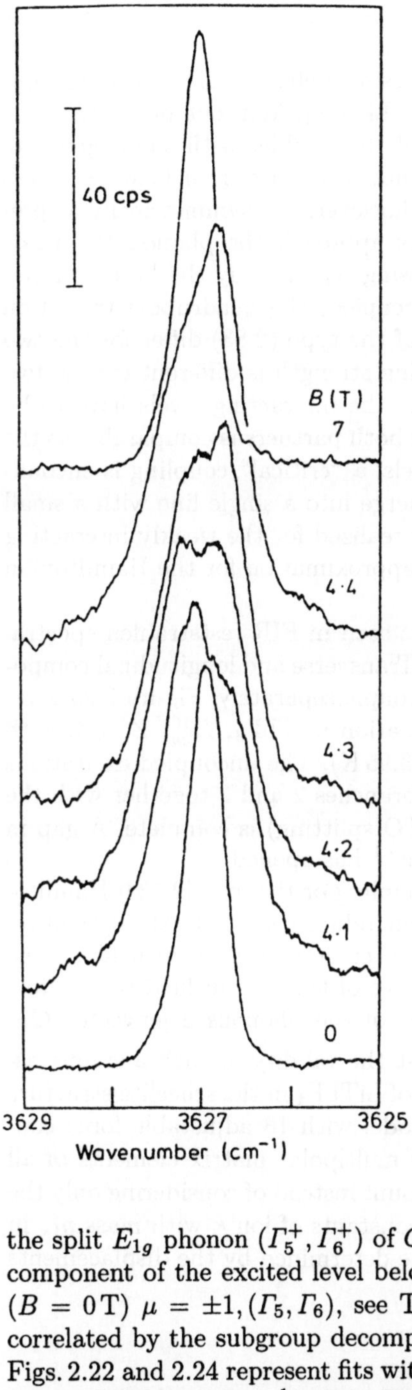

Fig. 2.23. Raman spectra in $z[xy, yy]x$ polarization of $Tb(OH)_3$ in the anti-crossing region between an E_{2g} phonon, $(OH)^-$ valence stretch, and a Zeeman level ($\mu = 2$) of the $^7F_4$ multiplet component. The scattering cross section of the electronic level outside the region of hybridization is undetectably low. From [2.143]

the split E_{1g} phonon (Γ_5^+, Γ_6^+) of C_{6h} (Fig. 2.21) interacts with one Zeeman component of the excited level belonging to the same rep at $\bar{\nu} = 234\,\mathrm{cm}^{-1}$, ($B = 0\,\mathrm{T}$) $\mu = \pm 1, (\Gamma_5, \Gamma_6)$, see Table 2.A.1. Both reps ($\Gamma_{5,6}^+ \leftrightarrow \Gamma_{5,6}$) are correlated by the subgroup decomposition of C_{6h} [2.50]. The solid curves in Figs. 2.22 and 2.24 represent fits with (2.87). In Fig. 2.22 exactly at resonance a gap of almost $10\,\mathrm{cm}^{-1}$ opens. Figure 2.23 demonstrates the resonance between an E_{2g} phonon in the OH^- valence stretch region of $Tb(OH)_3$ and a

Zeeman level ($\mu = 4$) of the $^7F_4$ multiplet. The OH^- ions are the constituents of the first coordination shell of the RE ions.

Figure 2.24 demonstrates the anticrossing behavior of a Zeeman doublet with a doubly degenerate phonon in $LiTbF_4$. Without magnetoelastic (m.e., [2.134]) interaction the crossings of the doublet with the degenerate phonon would occur at two different values of the external field. At small field strength, the phonon starts to split, however, not symmetrically as previously, because the Zeeman components approach the phonon frequency differently with growing field. In the crossing region, only the lower component of the electronic ground state is occupied, the quadrupole transition matrix elements occuring in expressions of the type (2.83) differ for the two phonon branches. Accordingly, the coupling strength is different for the two branches. In addition, the finite widths of the interacting levels have to be considered. If the sum of the linewidths of both partners is comparable to the repulsion of the nominally undamped levels, a "critical" coupling is encountered [2.147] where the two excitations merge into a single line with a small frequency renormalization. Such a case is realized for the weakly interacting branch in Fig. 2.24. In this situation the approximation for the Hamiltonian leading to (2.87) is not acceptable.

Similar anticrossing effects can also be seen in FIR reststrahlen spectra. Here an additional complication appears: Transverse and longitudinal components of both interacting polar partners couple separately. Figure 2.25 gives an example [2.148] of this interesting situation in TbF_3, D_{2h}^{16}, $Z = 4$, with a ferromagnetic phase transition at $T_c = 3.95\,\mathrm{K}$). The uncoupled excitations cross near $B = 5\,\mathrm{T}$, the hybridization of branches 2 and 3 together with the exchange of oscillator strength (i.e. LO–TO splitting) is complete. A gap in the region of the LO–TO-modes of $\approx 2\,\mathrm{cm}^{-1}$ has opened.

In a simple interpretation of the JT coupling or the m.e. [2.134] Hamiltonian (2.63) the coupling constant b_Γ was introduced as due to the dynamical crystal field, i.e. as the modulation of the crystalline electric field by the phonons or, more precisely, as the derivative of the crystal field parameters B_q^k, (2.70) with respect to the amplitude of the phonons eigenvector Q_Γ, $b_\Gamma = \frac{\partial B^k}{\partial Q_\Gamma}$, multipole order $k = 2$. To test the validity of such a simple assumption the optical phonon eigenvectors of $LiTbF_4$ in the scheelite structure have been calculated using a rigid ion model with 18 adjustable force constants and 3 ionic charges and by taking multipolar matrix elements of all orders ($k = 2, 4, 6; q = -k, \ldots, k$) into account instead of considering only the lowest order $k = 2$ [2.149]. The coupling constants of ion κ with mass m_κ in the lth unit cell with the phonon (Γ, p) are determined by the displacements $u_\alpha(l, \kappa)$ in the α direction, using

$$b_{\Gamma,p,\kappa}^{k,q} = \left.\frac{\partial B_\kappa^{k,q}}{\partial Q_{\Gamma,p}}\right|_{Q=0} \cdot \left(\frac{\hbar}{2\omega_{\Gamma,p}}\right)^{\frac{1}{2}}$$

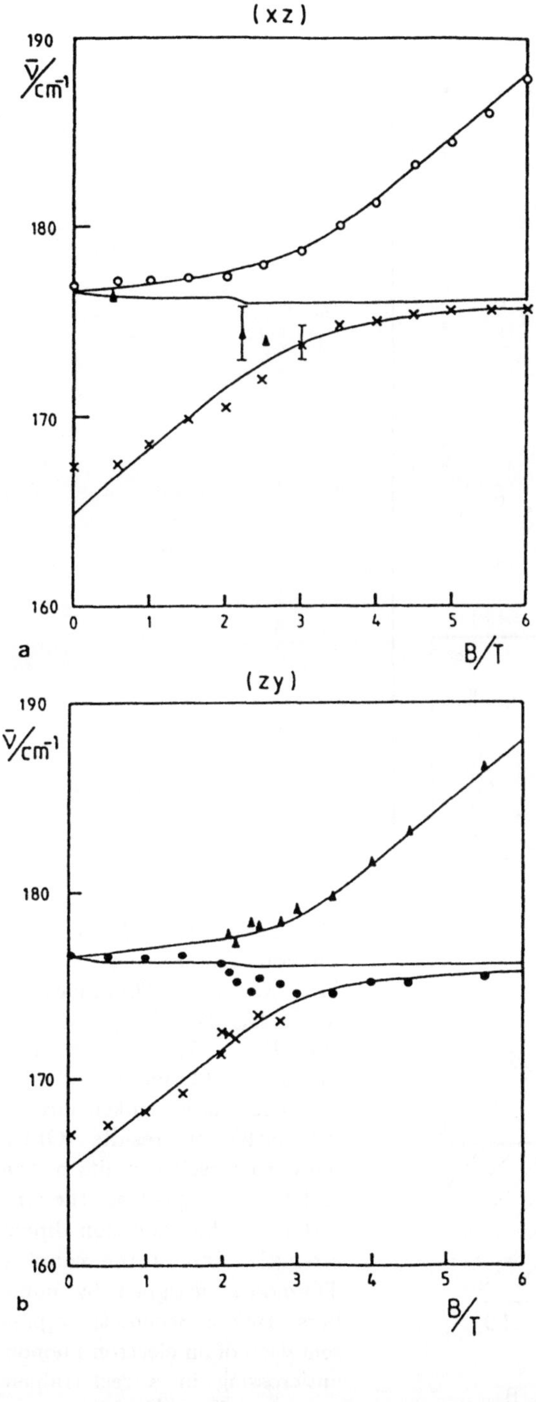

Fig. 2.24. Anticrossing region between an E_g phonon and two Zeeman components in $LiTbF_4$ with resonances of the uncoupled excitations at $\approx$ 2.15 T (critical coupling, small energy renormalization of the phonon branch, above) and $\approx$ 2.95 T (strong coupling, complete hybridization, influence of damping negligible, below). From [2.146]

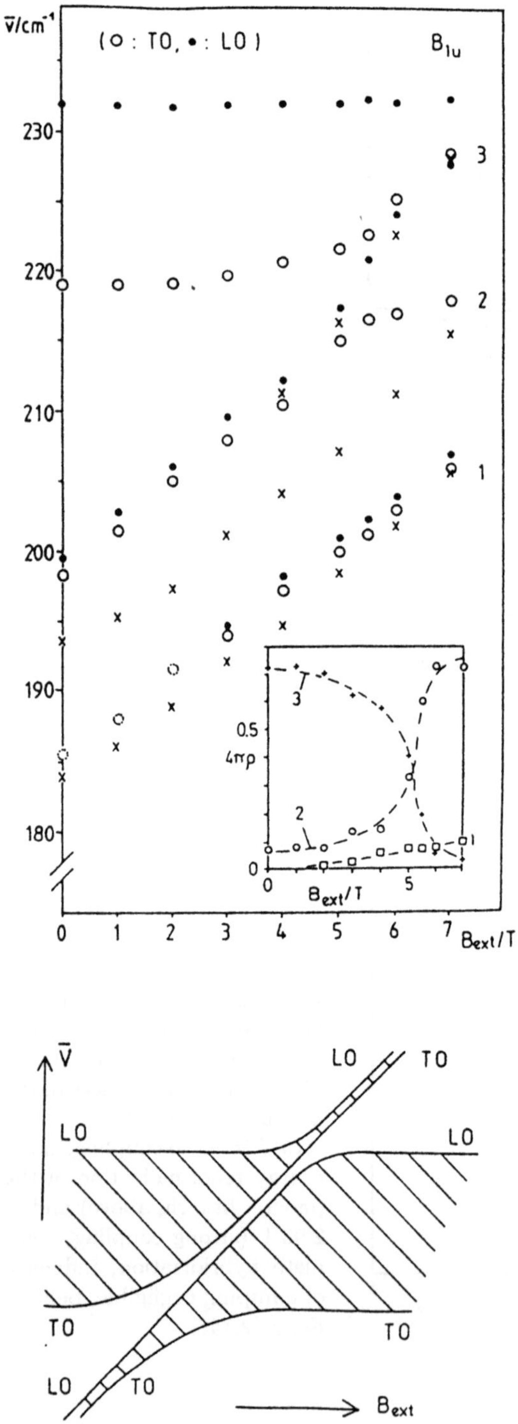

Fig. 2.25. *Above*: $4f$ electron-phonon anticrossings of polar LO- and TO-modes in $TbF_3, \boldsymbol{B}_{ext} \| x$, Voigt geometry, $T < T_c; \boldsymbol{E} \| z$. ○: TO-modes, •: LO-modes, ×: fluorescence data; broken circles: ir-transmission results. Other data from oscillator fits of the reststrahlen spectra, the insert gives the transition dipole strengths $4\pi\rho$ of the coupled TO-modes, assigned by numbers. *Below*: Schematic representation of an electron phonon anticrossing in a reststrahlen spectrum. From [2.148]

$$= \sum_{l,\kappa,\alpha} \frac{\partial B^{k,q}}{\partial u_\alpha \binom{l}{\kappa}} \Bigg|_{u_\alpha=0} \cdot e_\alpha^\kappa(p) \cdot \left(\frac{\hbar}{2\omega_{\Gamma,p} m_\kappa} \right)^{\frac{1}{2}} . \qquad (2.89)$$

Here $e_\alpha^\kappa(p)$ is the (κ, α) component of the phonon eigenvector p at the zone center. These coupling parameters $B_{\Gamma,p,\alpha}^{k,q}$ have been calculated explicitly in the point charge approximation which is a poor approximation for the calculation of the B_q^k because of the slow convergence of the lattice sums for B_0^2 and B_0^4. The $b_{\Gamma,p,\kappa}^{k,q}$ have been used to calculate the m.e. effects in $LiTbF_4$. In Table 2.11 the calculated results are compiled together with experimental values. The agreement is for most phonons surprisingly good (mostly within 30%) and can be improved in some cases by adjusting the phenomenlogical halfwidths of the coupled states. The performance of the point charge model in this context is improved over the case in which the B_q^k are calculated, because the dynamic part of the crystal field is essentially determined by the vibrational amplitudes which are small as compared to the interionic distances entering the static parameters.

Magnetoelastic effects have also been investigated in isomorphous $LiTmF_4$ as a function of temperature and hydrostatic pressure [2.150], see also [2.151], and compared with lattice dynamical calculations, similar to those discussed above, of the coupling constants. For the B_g phonons good agreement between theory and experiment is found.

The experimental findings on phonon-induced coherent electronic states (Frenkel excitons) in PrF_3 [2.137], [2.138] are surprising. In the single-ion crystal field model of localized excitations only 9 non-degenerate states of the non-Kramers ion are expected in PrF_3 in the ground state multiplet com-

Table 2.6. Experimental and calculated effects of magnetoelastic interaction in $LiTbF_4$; Phonon shift $\Delta\bar{\nu} = \bar{\nu}(T = 2\,\mathrm{K}) - \bar{\nu}(T = 150\,\mathrm{K})$. From [2.149]

Symm.	$\bar{\nu}_p$	Phonon shift $\Delta\bar{\nu}$		Magn. Phonon splittg.		Anticrossing		Remarks
	cm^{-1}	exp.	calc.	exp.	calc.	exp.	calc.	
A_g	129		−0.5					not observed
	269	0	0.1					-
	420	0	0.2					-
B_g	152	0	−0.6			10.8	11.7	-
	212	−2.5	−5.8*			6.0	11.8	*calc. with damp.: −1.6
	322	0	0.1					-
	363	−1.5	−0.1					-
	420	−1.6	0.0					-
E_g	126	0	0.6	1.5	1.9	13.6	14.8	strongly coupl. comp.
	126					?	3.9	weakly coupl. comp.
	176	0	0.8	0.5	0.1	5.0	4.9	strongly coupl. comp.
	176					1.0	6.9	weakly coupl. comp.
	326	0	1.4	9.4	7.4			
	360	0	0.3	≤ 1.0	2.5			
	451	0	0.2	≤ 1.0	1.4			

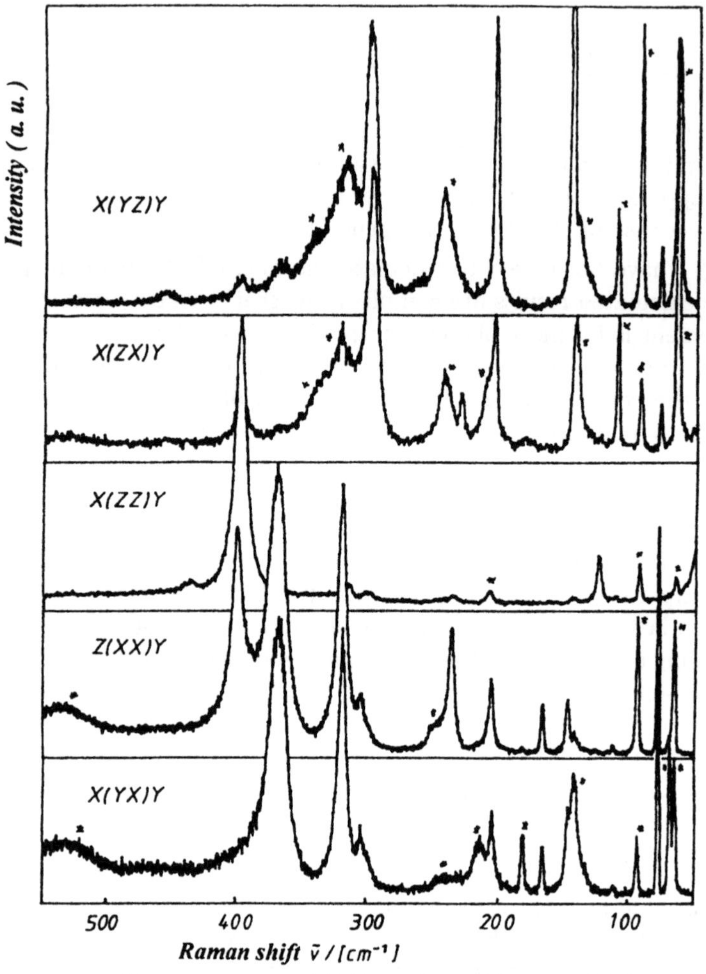

Fig. 2.26. Raman spectra of PrF_3 at $T = 2\,K$, $\lambda_l = 514.5\,nm$ in various polarizations. Electronic transitions within the $^3H_4$ ground state manifold are marked by asterisks. From [2.137]

ponent $^3H_4$. However, as Fig. 2.26 demonstrates, 16 transitions at different frequencies (including IR absorption and reflection data) are observed. The six equivalent Pr^{3+} ions (site symmetry C_2 with states A or B, respectively) generate six exciton bands which emerge from one single-ion transition. These exciton states transform according to the reps of the factor group [2.55] D_{3d}[17] and obey the pertinent selection rules with electronic degeneracies due to the intracell interactions of the Pr^{3+} ions. The observed Davidov splittings ($\leq 4\,cm^{-1}$, with D_{3d} symmetry labels) and the temperature dependence (inset) are reproduced in Fig. 2.27. Transitions to $^3H_5$ and to other higher-lying

[17] $A(\mu = 0) \to A_{1g}, A_{1u}, E_g, E_u$; $B(\mu = 1) \to A_{2g}, A_{2u}, E_g, E_u$

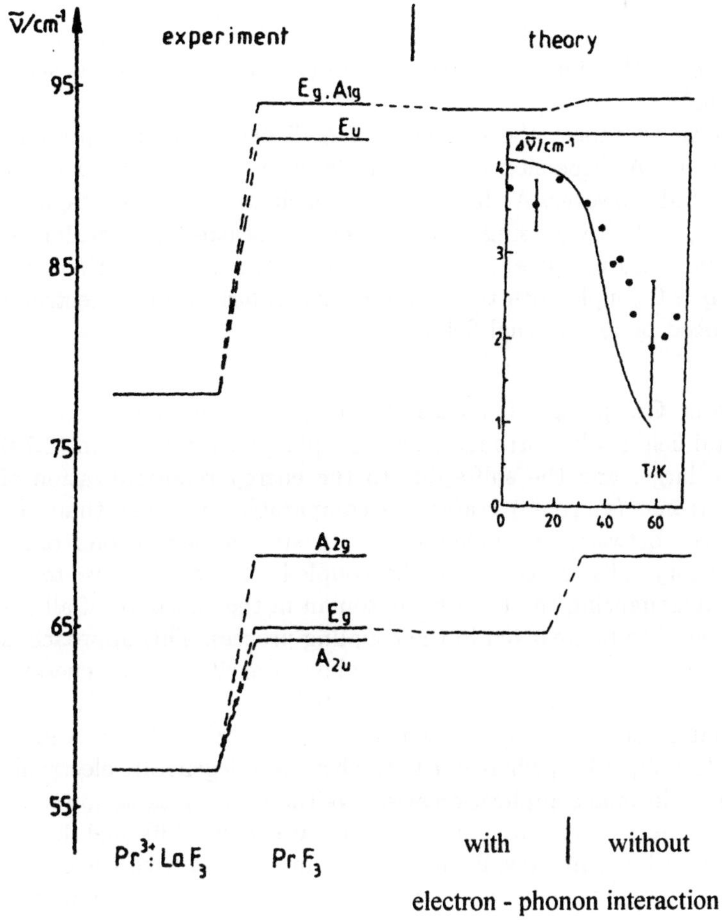

Fig. 2.27. Phonon mediated electronic Davydov splitting of the lowest crystal field levels in . *Left*: experimental values; *right*: calculated results for the E_g-states, using the Green's function method and adjusting 5 coupling constants b [see (2.79)] to all experimental values. *Insert*: Temperature dependence of the electronic $(E_g - A_{2g})$ splitting near $65\,\text{cm}^{-1}$ in PrF_3. *Solid line*: Theory. From [2.137]

multiplets, on the other hand, can be interpreted in the single-ion model, as usual. Other ion–ion coupling mechanisms (exchange, multipole–multipole interaction, etc.) can be calculated approximately and are found to be too small to produce the observed Davydov splittings.

The anomalous temperature dependence of an E_g phonon in PrF_3, its calculated asymptotic $(T \to \infty)$ frequency and its temperature shift in diamagnetic LaF_3, which is solely due to anharmonicity, are plotted. Most surprising is the (small) magnetic splitting of both the E_g phonon and the exciton states in Fig. 2.28. The phonon splitting is linear in B because PrF_3 is a van Vleck paramagnet (singlet ground state with field-induced magnetic moments). All m.e. effects in this compound (both in the phonon and in the exciton system) can be described by the simple JT Hamiltonian (2.70), with the same set of coupling constants.

It is tempting to expect the existence of $\boldsymbol{q}$ dispersion in the presence of excitonic effects. A dispersion of m.e. effects with wavevector $\boldsymbol{q}$ has not been demonstrated, however. A (hitherto unpublished) search for magnetic phonon splitting at finite $\boldsymbol{q}$ using inelastic neutron scattering experiments in a magnetic field did not give observable results. This indicates that only phonons with $\boldsymbol{q} \approx 0$ couple with the homogeneous polarization of electronic multipoles induced by an external field.

2.3.3.2 Strong Coupling. The case of strong m.e. coupling is realized, when the crystal field itself is strong, i.e. the coupling parameters b in (2.79) and (2.89), are large, and the shifts due to the energy renormalization of phonon or electronic (vibronic) states are comparable or larger than the energy differences between uncoupled states. In such a case second order perturbation theory fails. Mixtures of the coupled excitations have to be considered by diagonalizing the total Hamiltonian in the subspace of all excitations considered to be involved in the coupling process. This approach is also approximate, as electronic and (multi-) phonon excitations at energies far removed from the coupling region are neglected.

Such a situation seems to be realized in CeF_3 [2.142], [2.152][18] and in metallic $CeAl_2$ [2.153], [2.154] where due to the Kramers degenerate electronic ground state an additional complication arises. Without m.e. interaction some of the vibronic product states are fourfold degenerate. This additional degeneracy, not tolerated by symmetry, is lifted by the interaction with the consequence that the number of excited vibronic states exceeds the sum of expected uncoupled electronic and vibrational states. This excess of states, which is a reliable indication of strong coupling for the spectroscopist, is demonstrated by a simple example [2.153], [2.142], considering two Kramers doublets as ground- and excited states ($|\varphi_0^{\pm}\rangle$ and $|\varphi_1^{\pm}\rangle$ at energies $\varepsilon = 0$ and $\varepsilon_{\text{el.}} = \eta_1$ and an E-phonon $|\nu_{a,b}\rangle$, ($\varepsilon_{\text{ph.}} = \nu_1$) with ground state $|\nu_0\rangle$, $\eta_1 \approx \nu_1$ as uncoupled states. The product states are: $|\varphi_1^{\pm}\rangle\cdot|\nu_0\rangle$, ($\varepsilon_{\text{c}} = \eta_1$) and $|\varphi_0^{\pm}\rangle\cdot|\nu_{a,b}\rangle$, ($\varepsilon_{\text{c}} = \nu_1$). The latter state is clearly fourfold. A block-diagonalized secular equation for the coupled states is obtained, with M_{A}, M_B as the matrix elements of m.e. interaction:

[18] Electronic Raman transitions to the crystal field levels of the $^2F_{\frac{7}{2}}$ multiplet component around 2 500 cm^{-1} are outside the phonon range and show no anomalies [2.142].

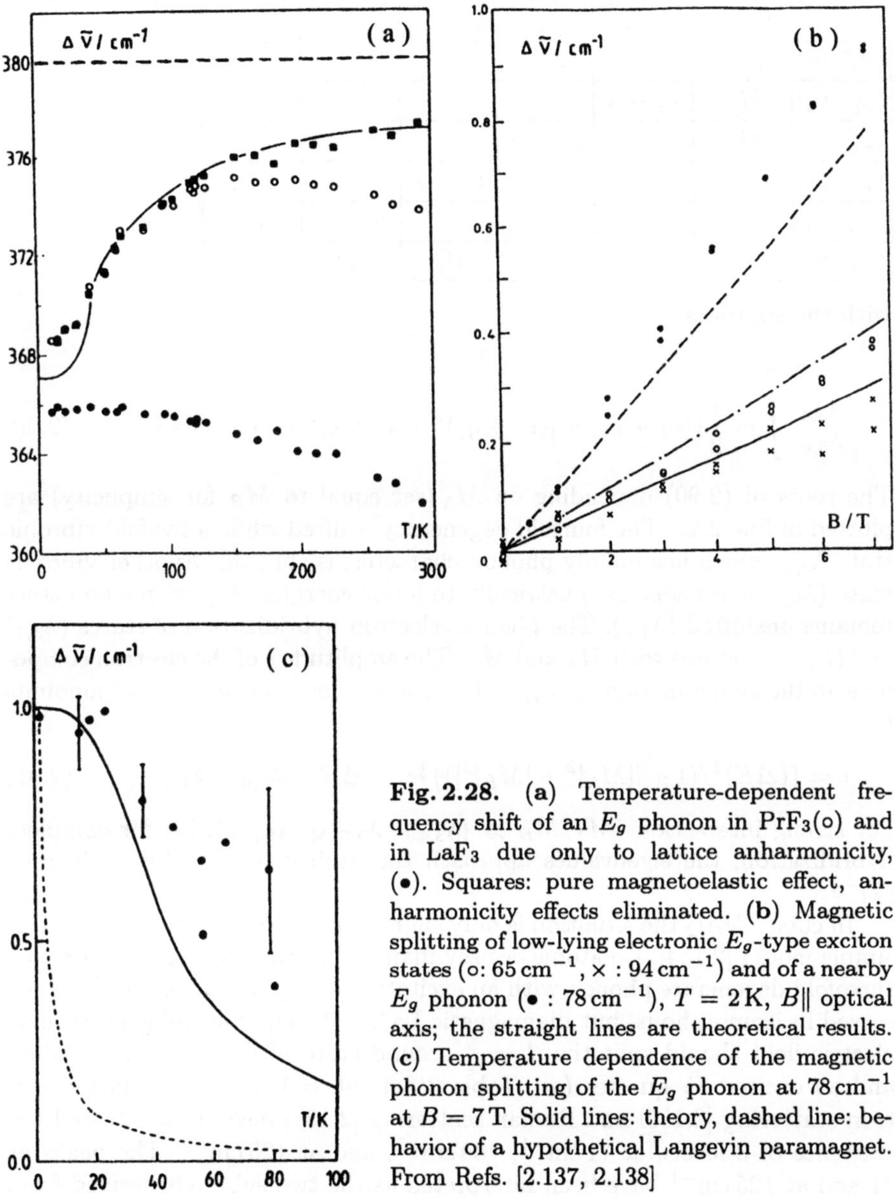

Fig. 2.28. **(a)** Temperature-dependent frequency shift of an E_g phonon in PrF_3(○) and in LaF_3 due only to lattice anharmonicity, (•). Squares: pure magnetoelastic effect, anharmonicity effects eliminated. **(b)** Magnetic splitting of low-lying electronic E_g-type exciton states (○: 65 cm^{-1}, × : 94 cm^{-1}) and of a nearby E_g phonon (• : 78 cm^{-1}), $T = 2$ K, $B \|$ optical axis; the straight lines are theoretical results. **(c)** Temperature dependence of the magnetic phonon splitting of the E_g phonon at 78 cm^{-1} at $B = 7$ T. Solid lines: theory, dashed line: behavior of a hypothetical Langevin paramagnet. From Refs. [2.137, 2.138]

	$\lvert\varphi_1^+, \nu_0\rangle$	$\lvert\varphi_0^-, \nu_a\rangle$	$\lvert\varphi_0^+, \nu_b\rangle$	$\lvert\varphi_0^-, \nu_b\rangle$	$\lvert\varphi_0^+, \nu_a\rangle$	$\lvert\varphi_1^-, \nu_0\rangle$
$\langle\varphi_1^+, \nu_0\rvert$	$\eta_1 - \lambda$	M_A	M_B			
$\langle\varphi_0^-, \nu_a\rvert$	M_A^*	$\nu_1 - \lambda$				
$\langle\varphi_0^+, \nu_b\rvert$	M_B^*		$\nu_1 - \lambda$			
$\langle\varphi_0^-, \nu_b\rvert$				$\nu_1 - \lambda$		M_B
$\langle\varphi_0^+, \nu_a\rvert$					$\nu_1 - \lambda$	M_A
$\langle\varphi_1^-, \nu_0\rvert$				M_B^*	M_A^*	$\eta_1 - \lambda$

with the six roots:

$$\lambda_{1,2} = \nu_1;$$
$$\left.\begin{matrix}\lambda_{3,4}\\ \lambda_{5,6}\end{matrix}\right\} = \frac{1}{2}\{(\nu_1 + \eta_1) \pm [(\nu_1 - \eta_1)^2 + 4(|M_A|^2 + |M_B|^2)]^{\frac{1}{2}}\}. \tag{2.90}$$

The roots of (2.90) depending on M_A (set equal to M_B for simplicity) are plotted in Fig. 2.29. The fourfold degeneracy is lifted while a twofold vibronic state $\lambda_{3,4}$, which has mainly phonon character, is split off. Another vibronic state ($\lambda_{5,6}$) is repelled symmetrically to lower energies. A pure phonon state remains unshifted ($\lambda_{1,2}$). The phonon–electron hybridization of states ($\lambda_{3,4}$) and ($\lambda_{5,6}$) increases with M_A and M_B: The amplitude c of the electron component in the eigenfunctions ($\lambda_{3,4}$) or the phonon component in ($\lambda_{5,6}$) amounts to:

$$c = \{(\Delta E)^2/(1 + [|M_A|^2 + |M_B|^2])\}^{\frac{1}{2}}; \qquad \Delta E = \lambda_{3,4} - \lambda_{1,2}. \tag{2.91}$$

For strong interaction ($M_A, M_B \gg |\lambda_{3,4} - \lambda_{5,6}|_{M_A,M_B=0}$), i.e. for complete hybridization, the eigenvalues approach the dashed lines in Fig. 2.29, $c \to (2)^{-\frac{1}{2}}$.

In cubic $CeAl_2$ the situation is only slightly more complicated than in this simple model [2.154]. Ce atoms occupy diamond lattice sites, giving rise to a threefold degenerate phonon with an excitation energy $\bar{\nu}_1 \approx 109\,\mathrm{cm}^{-1}$, as observed in isomorphous, but diamagnetic $LaAl_2$ [2.154]. The cubic crystalline electric field should split the $J = \frac{5}{2}$ ground state of Ce into a $\bar{\Gamma}_7$ doublet and an excited $\bar{\Gamma}_8$ quartet (see Table 2.A.1) at $\approx 100\,\mathrm{cm}^{-1}$. Inelastic neutron scattering [2.155] and Raman scattering [2.154] have demonstrated the existence of 3 levels, at $71\,\mathrm{cm}^{-1}$, $109\,\mathrm{cm}^{-1}$, and at $125\,\mathrm{cm}^{-1}$. The peaks at 71 and at $125\,\mathrm{cm}^{-1}$ have been interpreted as the two sublevels derived from the electronic $\bar{\Gamma}_8$ state and the level at $109\,\mathrm{cm}^{-1}$ corresponds to the unshifted phonon component $\lambda_{1,2}$ from (2.90). In Fig. 2.30 the temperature dependence of the Raman spectrum of $CeAl_2$ taken in backscattering geometry is shown. The upper level of the split quartet does not appear in the Raman spectrum, it is argued that its phononic admixture is small, hence the Raman cross section is small, too. Attempts to explain the structure in the neutron and Raman spectra by a dynamical Jahn–Teller effect in the $\bar{\Gamma}_8$ multiplet were unsuccessful [2.155].

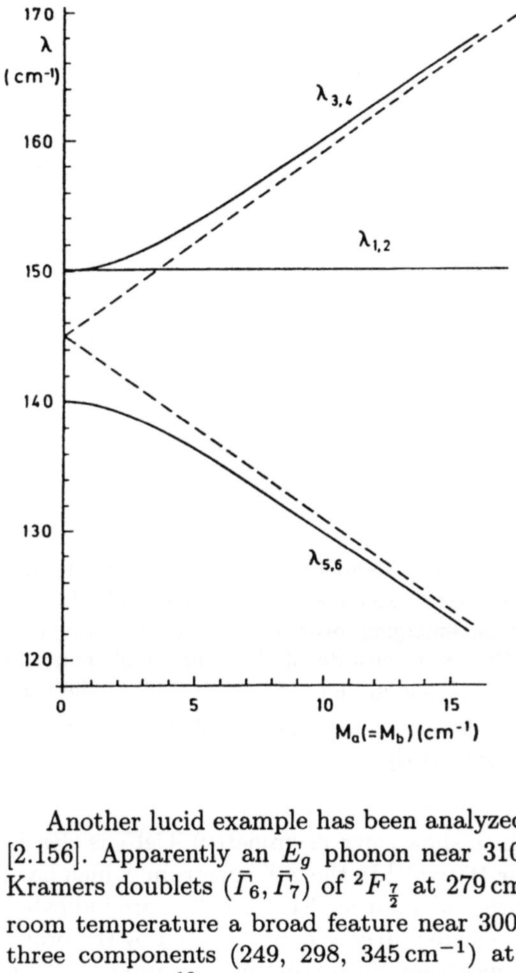

Fig. 2.29. Model calculations of vibronic interaction (strong coupling case, twofold degenerate electronic ground state) between an E-phonon ($\bar{\nu}_1 = 150\,\mathrm{cm}^{-1}$) and an electronic doublet ($\bar{\eta}_1 = 140\,\mathrm{cm}^{-1}$), depending on the size of magnetoelastic matrix elements $\mathrm{M_A}$, $\mathrm{M_B}$. Broken lines: Asymptotic behavior of vibronic energies for $M_A, M_B \gg |\lambda_{1,2} - \lambda_{5,6}|_{M_{A,B}=0}$. From [2.142]

Another lucid example has been analyzed in $\mathrm{YbPO_4}$($\mathrm{D_{4h}}$, site D_{2d}, $4f^{13}$) [2.156]. Apparently an E_g phonon near $310\,\mathrm{cm}^{-1}$ couples with two nearby Kramers doublets ($\bar{\Gamma}_6, \bar{\Gamma}_7$) of ${}^2F_{\frac{7}{2}}$ at $279\,\mathrm{cm}^{-1}, \bar{\Gamma}_6$, and at $298\,\mathrm{cm}^{-1}, \bar{\Gamma}_7$. At room temperature a broad feature near $300\,\mathrm{cm}^{-1}$ is found which splits into three components (249, 298, $345\,\mathrm{cm}^{-1}$) at 4.2 K. Using a model of three coupled modes[19], electron–phonon coupling constants of 37 and $24\,\mathrm{cm}^{-1}$ have been fitted. These values decrease with increasing temperature.

In a system like $\mathrm{CeF_3}$, however, with 11 E_u and 12 E_g phonons in the range of the two excited levels of ${}^2F_{\frac{5}{2}}$ (site symmetry $\approx D_{3h}$) and a strong crystal field [2.116] the situation becomes quite complicated [2.152]. Each E-phonon provides one additional vibronic state and shifts the energy of the interacting electronic level, a nondegenerate phonon only causes an energy renormalization of the coupled states. Higher order (two-phonon) states may also couple. In Fig. 2.31 the E_g Raman spectra (E_{1g} in D_{6h} approximation) of isomorphous $\mathrm{LaF_3}$, $\mathrm{CeF_3}$, $\mathrm{PrF_3}$ at 2 K are compiled; $\mathrm{NdF_3}$ which displays weak m.e. effects [2.157] behaves similarly to $\mathrm{PrF_3}$. While the spectra of $\mathrm{LaF_3}$ and $\mathrm{PrF_3}$ with weak coupling are closely related (except for the extra

[19] This model is not equivalent to (2.90) because the Kramers degeneracy of the ground state has not been taken into account.

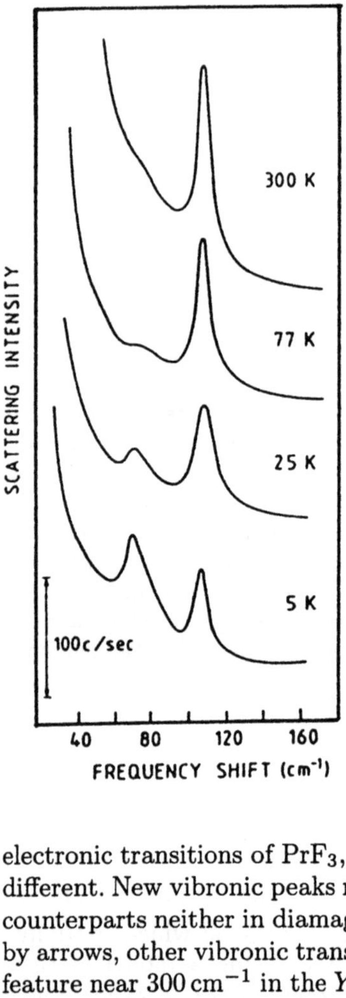

Fig. 2.30. Temperature dependence of the unpolarized Raman scattering of cubic $CeAl_2$. The peak emerging near $71\,cm^{-1}$ below 77 K is a vibronic eigenstate of the coupled phonon and crystal field systems ($4f^1$, $^2F_{\frac{5}{2}}$), while the peak at $109\,cm^{-1}$ is essentially of phononic character. From [2.154]

electronic transitions of PrF_3, which show some asymmetry), CeF_3 is clearly different. New vibronic peaks near $160\,cm^{-1}$ in the ZX spectrum, which have counterparts neither in diamagnetic LaF_3 nor in PrF_3 or NdF_3, are indicated by arrows, other vibronic transitions are unresolved and covered by the broad feature near $300\,cm^{-1}$ in the YZ spectrum. Such vibronic transitions are also observed in the A_{1g} spectrum but not in E_{2g}, because only the A_{1g} and the E_{1g} phonons allow pseudovector scattering (see Sect. 2.3.1). Most remarkable are the almost complete asymmetry ($YZ \leftrightarrow ZX$) of the CeF_3 spectra and the unusually large linewidths of these transitions as compared to the other compounds.

Pure electronic transitions within $^2F_{\frac{5}{2}}$, expected from the previous discussion, are not observed in CeF_3, in contrast to $CeCl_3$ [2.158], [2.140]. They should be expected near 160 and $280\,cm^{-1}$ [2.142], where the strongest vibronic transitions occur. All observed transitions obey the selection rules of the unit cell factor group D_{3d}, (approximately D_{6h}), but not those of the site symmetry C_2 of the RE ions, thus indicating an expected strong delocalization also of the vibronic states. The vibronic transitions exhibit characteristic asymmetric lineshapes in the wavenumber regions where vibronic transitions overlap. The low-intensity dips near $165\,cm^{-1}$ and $110\,cm^{-1}$ in the ZX spec-

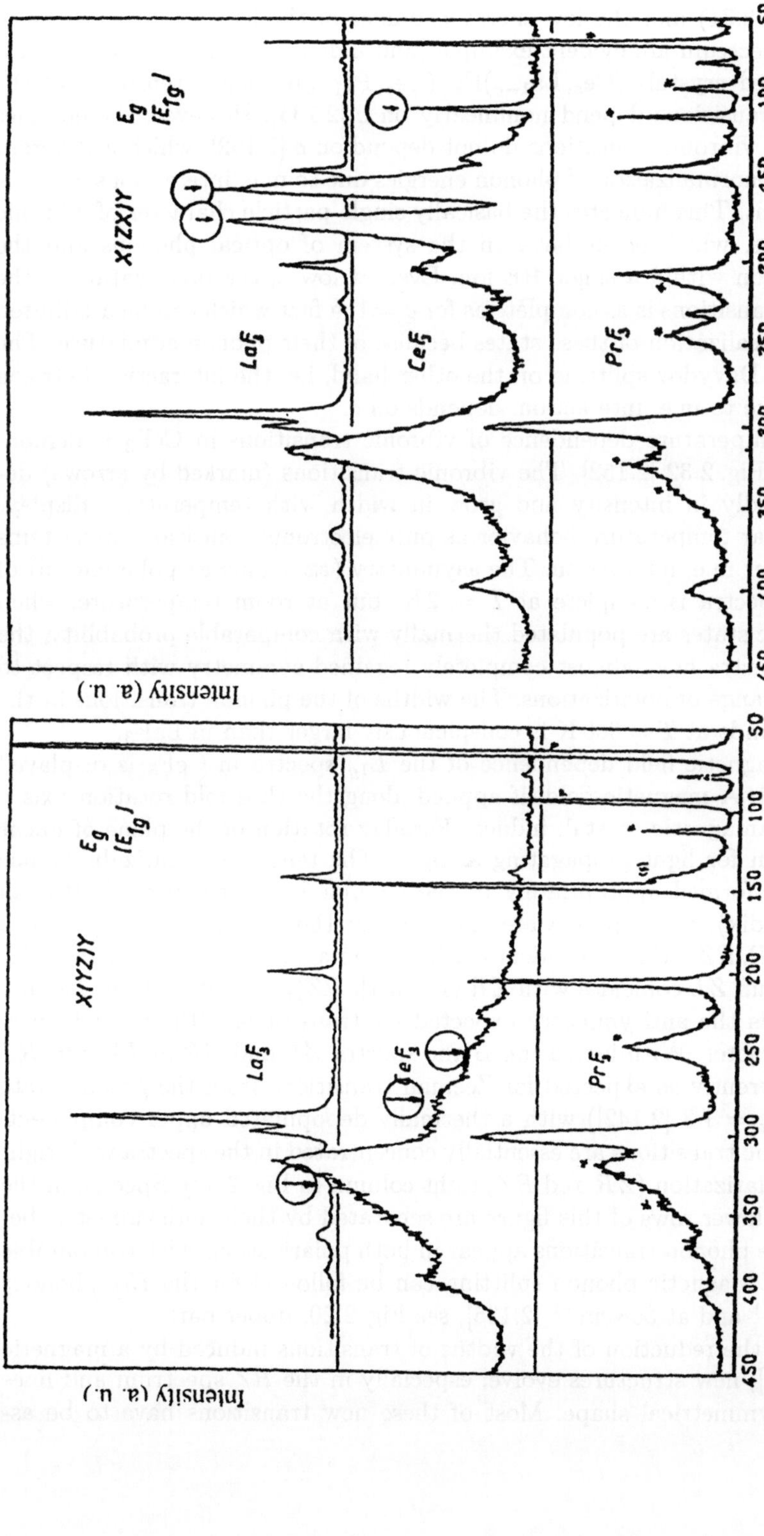

Fig. 2.31. Raman spectra of isomorphous LaF_3, CeF_3, PrF_3 in the tysonite structure; E_g modes (approx. E_{1g}) at orthogonal polarizations, T = 2 K. Arrows: Vibronic states in CeF_3, *: Electronic transitions in PrF_3. From [2.152]

trum (Fig. 2.31) and the low-energy shoulder of the $315\,\mathrm{cm}^{-1}$ transition in the YZ spectrum are evident examples and indicate Fano-like interferences.

In mixed crystals $(\mathrm{Ce}_c, \mathrm{La}_{1-c})\mathrm{F}_3$, $(\mathrm{Ce}_c, \mathrm{Pr}_{1-c})\mathrm{F}_3$ the intensities of the vibronic transitions depend nonlinearly on c [2.135]. However, the energies of the new vibronic transitions do not depend on c [2.152], which is different from the renormalization of phonon energies due to m.e. interactions which is c-dependent. This indicates the basically single-particle character of vibronic interactions, which occur between the system of optical phonons and the $4f$-transition within a single RE ion. Even at low c, the polarization of the vibronic transitions is as complete as for $c = 1$, a fact which can be attributed to the delocalization of these states because of their phonon admixture. The size of the Davydov splitting on the other hand, i.e. the interaction between RE ions due to m.e. interaction, depends on c.

The temperature dependence of vibronic transitions in CeF_3 is demonstrated in Fig. 2.32 [2.152]. The vibronic transitions (marked by arrows) decrease rapidly in intensity and grow in width with temperature, displaying a similar temperature behavior as pure electronic transitions in systems with smaller m.e. interaction. The asymmetry (zx versus yz polarization) of the E_{1g} spectra is complete at $T = 2\,\mathrm{K}$, but at room temperature, when all vibronic states are populated thermally with comparable probability, the phonon spectra have almost completely regained symmetry with respect to the interchange of polarizations. The widths of the phonon transitions in the $4f$ compounds at $T = 300\,\mathrm{K}$ is conspicuously larger than in LaF_3.

The magnetic field dependence of the E_{1g} spectra in CeF_3 is displayed in Fig. 2.33. A magnetic field, if applied along the threefold rotation axis z in the paramagnetic crystal, induces Faraday rotation of the plane of linear polarization for light propagating along z. The tensor of polarizability has to be transformed appropriately to the circular basis (equation (2.101) in the Appendix). Four spectra are expected for the orthogonal polarizations ZL, LZ, ZR, RZ, which are identical in a diamagnetic crystal. At $B = 0\,\mathrm{T}$ the spectrum ZL coincides with ZR (LZ with RZ), but differs from LZ and RZ. This is the antisymmetry expected for transitions with (partial) electronic character. With increasing B the spectra ZL and ZR or LZ and RZ evolve differently as expected for Zeeman transitions from the ground state doublet ($g_z = 1.3$ [2.142]) with a thermally depopulated upper component. The vibronic transitions are essentially concentrated in the spectra with right circular polarization (ZR and RZ, right column in Fig. 2.33). Spectra in the upper and lower rows of this figure are separated by their antisymmetric behavior. The phonon transitions appear in both polarizations with comparable intensities. Magnetic phonon splitting can be followed for the E_{1g} phonons at $203\,\mathrm{cm}^{-1}$ and at $308\,\mathrm{cm}^{-1}$ [2.135], see Fig. 2.20, upper part.

Due to the reduction of the widths of transitions induced by a magnetic field [2.133], new structures evolve, especially in the RZ spectrum and lines regain a symmetrical shape. Most of these new transitions have to be as-

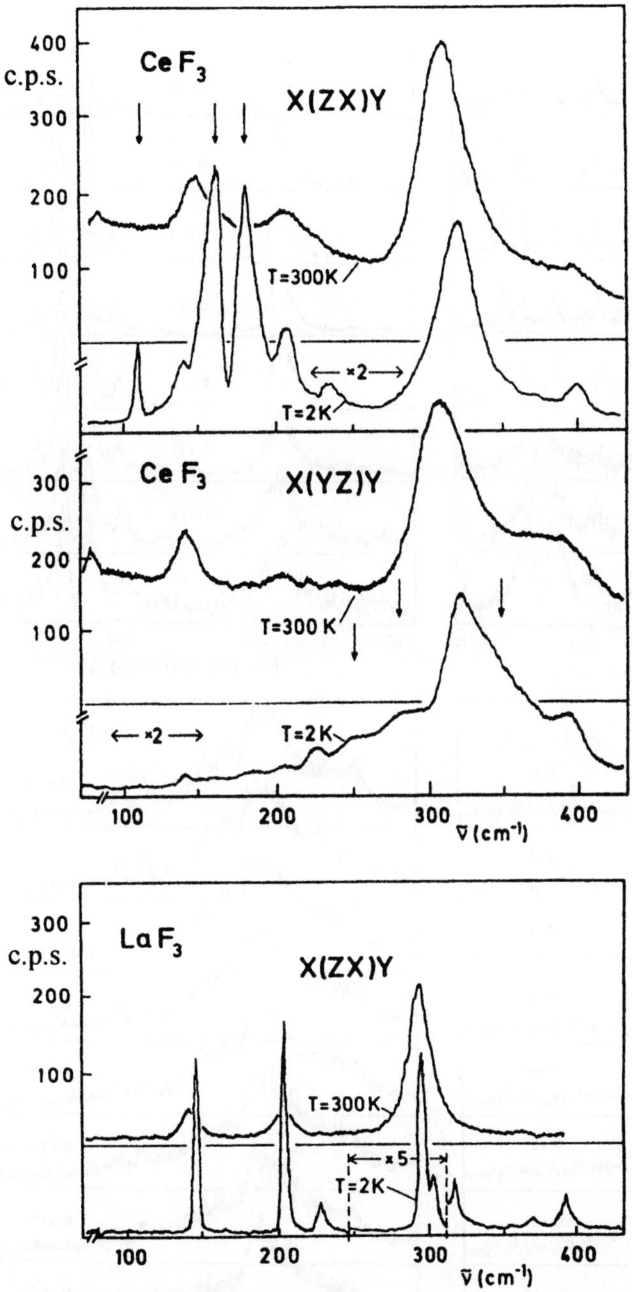

Fig. 2.32. Raman spectra of CeF_3 (zx and yz) and LaF_3 (zx, identical with yz) at $T = 2\,K$ and $T = 300\,K$, displaying the temperature dependence of antisymmetric scattering of vibronic transitions. The vibronics in CeF_3 are marked as in Fig. 2.31. From [2.142]

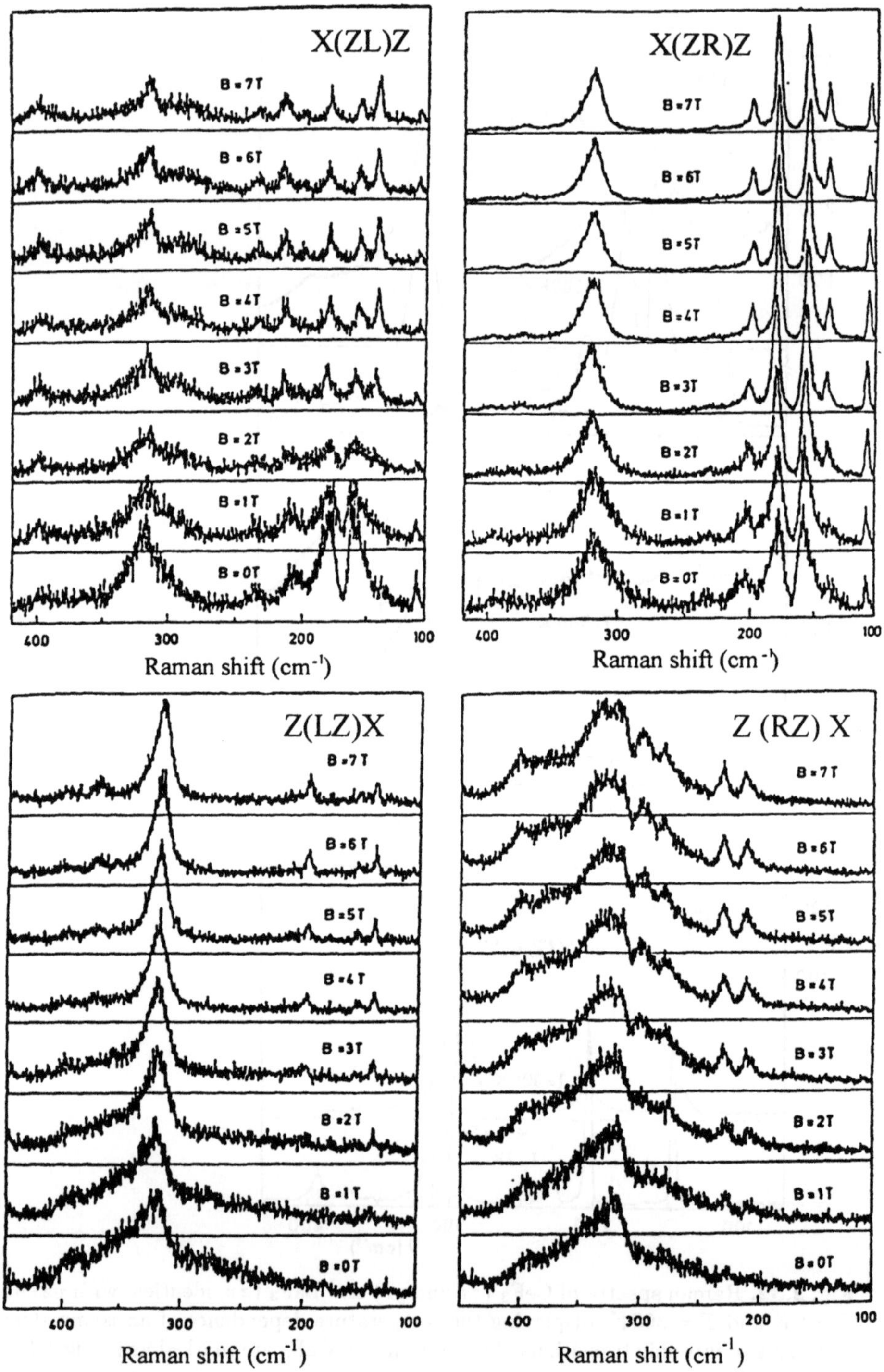

Fig. 2.33. Magnetic field dependence of E_{1g} Raman spectra of CeF_3, using orthogonal circular polarization (L: left (+), R: right (-), see footnote 6) of incident or scattered light, $B \| z$, $T = 2\,K$. From [2.152]

signed to vibronic transitions. It is important to note that the presence of the magnetic field allows us to separate the phonon transitions from the vibronic bands by their different polarization properties. The vibronic transitions start from the lower component of the ground state Kramers doublet, the phonons are not severely touched (except for the splitting) by the lifting of the electronic degeneracy. Even at saturating fields, however, the observed linewidths in CeF_3 are larger than in other isomorphous compounds (PrF_3, LaF_3).

Vibronic transitions can also be assigned by their polarization interchange in the anti-Stokes spectra: Lines which appear in the Stokes spectrum in ZX polarization, will be observed as XZ lines in the anti-Stokes spectrum, and vice versa (process odd with respect to time reversal). The unique polarization properties of vibronic transitions also exclude their assignment to Davydov components due to the formation of Frenkel excitons [2.137].

For the calculation of the vibronic energies in CeF_3, a more realistic extension of the previous model (2.90) has been used [2.152]. The excited crystal field doublets are coupled to all E_{1g} phonons observed (at 139, 203, 232, and 315 cm^{-1}). A subspace of 20 states is spanned, in which the Hamiltonian $\mathcal{H}_{\text{phon.}} + \mathcal{H}_{\text{cryst. field}} + \mathcal{H}_{\text{interact.}}$ is diagonalized. The matrix again factorizes in two identical 10×10 blocks (Kramers degeneracy). The off-diagonal elements have been selected to reproduce qualitatively the observed vibronic energies. Fitting procedures were not applied because of the severe approximations involved: The number of m.e. coupling constants per coupled phonon in CeF_3, as required by symmetry, amounts to 27 (5 if only quadrupolar coupling is considered) [2.149]. These sets of coupling constants are too large to be determined from experiment. The complicated sums over complex matrix elements (2.83) have been replaced simply by real numbers, a procedure justified by its success.

As previously discussed in the model (2.90), from each uncoupled E_{1g} phonon in CeF_3 one new phonon coupled to the electronic ground state and a vibronic state hybridized with an excited crystal field level of $^2F_{\frac{5}{2}}$ are derived. Another vibronic state arises from each excited electronic level. Each of these new states is twofold (Kramers) degenerate. The experimental results (Figs. 2.31–33) display more vibronic states than calculated in this simple model. This may be due to the neglect of coupling to two-phonon states in the model. The m.e. coupling constants which reproduce the data vary between 6.5 cm^{-1} and 36 cm^{-1}, a spread also observed in other systems [2.140, 2.137, 2.149, 2.156]. To generate the vibronic doublet near 165 cm^{-1} in CeF_3, a coupling constant between the phonon at 203 cm^{-1} and the crystal field state near 280 cm^{-1} of 36 cm^{-1} had to be chosen, the coupling strength of the same electronic level to the phonon at 139 cm^{-1} amounts to 7.5 cm^{-1}. This choice shifts the vibronic component of the phonon mode at 203 cm^{-1} towards lower energies close to the phonon state at 145 cm^{-1} (with a coupling strength of 6.5 cm^{-1} to this phonon mode).

The effect of a magnetic field $B\|z$ can also be reproduced within the framework of this simple model by introducing the field-dependent electronic transition energies from the lower, non-depleted Kramers component of the ground state to the uncoupled excited crystal field states and additional off-diagonal matrix elements of the m.e. interaction, which couple the two orthogonal components of an E_g phonon, into the (10×10) secular matrix. The m.e. interaction matrix elements have been selected to reproduce the observed magnetic phonon splittings in CeF_3 [2.135]. Magnetic splittings of vibronic levels have not been observed in experiment; only nonlinear saturating shifts with B in the E_{1g} spectra have (see Fig. 2.10 in [2.142]). These bands are interpreted as transitions to one component of a split pair of levels, the intensity of the other component is too low for observation as only $\Delta m = \pm 1$ differences can be bridged by Raman scattering with circular polarization, which does not occur in states transforming according to the reps of the factor group ($\approx D_{6h}$). If the reps of the site group (C_2) would apply, both components should show up in the spectra. This supports the model of delocalized vibronic states in the strong coupling case.

The Raman intensities of vibronic transitions are also reproduced by the model. Extending the method of (2.88) to the present situation in CeF_3 by using the eigenvectors determined from the (10×10) secular matrix, and adjusting the excitation amplitudes P and empirical halfwidths to the experimental results, the simulated spectra of CeF_3 in Fig. 2.34 have been calculated. In the case of the spectra at finite field B, these halfwidths have been reduced ad hoc by 50%. The agreement with the experimental data (Fig. 2.33, ZR, upper right) is quite satisfying, the Fano anomaly near $165\,\mathrm{cm}^{-1}$ is reproduced [2.152]. Attempts to calculate the observed complete asymmetry of vibronic transitions in the E_{1g} Raman spectra from the crystal field eigenfunctions using the Judd–Ofelt approach [2.60], [2.61] have not been successful so far.

In conclusion, vibronic coupling generates additional transitions in infrared and Raman spectra which lie beyond the simple single-ion model. Only crude theoretical approximations exist at present, more experimental and theoretical work is needed in this field to answer the many questions which still remain open.

2.4 Applications

2.4.1 Crystal Field Levels in High-T_c Superconductors

Inelastic scattering of light is an important method of research in the vast field of high-T_c superconductivity. Most Raman investigations concentrate on scattering processes from itinerant charge carriers of electron or hole character or due to magnetic fluctuations in the Cu–O planes of cuprates, and are beyond the scope of this chapter. Mixed oxides of copper and rare earths ($(RE)_2CuO_4$ and structures of the $YBa_2Cu_3O_7$-type, where Y^{3+} is replaced

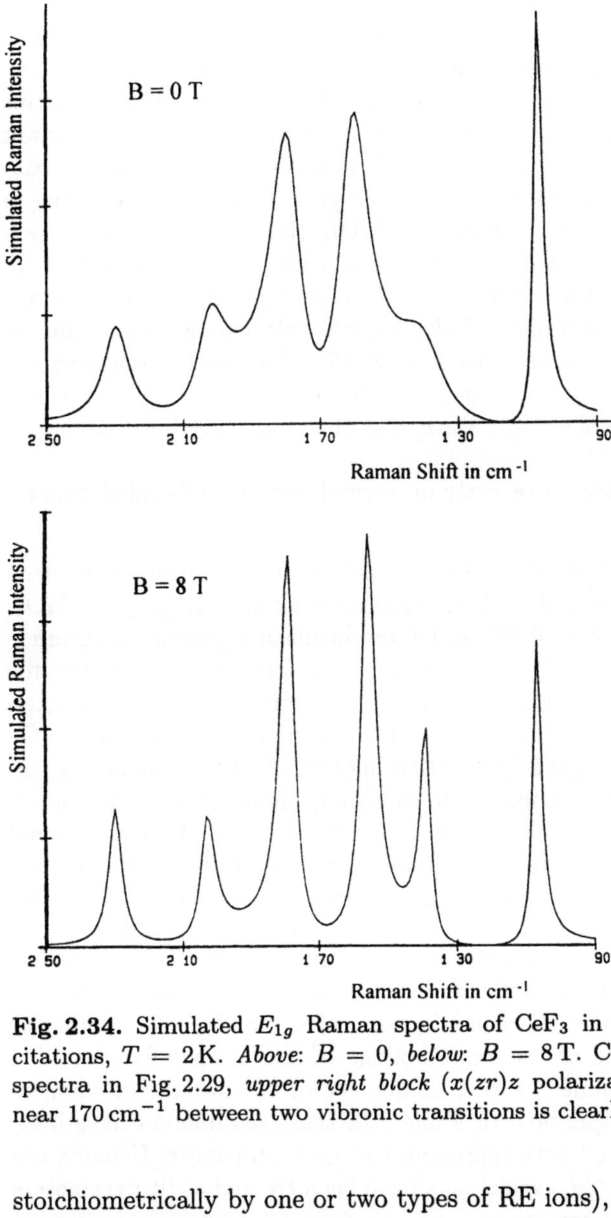

Fig. 2.34. Simulated E_{1g} Raman spectra of CeF_3 in the region of vibronic excitations, $T = 2\,K$. *Above*: $B = 0$, *below*: $B = 8\,T$. Compare with experimental spectra in Fig. 2.29, *upper right block* ($x(zr)z$ polarization). The Fano anomaly near $170\,cm^{-1}$ between two vibronic transitions is clearly evident. From [2.152]

stoichiometrically by one or two types of RE ions), on the other hand, form an important subgroup of high-T_c compounds. We concentrate on these materials where the electrons in the $4f$ shell serve as sensitive local probes for such features as small crystal-field or structural changes on doping, e.g., of the oxygen content, monitoring the charge transfer in the CuO_2 planes or between the planes and the CuO chains, or where they test a number of important vibrational, electronic, and magnetic interaction mechanisms.

In this endeavor, light scattering from crystal field excitations competes severely with inelastic neutron scattering (INS), because the superconducting compounds are strongly opaque and levels of scattered light are very low. Recent instrumental progress with cooled CCD detectors has opened this area for optical spectroscopy but both techniques have provided important information. General reviews on the class of materials of concern in our context are given in [2.159]. A review of recent progress by Raman scattering in high-T_c superconductors can be found in [2.160], the study of magnetic excitations (single-ion, cluster and collective excitations) in RE-based high-T_c superconductors by INS has been reviewed in [2.161]. The complementary advantages of Raman scattering and INS crystal field excitations studies in HTS superconductors are summarized in [2.167]. The most prominent advantage of Raman scattering is that CEF excitations involving different J multiplets can be easily observed. With INS they can only be observed in special cases [2.190].

Research has concentrated recently on several topics, as detailed below.

2.4.1.1 Crystal Field Studies. Crystal-field (CF) transitions have been studied in the electron-doped high-T_c superconductors $(RE)_{2-x}Ce_xCuO_4$, (RE = Pr, Nd, or Sm; $x \approx 0.15$) and their insulating parent compounds ($x = 0$) [2.18], [2.162]– [2.169], e.g., in order to understand thermodynamic functions such as magnetic susceptibility and specific heat. The CF excitations can also be used to investigate the charge transfer to the CuO_2 planes following electron doping by substituting $Ce_x^{(4+)}$, or to register oxygen-vacancy-induced charge redistribution in the CuO_2 planes of nonstoichiometric compounds. Furthermore, they allow the observation of two-dimensional percolative networks generating superconductivity, they monitor magnetic fluctuations and the opening of a gap in the spin excitations of the CuO_2 planes. In isomorphic Nd_2CuO_4, Pr_2CuO_4 and in Sm_2CuO_4 Raman transitions to excited CF states within the ground state multiplet and to higher multiplets have been studied and the symmetries of the observed transitions determined using the Raman selection rules for C_{4v} site symmetry (Table 2.A.1). Hence, one of the shortcomings of INS from polycrystalline samples has beeen overcome[20] and a unique set of CF parameters can be determined for these compounds. In Table 2.12 these parameters have been compiled and are compared with representative neutron results. Usually, the free ion energies of the $^{2S+1}L_J$ multiplets have been treated as fit parameters

[20] To determine the symmetry types of CF levels, neutron scattering intensities have to be calculated from CF eigenfunctions and compared with experiment. Crystal field transitions can exhibit different dependences on Q (momentum transfer) in INS, according on their types of symmetry [2.170]. This has not yet been fully evaluated to provide an additional tool for assigning transitions, see also Ref. [2.171].

Table 2.7. CF parameters B_q^k (Wybourne notation) for Nd^{3+}, Pr^{3+}, Sm^{3+} in $(RE)_2CuO_4$ in cm^{-1}, from Raman data (RS), compared with results from inelastic neutron scattering in powder samples (INS); see also [2.167]

RE	Nd_2CuO_4		Pr_2CuO_4		Sm_2CuO_4	
Source	RS	INS	RS	INS	RS	INS
Ref.	[2.163]	[2.168]	[2.166]	[2.168]	[2.164]	-
B_0^2	−327	−226	−235	−226	−329	-
B_0^4	−2264	−2121	−2287	−2428	−1524	-
B_0^6	215	274	32	210	239	-
B_4^4	1649	1605	1864	1839	1662	-
B_4^6	1477	1476	1519	1807	1345	-

Table 2.8. CF parameters B_q^k (Wybourne notation) for Nd^{3+} and Pr^{3+} in the n-type superconductors $(RE)_{1.85}Ce_{0.15}CuO_4$ in cm^{-1} from Raman data (RS) for the three different sites (I, II, III) and from inelastic neutron scattering in powder samples (INS) for parameter sets (I) and (II)

RE	$Nd_{1.85}Ce_{0.15}CuO_4$					$Pr_{1.85}Ce_{0.15}CuO_4$			
Site	I	II	III	(I)	(II)	I	II	IIIa	IIIb
Source	RS	RS	RS	INS	INS	RS	RS	RS	RS
Ref.	[2.169]			[2.168]		[2.162]			
B_0^2	556	−175	−420	218	−645	−242	−137	−282	−234
B_0^4	−2390	−2374	−2497	−2315	−1863	−2299	−2395	−2016	−2218
B_0^6	421	251	128	484	516	202	331	153	169
B_4^4	1673	1661	1889	1774	1911	1839	1839	1839	1839
B_4^6	1387	1472		1411	1403	1807	1807	1807	1807

and J mixing within the entire set of observed multiplets has been considered. More data can be found in the references quoted.[21]

In $Nd_{1.85}Ce_{0.15}CuO_4$ Raman experiments [2.169] indicate that all CF levels in this compound occur in triplets around the CF excitations of Nd_2CuO_4, suggesting the presence of three inequivalent sites associated with one unperturbed and two perturbed Nd^{3+} ion sites as a result of cerium doping. This is illustrated in Fig. 2.35. The CF spectrum corresponding to the central peaks

[21] In the literature there exist various definitions of crystal field parameters $B_q^k = B_{k,q}$ and conversion tables between the systems, sometimes erroneous, are found; see Refs. [2.29] and [2.30] for details. In this review the crystal field Hamiltonian is written in the *Wybourne notation*: $\mathcal{H}_{CF} = \sum_{k,q,i} B_{k,q} C_{k,q}(\vartheta_i, \varphi_i)$. $C_{k,q} = \left(\frac{4\pi}{2k+1}\right)^{\frac{1}{2}} \cdot Y_{k,q}$. Here $C_{k,q}$ is the tensorial operator, $Y_{k,q}$ are the spherical harmonics. ϑ_i, φ_i are the spherical coordinates of the ith electron in the unfilled shell.

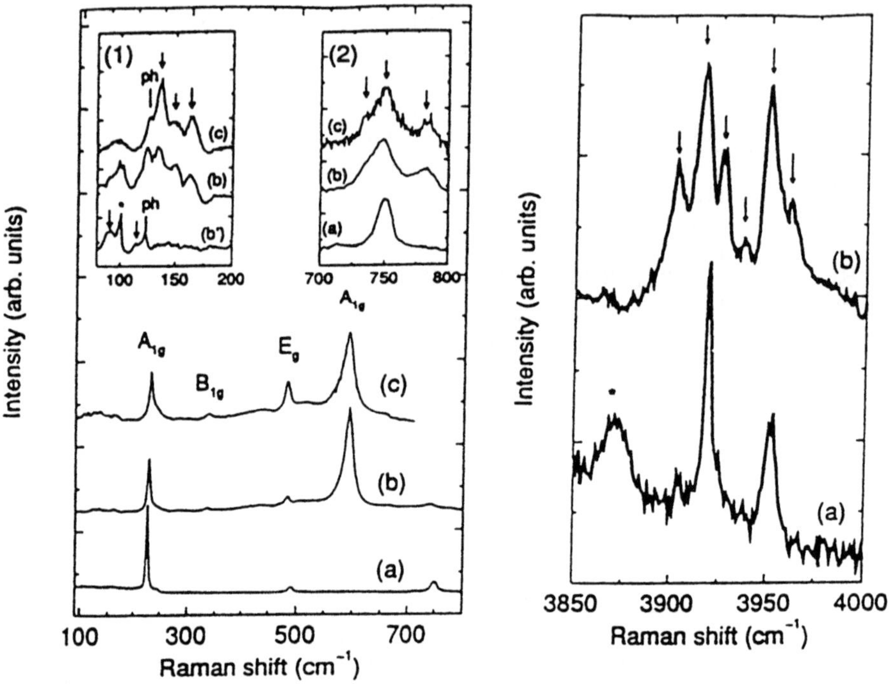

Fig. 2.35. (*Left*): Raman CF and phonon excitations in Nd_2CuO_4 (a) and in $Nd_{1.85}Ce_{0.15}CuO_4$ (b), (b'), and (c); $T = 18\,K$. (a), (b), (c): $y[z{x \atop z}]\bar{y}$; (b'): $y[x{x \atop z}]\bar{y}$. The insets mark with arrows the ground state CF excitations of three inequivalent sites. ph and * indicate phonons and plasma lines. (*Right*): CF transitions around 3900 cm^{-1} in Nd_2CuO_4 (a) and in $Nd_{1.85}Ce_{0.15}CuO_4$ (b), where again the inequivalent Nd^{3+} sites are marked. $*$: luminescence band. From [2.169]

in the spectra of $Nd_{1.85}Ce_{0.15}CuO_4$ is ascribed to the unperturbed site because of its similarity with Nd_2CuO_4, whereas the satellite transitions correspond to the perturbed Nd^{3+} sites. In Table 2.13 the CF parameters derived are compiled, the equivalence between site II and the Nd_2CuO_4 CF parameters suggests that site II corresponds to the unperturbed Nd^{3+}. In the same material, splitting of the CF states of the Kramers degenerate ground state and the first excited multiplet components have been observed [2.169]. This splitting suggests that there are regions in $Nd_{1.85}Ce_{0.15}CuO_4$ where magnetic order persists, see also [2.172]. CF doublets, with about 5 cm^{-1} separation between their components, have also been observed in pure Nd_2CuO_4 for $T \leq 150\,K$ (Fig. 2.36, [2.173]). At these elevated temperatures a magnetic origin of the splitting may seem to be rather unlikely; the splitting may therefore be attributed to a Davydov splitting ($Z = 2$ ions per unit cell, see Sect. 2.2.7). The Davydov assignment has, however been ruled out in [Dufour et al., Phys. Rev. Lett. **51**, 1053 (1995)]. The coupling mechanism is different from the previously discussed phonon or exchange mediated interactions,

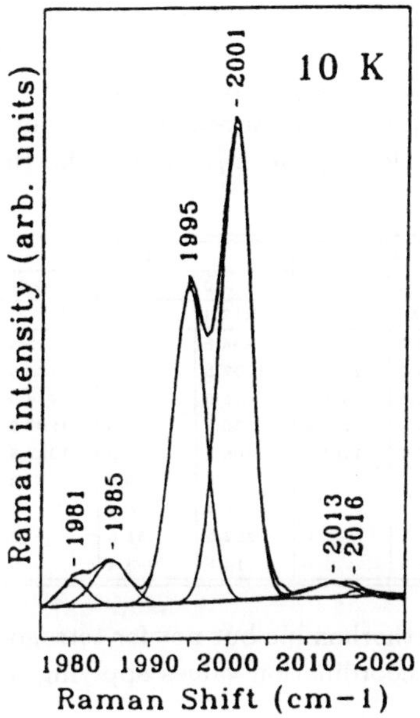

Fig. 2.36. Doublet structure and deconvolution of the $^4I_{\frac{9}{2}} \rightarrow {}^4I_{\frac{11}{2}}$ Raman lines in Nd_2CuO_4, $T = 10\,K$. From [2.173]

since all CF levels are subject to the coupling. The authors of [2.173] refer to an overlap of the $4f^3$ wavefunction with the strongly polarizable O^{2-} p orbitals as a possible source of this effect. In $Pr_{1.85}Ce_{0.15}CuO_4$ [2.162] the three Pr^{3+} sites have been characterized in more detail. One site is associated with charge transfer due to the replacement $Pr^{3+} \rightarrow Ce^{4+}$. It is also claimed that site III may be twofold (IIIa and IIIb). In Pr_2CuO_4 and in $(Pr, Ce)_2CuO_4$, where the tetragonal $(RE)_2CuO_4$ structure is at its stability limit, a broad phonon-like feature in the Raman spectrum (A^* mode) near $590\,cm^{-1}$ has been observed and ascribed to the presence of some degree of disorder in the oxygen sublattice induced by the structural instability [2.174].

The neutron results in Table 2.13 compare favorably with the Raman values, however the lower energy resolution of INS only justifies a model with two different CF environments. The CF parameters based on this model are marked as (I) and (II), but cannot be correlated with the three sites derived from the Raman work. Neutron-powder diffraction [2.172] has been applied to study the short-range atomic structure of $Nd_{2-x}Ce_xCuO_{4-y}$, $x = 0.165$ and 0.2. The local structure in the CuO_2 planes is found to be spatially inhomogeneous (the planes are buckled) with two types of local regions (domains sized $\approx 6\,Å$), one heavily distorted the other relatively undistorted, giving rise to the subtly different Nd^{3+} environments.

In Table 2.14 the CF parameters of $(RE)Ba_2Cu_3O_{7-\delta}$, $(0 \leq \delta \leq 1$, RE $= Nd^{3+}$, Pr^{3+}, Eu^{3+}, Gd^{3+}, Er^{3+}, $Ho^{3+})$, site symmetry of RE: D_{2h} for $\delta \leq 0.6$, D_{4h} for $\delta \geq 0.6$, and derived solely from neutron data, have been listed for comparison [2.170, 2.175, 2.176, 2.177, 2.178, 2.179, 2.180]. In these compounds, for δ increasing from 0 to 1, structural transitions, a strong de-

Table 2.9. CF parameters B_q^k (Wybourne notation) from inelastic neutron scattering in powder samples for $Nd^{3+}, Pr^{3+}, Eu^{3+}, Gd^{3+}, Er^{3+}, Ho^{3+}$ in $(RE)Ba_2Cu_3O_{7-\delta}$ in cm^{-1}. The RE site symmetry is orthorhombic for $\delta \leq 0.6$, and tetragonal for $\delta \geq 0.6$. The experimental uncertainties as given in the original references amount to $\leq 1\%$ for $B_0^4, B_0^6, B_4^6, \approx 5\%$ for B_0^2, and $\geq 10\%$ for $B_2^2, B_2^4, B_2^6, B_6^6$. For the Ho compound; see also [2.175]

RE	Nd			Pr	Eu	*Gd*	Er		Ho		
δ	0.02	0.98	0	0	0	0	0.02	0.91	0	−0.05	0.89
Ref.	[2.178][a]		[2.176], [2.177]		[2.170]		[2.180][a]		[2.177]	[2.179][a]	
B_0^2	592.0	564.6	416.8	451.6	790.4	822.6	224.7	101.6	434.8	283.9	112.9
B_0^4	−2819.5	−2980.8	−2712.4	−2773.1			−2080.7	−2168.5	−1907.5	−2193.7	−2180.9
B_0^6	539.4	556.2	620.7	786.6			473.6	460.7	471.6	512.3	474.9
B_4^4	1603.3	1678.1	1669.4	1491.3			1209.2	1205.4	1050.1	1243.9	1297.2
B_4^6	2003.7	1965.8	1978.2	2612.5			1202.4	1201.0	1305.1	1339.3	1342.8
B_2^2	29.0	0	148.4	160.8	32.9	72.4	76.7	0	76.6	79.0	0
B_2^4	84.7	0	11.7	13.8			104.7	0	−297.4	36.7	0
B_2^6	−4.8	0	−268.6	−340.4			−7.2	0	−252.6	−31.5	0
B_6^6	6.4	0	82.6	104.7			5.4	0	−14.8	−3.4	0

[a]) The parameters $B_2^2, B_2^4, B_2^6, B_6^6$, occurring for orthorhombic but not for tetragonal site symmetry, were fixed at their geometrical coordination values applying to the point charge model.

crease in the superconducting transition temperature $T_c \to 0$, and a metal-to-semiconductor transition occurs. The RE ions are sandwiched between the two superconducting CuO_2 planes. At $T \leq 3\,K$ transitions to two- and three-dimensional antiferromagnetic order in the RE sublattice have been found. In (RE)1, 2, 3, 7 the RE site can be occupied by all RE's and Cm, except Ce and Tb. Pr1, 2, 3, 7 is not superconducting[22] and displays very large intrinsic widths of the neutron peaks. Ce^{3+} and Pr^{3+} do not fit or will only fit with difficulties into the (RE)1, 2, 3, 7 structure, Pr^{3+} may also occupy the Ba site, with disorder as a result. Based on this CF analysis there is no reason for assuming either Pr^{4+} or a mixed valence for the Pr ion in this compound.

Significant changes have been observed in the CF parameters as a function of oxygen content δ, which cannot be related to structural changes alone [2.184a,b]. These changes have been interpreted, however, by introducing the concept of charge transfer from the Cu–O chains to the CuO_2 planes. CF parameters for the limiting values 0 and 1 of δ have been given in Table 2.14, numbers for intermediate δ values can be found in the references. Obviously, the parameters $B_0^2, B_2^2, B_0^4, B_4^4, B_0^6$ probe this influence. For RE = Ho^{3+}, for example, an electronic charge transfer of $0.08e$ per oxygen is derived for δ going from 0 to 1 upon oxygen reduction. This transfer compensates the hole concentration in the planes, hence giving immediately an interpretation of the reduction of T_c, i.e. the suppression of superconductivity, with δ. Similar results were found for other RE ions [2.178, 2.179, 2.180].

[22] For a recent report of superconductivity in $PrBa_2Cu_3O_7$ see K. Oka, Z. Zou, J. Ye, Physica C **300**, 200 (1998)

CF parameters for $TmBa_2Cu_4O_8$ have been reported in [2.181], as determined by various methods, and have been compared with values for the isomorphic Ho compound [2.179]. Observed discrepancies between both compounds for the rank-2 CF parameters (which are in general the parameters most sensitive to small changes in the crystalline environment [2.29]) are considered as genuine and not due to the specific conditions of the various spectroscopic techniques.

Unfortunately, only a few Raman transitions have been observed in (RE)1,2,3,7 which do not allow to determine the B_q^k reliably [2.184a,b,c]. The Raman transitions observed in this series of compounds are generally close to resonance with a phonon of identical symmetry, borrowing their intensity from this excitation via the coupling process (see Sect. 2.3.3.2 and the next subsection), while in the $(RE)_2CuO_4$ compounds the intensities of the Raman transitions between CF levels are solely of electronic origin. The CF parameters of all the ions in Table 2.14 have been found to obey reasonably well scaling relations with the RE atomic number as derived from Hartree–Fock calculations of the atomic states. Using the results of the CF analysis the anisotropic, T-dependent magnetic suceptibilities have been calculated applying the Van Vleck formalism [2.182], [2.183]. They agree surprisingly well with the experimental data.

2.4.1.2 Effects of $4f$ Electron–Phonon Coupling in High-T_c Superconductors. Because of the high density of optical phonons on the energy scale in high-T_c RE superconductors with many ions per unit cell, chances are that resonances between phonons and electronic intra multiplet transitions of the same symmetry occur; they may complicate the appearance of both Raman and INS spectra. An interesting example of this type has been studied in detail for $NdBa_2Cu_3^{16}O_7$ single crystals [2.184], [2.185]. In Fig. 2.37 Raman spectra are shown, displaying a double peak structure near $300\,cm^{-1}$ which is not observed in other isomorphic (RE)1,2,3,7 compounds. A similar structure has been observed in the same material by INS [2.186]. The experimental results can be summarized as follows: Both components of the doublet have B_{1g} symmetry. The splitting of the doublet increases with decreasing temperature (Fig. 2.38). The isotopic substitution of ^{16}O by ^{18}O results again in a temperature dependent double peak, however with a different intensity ratio of the two components (right panel of Fig. 2.38). The B_{1g} phonon involved can be identified as the out-of-phase motion of the O(2)–O(3) ions (plane oxygen) which strongly distorts the CF around the Nd^{3+} ions. The phonon frequency is lowered upon isotopic substitution ($^{16}O \rightarrow {}^{18}O$) roughly by a factor of $\sqrt{16/18}$.

Obviously, this is an example of strong coupling as discussed in Sect. 2.3.3.2. However, the situation is simpler here because the phonon is nondegenerate. At low temperatures, a 2×2 matrix, as deduced from (2.90), is sufficient for a calculation of the renormalized phonon and CF frequencies and coupled eigenfunctions and of the spectral function (2.87) used for a fit of the experimental data (Fig. 2.38). The following uncoupled phonon frequencies have been derived from the measured Raman spectra: $\omega_{\text{ph}} = 308\,\text{cm}^{-1}, \omega_{\text{CF}} = 304\,\text{cm}^{-1}$ for the ^{16}O compound, $\omega_{\text{ph}} = 290\,\text{cm}^{-1}$ for the ^{18}O compound, the coupling constant Δ_0 was found to be $35 \pm 3\,\text{cm}^{-1}$ in both cases. Additional transitions involving phonon excited states have to be considered in the coupling matrix only at elevated temperatures, together with their thermal occupation factors and phenomenological damping terms (here $\Gamma = 10\,\text{cm}^{-1}$). This fact increases the dimension of that matrix. For a correct reproduction of the observed temperature dependence of the coupling phenomena (Fig. 2.34), a variation of Δ with T according to: $\tilde{\Delta} = \Delta_0[1 - (\kappa T)^2]$ has to be introduced phenomenologically, thus yielding another fit parameter $\kappa = 2.6 \cdot 10^{-3}\,\text{K}^{-1}$, which is assumed to be proportional to $\langle u_{\text{ph}} \rangle^4 \sim T^2$ because of anharmonic effects. In $NdBa_2Cu_3\,^{16}O_6$ essentially the same phenomena have been observed.

A similar double-peak feature has been obtained in the single-crystal Raman spectra of insulating $Pb_2Sr_2NdCu_3O_{8+\delta}$ [2.187]. In this reference, a re-

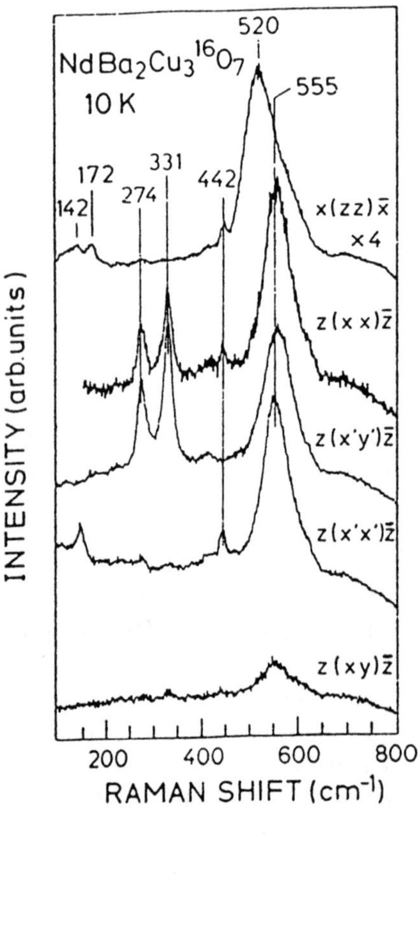

Fig. 2.37. Polarized Raman spectra ($T = 10\,\text{K}$) of a $NdBa_2Cu_3\,^{16}O_7$ single crystal taken in backscattering geometry. The double peak structure at $274\,\text{cm}^{-1}$ and at $331\,\text{cm}^{-1}$ is due to the coupled excitation involving a B_{1g} phonon (factor group symmetry approximately D_{4h}) and a Nd^{3+} crystal field excitation. All other Raman lines are due to phonon transitions. From [2.184]

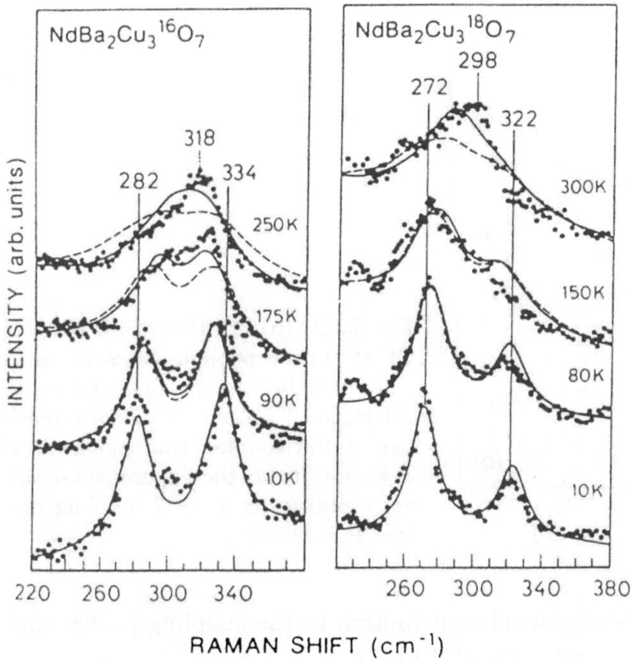

Fig. 2.38. Temperature dependence of the Raman spectra of $NdBa_2Cu_3\,^{16}O_7$ (*left*) and of $NdBa_2Cu_3\,^{18}O_7$ ceramic samples (*right*). The solid and dashed curves represent fits to the experimental data (dots) with and without anharmonic contributions, respectively. From [2.184]

fined discussion of the coupling model for $NdBa_2Cu_3O_7$ is also given. The structure of this material is rather similar to that of $NdBa_2Cu_3O_7$. Near a strong phonon of B_{1g} type at $275\,cm^{-1}$, it shows a peak around $335\,cm^{-1}$, of the same symmetry, which does not have pure phonon character. It can be attributed to an electronic transition in the $4f^3$ configuration, again borrowing its intensity from the nearby phonon. The analysis of this coupling phenomenon along the lines discussed above results in decoupled frequencies $(T \approx 10\,K) : \omega_{ph}(^{16}O) = 287 \pm 3\,cm^{-1}, \omega_{CF} = 329 \pm 3\,cm^{-1}, \tilde{\Delta} = \Delta_0[1-(\kappa T)]$, $\Delta_0 = 23 \pm 3\,cm^{-1}$, $\kappa = 2.8 \times 10^{-3}\,K^{-1}$; $\omega_{ph}(^{18}O) = 282 \pm 3\,cm^{-1}$, (ceramic sample), $\omega_{CF} = 315 \pm 3\,cm^{-1}$, $\Delta = 25 \pm 3\,cm^{-1}$. At $T > 100\,K$, again the thermally excited phonons have to be considered in the calculations together with the CF transitions. Here the phenomenological coupling constant $\tilde{\Delta}$ depends linearly on temperature. Phonon dispersion, which was neglected previously, was invoked for this result. The coupling constant for $Pb_2Sr_2NdCu_3O_{8+\delta}$ could be approximated rather well by using the appropriately scaled CF parameters of $NdBa_2Cu_3O_6$. The electron–phonon coupling constant b [see (2.79) and (2.89)] has been calculated for the Nd^{3+} crystal field levels in $NdBa_2Cu_3O_{7-\delta}$ using a point charge model. Charge transfer between two

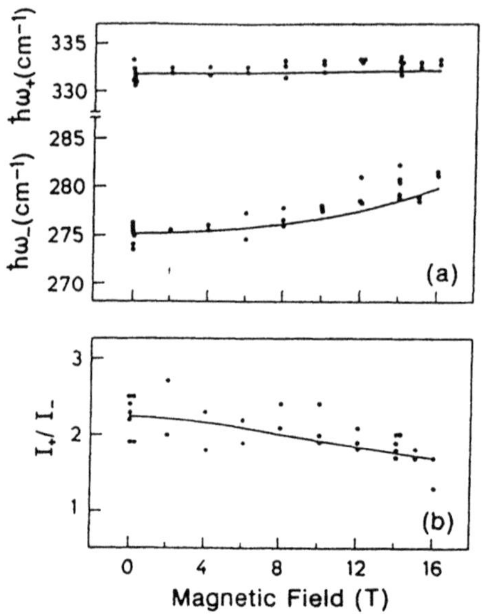

Fig. 2.39. (**a**) Position of the coupled CF-excitation phonon states vs. magnetic field $B \perp$ to the CuO_2 planes in $NdBa_2Cu_3{}^{16}O_7$. (**b**) Ratio of intensities of the coupled transitions. Solid lines are fits to the experimental values (dots) using a 2×2 coupling matrix. From [2.185]

oxygen atoms yields an additional contribution to the coupling [2.188] that is found to agree with the experimental values.

The application of a magnetic field in $NdBa_2Cu_3O_7$ raises Kramers degeneracy in Nd^{3+} and the resonance condition with respect to the optical phonon can be tuned through [2.185]. Results for fields up to a field of 16 T perpendicular to the CuO_2 planes are shown in Fig. 2.39. A nonlinear shift of the lower component, of $5.6\,\mathrm{cm}^{-1}$ (compare with Fig. 2.29, strong coupling), is observed towards higher frequencies, with the higher-energy feature almost unaffected by the field. The intensity ratio of the two components changes with B in favor of the lower component, which borrows an increasing amount of intensity from its partner at higher frequency. The lower-energy structure also exhibits a pronounced broadening with a quadratic dependence on B. The convincing fits displayed in Figs. 2.38 and 2.39 could be achieved only when CF eigenfunctions with contributions from the multiplet component ${}^4I_{\frac{9}{2}}$ next to ${}^4I_{\frac{11}{2}}$ were applied in the calculations.

An external pressure will also modify the frequencies of the coupled modes and the coupling parameters: In Fig. 2.40a the low-temperature Raman spectra of $NdBa_2Cu_3O_{6.8}$ single crystals are shown at different, approximately hydrostatic, pressures. They display a clear shift of the coupled modes (both CF state and B_{1g} phonon, $\frac{\mathrm{d}\,\omega_{\mathrm{ph}}}{\mathrm{d}\,P} = 2.80\,\mathrm{cm}^{-1}/\mathrm{GPa}$, $\frac{\mathrm{d}\,\omega_{\mathrm{CF}}}{\mathrm{d}\,P} = 2.46\,\mathrm{cm}^{-1}/\mathrm{GPa}$) to higher energies and also of other uncoupled phonons (Fig. 2.40b). The coupling constant $\Delta = 28.1\,\mathrm{cm}^{-1}$ is also pressure dependent: $\frac{\mathrm{d}\,\Delta}{\mathrm{d}\,P} = 0.26\,\mathrm{cm}^{-1}/\mathrm{GPa}$. Point charge model calculations have yielded a pressure coefficient for the CF level and for Δ in quantitative agreement with the experiment [2.189]. This finding again supports the empirical result

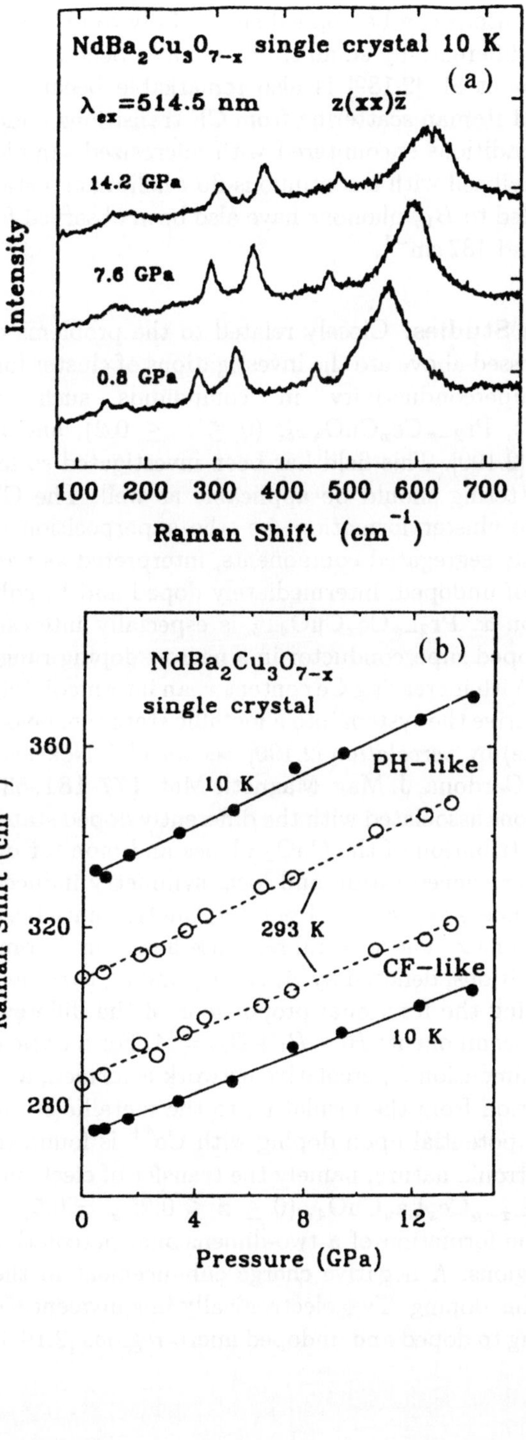

Fig. 2.40. (a) Low-temperature Raman spectra of single-crystalline $NdBa_2Cu_3O_{6.8}$ at different (≈ hydrostatic) pressures produced in a diamond-anvil cell, $z(xx)\bar{z}$. (b) Pressure dependence of Raman wavenumbers in the region of the coupled CF and phonon transitions. Experimental errors: $\pm 2\,cm^{-1}$ at 10 K, $\pm 6\,cm^{-1}$ at 293 K. From [2.189]

that point charge model calculations can be applied successfully to problems where the charges experience differentially small shifts, see also Sect. 2.3.3.1. The experiment of *Goncharov* et al. [2.189] is also remarkable because it proves the general feasibility of Raman scattering from CF transitions under the experimentally difficult conditions encountered with microsized samples in a high-pressure diamond-anvil cell with a continuous-flow helium cryostat. Recently, CF transitions coupled to B_{1g} phonons have also been observed for $SmBa_2Cu_3O_7$ [2.184c] at 90 and 187 cm^{-1}.

2.4.1.3 Phase Separation Studies. Closely related to the problems of crystal field spectroscopy discussed above are the investigations of cluster formation and percolative superconductivity in compounds such as $(RE)Ba_2Cu_3O_x$, $(6 < x < 7)$, $Pr_{2-x}Ce_xCuO_{4-\delta}$, $(0 \leq x \leq 0.2)$, and in superoxygenated $La_2CuO_{4+\delta}$ [2.190]. This field has been investigated so far only by INS, but Raman scattering should be applicable as well. The CF is a local probe to investigate cluster formation, i.e. the superposition of at least three different, spatially segregated components, interpreted as partially coexisting local regions of undoped, intermediately doped and heavily doped character, depending on x. $Pr_{2-x}Ce_xCuO_{4-\delta}$ is especially interesting because it is an electron-doped superconductor in a narrow doping range $(0.14 \leq x \leq 0.17), T_c \leq 25$ K. With increasing Ce content x, an intermediately doped state evolves which will drive the system into a metallic state (i.e. below T_c into a superconducting state) by *percolation* [2.190], see also [V. Nekvasil, S. Jandl, T. Strach, T. Ruf, M. Cardona, J. Mag. Magnetic Mat. **177-181**, 535 (1998)]. The electronic excitations associated with the differently doped states are indicative of the charge distribution of the CuO_2 planes and monitor directly the changes of the carrier concentration and local symmetry induced by doping. In Fig. 2.41 neutron energy spectra of $Pr_{2-x}Ce_xCuO_{4-\delta}$ have been displayed for six Ce concentrations x. The spectral response has been decomposed into four individual transitions, denoted by A, B_1, B_2 and C, that vary differently with x, thus reflecting the fractional proportions of the different cluster types. For $x \approx 0.14$, the components $B(= B_1 + B_2) = A$. For a critical volume fraction of 50% a two-dimensional percolative network is formed, and the system undergoes a transition from the insulating to the metallic state. The main influence on the CF potential upon doping with Ce^{4+} is found to be not of structural but of electronic nature, namely the transfer of electrons into the CuO_2 planes. In $Nd_{2-x-y}Ce_xLa_yCuO_4$, $(0 \leq x \leq 0.2;\ y = 0.5, 1)$ evidence has been found for the formation of a two-dimensional percolative network of Ce-doped micro-regions. A negative charge enhancement in the CuO_2 planes is derived upon this doping. Two electronically inequivalent Ce sites are identified corresponding to doped and undoped micro-regions [2.191].

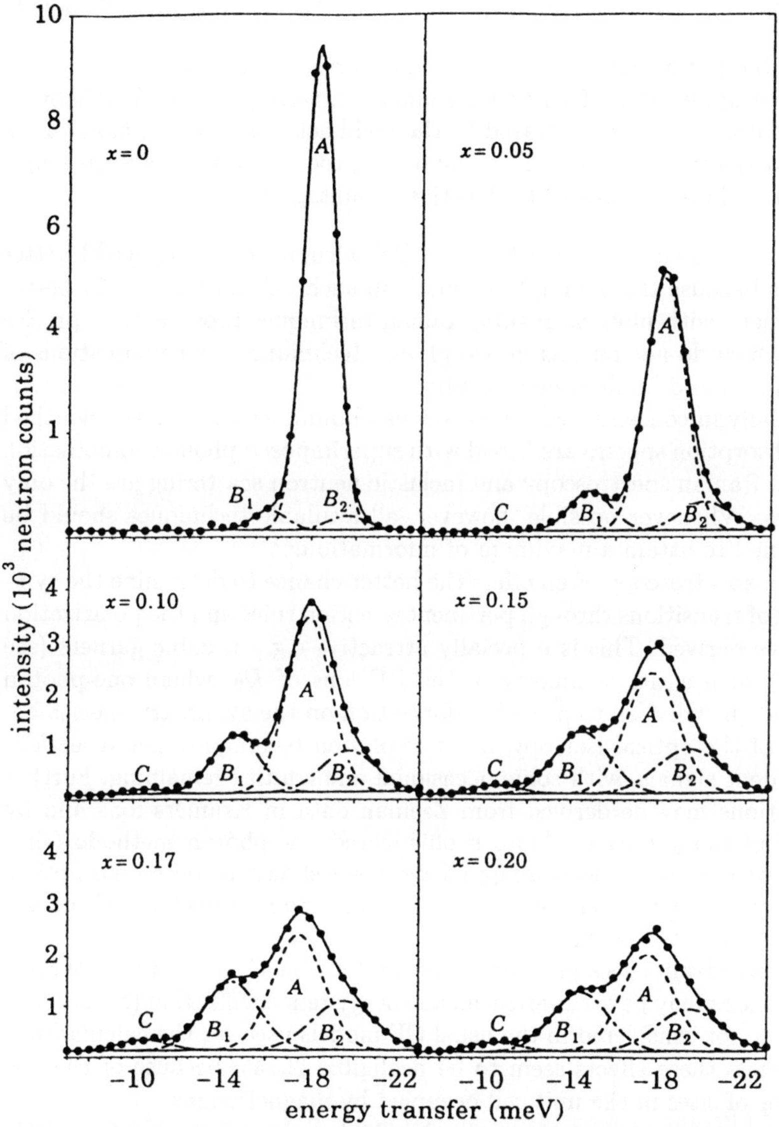

Fig. 2.41. Inelastic neutron scattering spectra from $Pr_{2-x}Ce_xCuO_{4-\delta}$, $0 \le x \le 0.20$; $T = 10\,K$, momentum transfer $Q = 1.8\,Å^{-1}$. The material is superconducting for $0.14 \le x \le 0.17$. The broken lines are fit results. From [2.190]

2.4.2 Crystal Field Studies in Ionic and Organometallic Crystals

Although the parametrization of crystal fields based on Raman experiments was the first application of electronic Raman scattering, this method is by no means obsolete, as is demonstrated by the vivid activities in the field of high-T_c superconductivity. For insulating materials, too, the Raman spectroscopic method still offers a number of attractive advantages:

1. For low lying electronic levels, when CF calculations mostly yield better results because the assumptions made in such calculations are in better agreement with physical reality, Raman techniques provide more precise data which do not rely, as in one-photon techniques, on combinations of absorption and luminescence results.
2. Especially in concentrated systems, when luminescence is often quenched and absorption spectra are broad with superimposed phonon combination bands, Raman spectroscopy and inelastic neutron scattering are the only choices. Whenever possible, however, all available techniques should be combined to attain a maximum of information.[23]
3. Raman spectroscopy often offers the better chance to determine the symmetry of transitions through pertinent selection rules and the polarization features derived. This is especially attractive e.g., in cubic garnets (see below) with a site symmetry of the RE ions of D_2, where one-photon methods usually fail to provide information on the symmetry species because of the optical isotropy, but two-photon techniques give direction-dependent signals, with some occasional ambiguities remaining. Further indications may be derived from Zeeman data in Kramers ions and by studying the g tensors. More sophisticated one-photon methods (site-selective polarized spectroscopy) have been shown to overcome optical isotropy and have been applied to determine the symmetries of the CF levels in garnets (see below).
4. For these advantages a price has to be paid: CF levels in the range of optical phonons may have experienced an energy renormalization (Sect. 2.3.3) which is not considered in the usual CF models used for the calculations. In garnets these effects seem to be negligible [2.145] because of the low number of sites in the unit cell occupied by magnetic ions.
5. Raman spectroscopy competes in this field with inelastic neutron scattering which has its domain in strongly absorbing or metallic systems.

Crystal fields in *cubic garnets*, for example, have been studied in great detail with Raman methods [2.192,2.193,2.194,2.195], with one-photon methods

[23] Some care has to be taken when data from samples with different concentrations c of the $4f$ ion are compared, as the CF levels may depend strongly on c. In DyAlG and (Y,Dy)AlG (see below), these differences amount up to $60\,\mathrm{cm}^{-1}$, [2.195]. See also Table 2.15 for a comparison of CF parameters of TbAlG and (Y,Tb)AlG.

in [2.196] (and in refs. therein), see also [2.29]. RE garnets are mixed metal oxides and have the general formula $3(\mathrm{RE})_2\mathrm{O}_3 \cdot 5(\mathrm{ME})_2\mathrm{O}_3$; ME = Al, Ga, or Fe. Their space group is $\mathrm{O_h^{10}}$, the unit cell has 6 equivalent RE sites with D_2 symmetry. These sites are however magnetically inequivalent with respect to the crystallographic unit cell and only one of the local axes (ξ, η, ζ) of the sites coincides with one axis of the cubic unit cell (axes: $\bar{x}, \bar{y}, \bar{z}$) [2.29]. The local sites can be grouped into 3 pairs of two sites each, each pair has two local axes, say (ζ, η), orthogonally crossed to each other and oriented in one of the planes $(\bar{x}, 0, 0), (0, \bar{y}, 0), (0, 0, \bar{z})$ under 45° with respect to $\bar{x}, \bar{y}$, or $\bar{z}$). The third local axis (ξ) is oriented either $\|\bar{x}, \|\bar{y}$, or $\|\bar{z}$.

Next we have to correlate the independent single-ion scattering processes occurring in D_2 site symmetry with the observed Raman response $[(\alpha_{\bar{x},\bar{y}})^{\Gamma_j}]^2$ of the cubic crystal belonging to the rep Γ_j of the unit cell group [2.193], [2.195]. There has been no interionic coupling detected in the garnets, hence the intensities provided by the different sites $[(\alpha_{\xi,\eta})_i^{\Gamma_{j'}}]^2$ have to superimposed, considering their directional anisotropy correctly, i.e.

$$[(\alpha_{\bar{x},\bar{y}})^{\Gamma_j}]^2 = \sum_{\text{sites}(i)} [R_i^{-1} (\alpha_{\xi,\eta})_i^{\Gamma_{j'}} R_i]^2 \tag{2.92}$$

Here the R_i are the rotational matrices which transform the scattering tensor $(\alpha_{\xi,\eta})_i^{\Gamma_{j'}}$ of rep $\Gamma_{j'}$ from the various local coordinate systems $(\xi, \eta, \zeta)_i$ into the unit cell system $(\bar{x}, \bar{y}, \bar{z})$. If the specific form of $[\alpha_{\bar{x},\bar{y}}]$ in terms of the tensor components of $(\alpha_{\xi,\eta})_i$ is not of importance, the correlation between the single-ion reps (for garnets: $\bar{\Gamma}_5$ for Kramers ions, $\Gamma_1 - \Gamma_4, (A, B_1, B_2, B_3)$ for non-Kramers ions, Table 2.A.1) and the reps $\Gamma_1^+ - \Gamma_5^+, A_{1g}, A_{2g}, E_g, F_{1g}, F_{2g}$ of the unit-cell scattering tensor $[\alpha_{\bar{x},\bar{y}}]$ will give the reps of the latter labeling the tensor components causing symmetric and antisymmetric scattering. In the case of the cubic garnets the rep A (symmetric scattering) of D_2 correlates with A_{1g}, E_g, F_{2g} of O_h, while B_1, B_2, B_3 (both symmetric and antisymmetric scattering) correlate with all reps of O_h except A. Hence symmetric electronic scattering will be observed in A_{1g}, E_g, F_{2g} of O_h, while antisymmetric scattering will concentrate in F_{1g} [2.193].

In a specific study of CF parameters in the garnets $3\mathrm{Tb_2O_3} \cdot 5\mathrm{Al_2O_3}$ (TbAlG) and in $3\mathrm{Eu_2O_3} \cdot 5\mathrm{Ga_2O_3}$ (EuGaG) [2.192], electronic Raman spectroscopy has been combined with fluorescence and infrared absorption spectroscopy. Only data from the ^7F_J multiplets $(0 \le J \le 6)$ have been considered. This simplifies considerably the calculations of the CF energies, since $\mathcal{H}_{\mathrm{CF}}$ is diagonal in S. In other RE ions, more complicated calculations with eigenfunctions in the intermediate coupling approximation (as obtained by diagonalizing simultaneously the Coulomb repulsion and the spin–orbit coupling of the $4f$ electrons) have to be performed [2.196].[24]

[24] The literature on crystal field theory and calculations for specific lattices and ions is abundant, see [2.29, 2.42, 2.197, 2.198] for details.

The crystal field Hamiltonian used for the garnets is (see footnote 15):

$$\mathcal{H}_{\rm CF} = \sum_{k,q} B_q^k C_{k,q}; \qquad k = 2,4,6;\ q = 0,2,4,6;\ q \le k. \tag{2.93}$$

The actual calculations are performed in two steps: First the matrix elements of (2.93) have to be calculated with standard techniques in a suitable basis (LS eigenfunctions of the lowest multiplet or intermediate coupling functions), using approximate values for the CF parameters, e.g., values determined previously for a different ion in the same lattice. After establishing a correlation between the calculated wavenumbers of CF levels and types of symmetry with the experimental data, a least-squares fit will minimize the remaining deviations and produce the final parameter values. In Fig. 2.42 the observed energies of CF levels of TbAlG and EuGaG are compared with calculated values according to this procedure. The remaining deviations are typical for this kind of analysis and are mainly caused by shortcomings of the model which is based on the assumption of point charges. In the real crystal this assumption is of course not met at all, overlap and covalency contributions as well as linear shielding effects are lumped together in the phenomenological CF parameters [2.198]. In Table 2.15 the parameter values of TbAlG and EuGaG have been compiled and compared with values of isomorphic $(Y_{1-c}Dy_c)$AlG, $c = 0.1$ or 0.01 [2.196]. The garnets have a strong CF, as becomes evident by comparison of the values in Table 2.15 with tabulated CF parameters, e.g., for the hexagonal $RECl_3$ or ethylsulphates [2.29]. The cubic parameters B_0^4, B_4^4, B_4^6 are especially prominent, giving some justification to the cubic approximation of the CF sometimes in use with garnets.

Recently a one-photon technique has been developed and applied to garnets: the site-selective polarized excitation and fluorescence spectroscopy [2.199]. Here a tunable laser of definite polarization and beam direction relative to the crystal axes excites the ions on the six differently oriented sites with a certain probability depending on the type (electric or magnetic) of the local transition dipoles and their orientation. This site-selective excitation

Table 2.10. CF parameters B_q^k in cm^{-1} (Wybourne notation) from optical spectroscopy in $3Tb_2O_3 \cdot 5Al_2O_3$ (TbAlG) and $3Eu_2O_3 \cdot 5Ga_2O_3$ (EuGaG); from [2.192]. The values for $3(Y_{0.99}Dy_{0.01})_2O_3 \cdot 5Al_2O_3$ ((Y,Dy)AlG) from [2.196] are not based on Raman results. The results for (Y,Tb)AlG and (Y,Eu)AlG have been obtained with the method of site-selective spectroscopy [2.199]

Substance	B_0^2	B_0^4	B_0^6	B_2^2	B_2^4	B_4^4	B_2^6	B_4^6	B_6^6
TbAlG	−320	−2328	704	319	278	895	−253	1073	−161
EuGaG	−160	−2104	848	194	253	950	−144	1196	−185
(Y,Dy)AlG	−340	−2336	656	273	304	1000	−222	1065	−81
(Y,Tb)AlG	−386	−2427	664	203	257	1031	−192	1129	−98
(Y,Eu)AlG	−308	−2270	1090	302	256	1076	−324	1538	−147

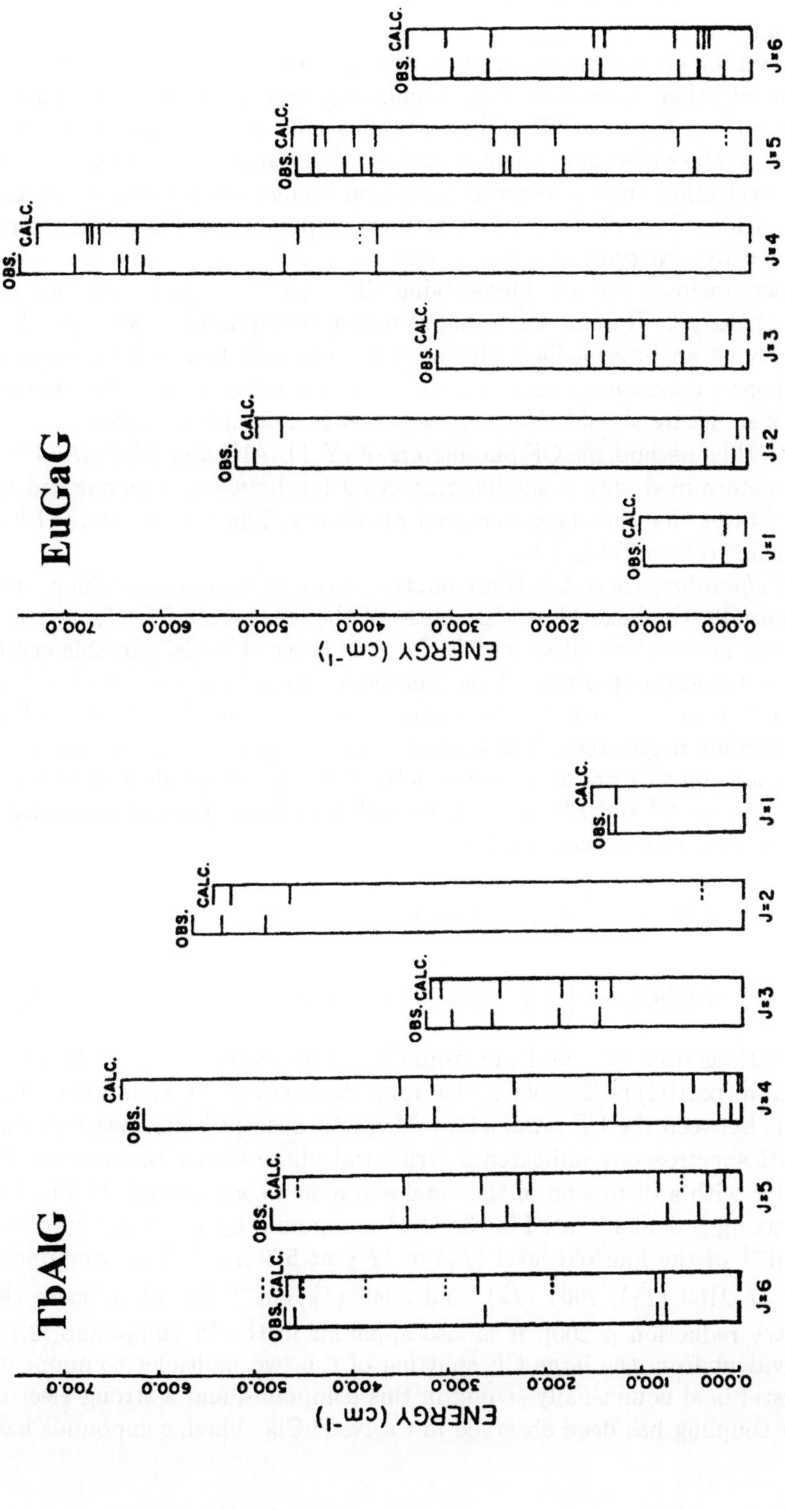

Fig. 2.42. Experimental (OBS.) and calculated (CALC.) CF levels in the ^7F-multiplets of TbAlG (*left*) and EuGaG (*right*), from [2.192]

results in an anisotropic fluorescence emission of the crystal with intensities depending on the orientations of polarizer and analyzer. These orientations are chosen for an optimized selection of a specific site and a maximum suppression of other unwanted sites. Combining various relative orientations, the transitions between CF states as well as the states themselves can be assigned to the different symmetry species. The method has to avoid migration of excitation energy between transition dipoles of different orientation. This is achieved in dilute crystals at low temperatures. The experiment is performed by observing the fluorescence at a fixed wavelength, the excitation laser sweeps across the higher lying CF levels. This procedure has some formal analogy to Raman scattering as it is a "two-photon" technique, however without an intermediate virtual state but with two real states where multiphonon relaxation starts or ends. Such a selection of specific sites in a complex structure should also be possible with Raman techniques.

With this method the CF parameters of (Y,Tb)AlG and (Y,Eu)AlG have been redetermined with a smaller rms deviation between observed and calculated states than has been achieved previously. The values obtained have been included into Table 2.15.

The *elpasolites* (Sect. 2.2.4) are another series of compounds widely studied, especially the hexachloro-elpasolites of the form $A_2B(RE)Cl_6$, where A and B are monovalent alkali metals. In most cases $A = Cs^+$. In this crystal structure (spacegroup $Fm3m$) the lanthanide ions from La^{3+} to Lu^{3+} are embedded in an octahedral coordination of 6 Cl^- ions with three- or fourfold electronic degeneracy. The material also accepts transition metals and actinide ions on this site. In this symmetry the only independent crystal field parameters are B_0^4 and B_0^6, since B_4^4 and B_4^6 are usually fixed at values taken from the point charge model [2.197]:

$$\begin{aligned}\mathcal{H}_{\mathrm{CF}} = \; & B_0^4\left[C_{4,0} + \left(\frac{5}{14}\right)^{\frac{1}{2}} \cdot (C_{4,4} + C_{4,-4})\right] \\ & + B_0^6\left[C_{6,0} - \left(\frac{7}{2}\right)^{\frac{1}{2}} \cdot (C_{6,4} + C_{6,-4})\right]. \end{aligned} \tag{2.94}$$

Deviations from this model or from the exact symmetry can be easily detected as a (partial) raising of the electronic degeneracy and violation of these relations between the CF parameters. Magnetic susceptibility measurements and ESR spectroscopy indicated a structural phase transition occuring below 160 K with a distortion of the octahedron as a consequence. In Fig. 2.43 this splitting is demonstrated in the Ce^{3+} compound by an energy difference of $15\,\mathrm{cm}^{-1}$ of the fourfold level ($\bar{\Gamma}_8$) of $^2F_{\frac{5}{2}}$ at $570\,\mathrm{cm}^{-1}$. The components of $^2F_{\frac{7}{2}}$ at 2160 ($\bar{\Gamma}_7$), 2661 ($\bar{\Gamma}_8$), and 3048 ($\bar{\Gamma}_6$) cm^{-1} do not indicate this symmetry reduction [2.200]. It is also apparent in the Pr compound [2.77]. As is evident from the large CF splitting of the two multiplet components, the crystal field is unusually strong in this compound and a strong electron phonon coupling has been observed in $Cs_2NaYbCl_6$. These compounds have also served as another test for the theories of intensities of electronic Raman scattering (Sect. 2.2.4): The mean ratio of F_1/F_2 is 0.22, close to the theoretical value of 0.25 [2.73], by assuming that the $4f^{n-1}5d^1$ configuration is the only intermediate configuration with an average energy of $10^5\,\mathrm{cm}^{-1}$ [2.77].

The CF parameters have been determined for some of the RE ions, with some remaining scatter in the numbers for B_0^4 and B_0^6 (in cm^{-1}, Wybourne notation) [2.200], [2.201]: Ce^{3+}: 2048 (2104); 272 (140.8). Pr^{3+}: 2168; 272. Nd^{3+}: 1792; 287. Eu^{3+}: 2432; 51.2. Ho^{3+}: 1664; 187. Yb^{3+}: 1368; 25.6. Values for B_0^2, due to the low-symmetry distortion, are approximately $10\,\mathrm{cm}^{-1}$.

In hexafluoroelpasolites the EuF_6^{3-} complex has been studied in $Cs_2NaY_{1-x}Eu_xF_6$ by one-photon techniques [2.202] at low Eu^{3+} concentrations ($0.001 \leq x \leq 0.2$). At low concentration no distortion from the octahedral symmetry was evident for the Eu^{3+} site, but a small splitting of magnetic dipole transitions was detected for $x = 0.2$ below 20 K. The fourth- and sixth-rank CF parameters were about 1.6 times larger than for $Cs_2NaEuCl_6$, the values are: $B_0^4 = 3138\,\mathrm{cm}^{-1}$, $B_0^6 = 382\,\mathrm{cm}^{-1}$, for $Cs_2NaYCl_6 : Eu$: $B_4^0 = 1928\,\mathrm{cm}^{-1}$, $B_0^6 = 247\,\mathrm{cm}^{-1}$.

Another interesting example is the *perovskite structure* $NdAlO_3$, which has D_{3d}^6 symmetry at room temperature but undergoes a trigonal-to-cubic (perovskite) transition at high temperature [2.203]. The site symmetry of Nd^{3+} is D_3, however the deviation from cubic (O_h) symmetry is very small: All the Γ_8 levels of the cubic CF field split into pairs of closely spaced levels of D_3 symmetry types, $\Gamma_4, \Gamma_{5,6}$ (Table 2.A.1) under the small trigonal distor-

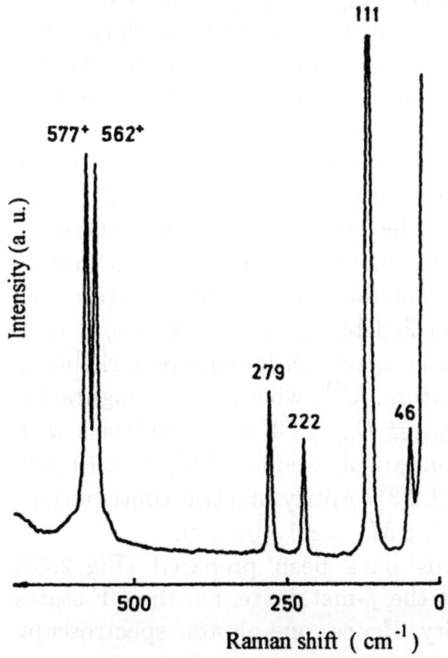

Fig. 2.43. Raman-spectrum of $Cs_2NaCe^{3+}Cl_6$ (elpasolite) at ≈ 35 K. The bands marked by an $^+$ are electronic transitions, from [2.200]

tion, which amounts to energy differences $\leq 65\,\mathrm{cm}^{-1}$ for the lower multiplet components ${}^4\mathrm{I}_{\frac{9}{2}}, {}^4\mathrm{I}_{\frac{11}{2}}$ of Nd^{3+}, as compared to differences of several hundred cm^{-1} due to the cubic contribution. CF parameters for the trigonal field have been determined from fits including J mixing: $B_0^2 = -530.4\,\mathrm{cm}^{-1}, B_0^4 = 490.3\,\mathrm{cm}^{-1}, B_3^4 = -449.4\,\mathrm{cm}^{-1}, B_0^6 = -1646.9\,\mathrm{cm}^{-1}, B_3^6 = -992.0\,\mathrm{cm}^{-1}, B_6^6 = -1026.6\,\mathrm{cm}^{-1}$ with a rms deviation of $6.6\,\mathrm{cm}^{-1}$. The observed ratios B_3^4/B_0^4, B_3^6/B_0^6, and B_6^6/B_0^6 are very close to the values expected for the cubic approximation (−1.195, 0.605, 0.634) where B_0^2 should be zero. Observed g-factors of the lowest CF levels are only in qualitative agreement with values calculated from CF eigenfunctions, a fact which is attributed to effects of the $4f$ electron–phonon interaction. Calculations of the Raman intensities based on the Judd–Ofelt approximation [2.60], [2.61], (Sect. 2.2.4) do not agree satisfactorily with experimental data, which is not surprising.

Organometallic complexes with f-elements and studies of coordination effects and ligand fields are a broad field of activities, especially for chemists. The literature is voluminous (see [2.204]) and certainly cannot be covered adequately in this review. We only want to give some flavor of this area by citing a single example [2.205], the sandwich complexes of f-elements, which have been known for 30 years. The homologue of ferrocene, uranocene (bis(cyclooctatetraenyl)uranium(IV), $(C_8H_8)_2U$, $U(COT)_2$ was the first compound synthesized; others, e.g., $Ce(COT)_2$, followed. These molecules are formed by two planar aromatic (C_8H_8) rings which are parallel and eclipsed, sandwiching the central metal ion with point symmetry C_i. The molecular symmetry is D_{8h}, the crystal displays the space group C_{2h}^5 [2.206]. The (C_8H_8) ring diameter (C–C distance) is 0.365 nm, the two planes are 0.386 nm apart. The large U(IV) ion (≈ 0.2 nm diameter) is framed by two gloriole-like rings. The lanthanide compounds were found to be highly ionic in comparison to the actinide complexes, which exhibit strong covalent contributions from metal f- and d orbitals to the metal–ligand interaction.

Uranocene displays moderately intense visible absorption bands, which have been attributed to charge transfer transitions from the $\mathrm{COT}^{2-}\pi$ orbitals to the uranium f-orbitals [2.207]. The resonance Raman spectra depicted in Fig. 2.44 show an electronic Raman line at $466\,\mathrm{cm}^{-1}$ with an unusual depolarization behavior, which is due to contributions of antisymmetric scattering. Its electronic nature has been verified by different methods, it does not shift under deuteration. This resonance spectrum is truly remarkable as it is taken in solution at room temperature. $\mathrm{U^{IV}}$ with a f^2 configuration has a ${}^3H_4$ ground state, split by the $D_{8h}(\equiv D_{\infty h})$ CF into five levels with $M_J = 0, \pm1, \pm2, \pm3, \pm4$. Raman transitions are allowed for $\Delta M_J = 0, \pm1, \pm2$, i.e. $\Delta M_\mathrm{S} = 0$, $\Delta M_\mathrm{L} = 0, \pm1, \pm2$ (Sect. 2.2.3). Antisymmetric contributions are possible only for $\Delta M_J = 0, (xy - yx), \Delta M_J = \pm1, (yz, zx)$.

Recently, half-sandwiched compounds have been prepared (Fig. 2.45) which do not have an inversion center at the f-metal site, i.e. the CF states are not characterized by a definite parity. Hence, one-photon spectroscopy

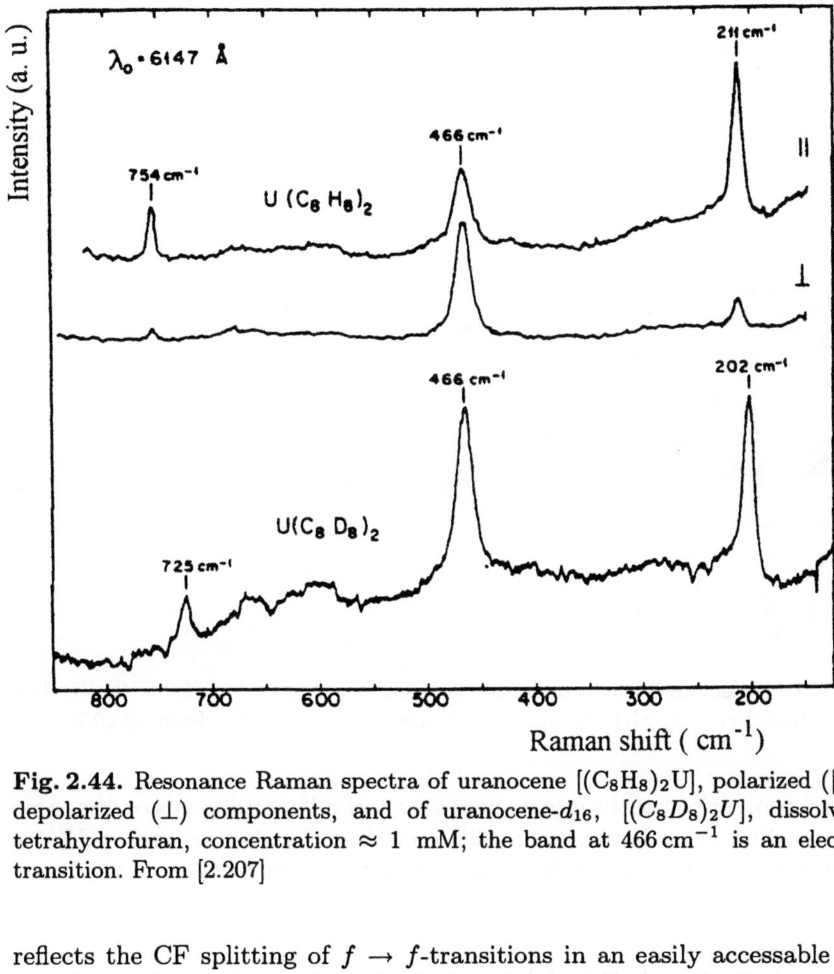

Fig. 2.44. Resonance Raman spectra of uranocene [$(C_8H_8)_2U$], polarized (||) and depolarized ($\perp$) components, and of uranocene-d_{16}, [$(C_8D_8)_2U$], dissolved in tetrahydrofuran, concentration $\approx$ 1 mM; the band at $466\,\mathrm{cm}^{-1}$ is an electronic transition. From [2.207]

reflects the CF splitting of $f \to f$-transitions in an easily accessable manner. It has been shown that the hydrotris(pyrazol) ligands (lower part of Fig. 2.45) contribute to the total CF only in higher order, i.e. the CF effects observed are essentially due to the single (COT) aromatic ring [2.205] with a site symmetry of C_{8v}. The Pr^{3+} compound ($Pr[COT]^+$) has been investigated in detail. The twofold degeneracy of the CF states in C_{8v} symmetry is partially raised by the weak trigonal perturbation due to the pyrazol ligands with an energy difference of $43\,\mathrm{cm}^{-1}$ at maximum. A Hamilton operator for single $4f$-electron states is diagonalized and the parameters fitted to the observed electronic absorption and emission bands, besides the CF terms the intraatomic Coulomb interaction (Slater parameters F^2, F^4, F^6) and the spin–orbit interaction are included (spin–orbit coupling constant ζ) with the pertinent parameters as additional fit variables. The following CF parameter have been obtained: $B_0^2 = -21\,\mathrm{cm}^{-1}, B_0^4 = -3732\,\mathrm{cm}^{-1}, B_0^6 = 254\,\mathrm{cm}^{-1}$. Other fit parameter were: $F^2 = 66238\,\mathrm{cm}^{-1}, \zeta = 748\,\mathrm{cm}^{-1}$.

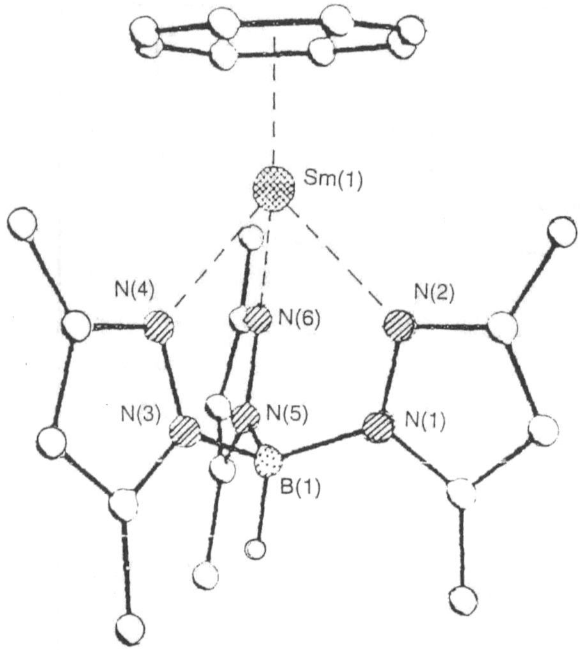

Fig. 2.45. Molecular structure of the half sandwich complex (COT)Sm[Hydro-tris(3,5-dimethyl-pyrazol-1-yl)borato]$^-$. Protons are not shown. From [2.205]

For a comparison of the crystal field effects of different ligands on the same f-ion, on different f-ions, even in different sites a scalar parameter N_v for the "strength" of the crystal field has come into use in chemical applications [2.208]:

$$N_v = \left[\sum_{k,q} (B_q^k)^2 \left(\frac{4\pi}{2k+1} \right) \right]^{\frac{1}{2}} . \tag{2.95}$$

This relation was derived under some ad hoc assumptions and found to be applicable only to parameters derived from electron configurations with $J \geq \frac{k}{2}$. Its unreflected use appears questionable. For (Pr[COT]$^+$): $N_v = 4417\,\mathrm{cm}^{-1}$, for the double sandwich the CFs are superposed in a crude approximation and $N_v = 8834\,\mathrm{cm}^{-1}$ is obtained. Values from model calculations were found for Ce[COT]$_2$) : $20344\,\mathrm{cm}^{-1}$, U[COT]$_2$: $34385\,\mathrm{cm}^{-1}$ or $32734\,\mathrm{cm}^{-1}$ (see [2.205]). For the cyclopentadienyl ligand (Cp) smaller N_v values are found experimentally: $Cp_3Pr \cdot NCCH_3$: $3555\,\mathrm{cm}^{-1}$; Cp_3UCl : $8242\,\mathrm{cm}^{-1}$, $[Cp_3U(NCBH_3)_2]^-$: $10046\,\mathrm{cm}^{-1}$. Clearly this situation is unsatisfactory.

The sandwich compounds cerocene, uranocene etc. have also attracted much attention in theoretical chemistry, because of their fundamental importance for stoichiometry and catalytic reactivity. Numerous sophisticated, large-scale calculations have been performed [2.209] to elucidate the ligand–(f-system) interaction. The results obtained differ markedly from the simple ionic model sketched above. Here the calculations are based on the assumption that the lowest f-state has to have a lower energy than the highest occupied molecular orbital of the ligand system. Hence a hopping of electrons between the two subsystems occurs, analogous to that in metallic cerium Kondo systems. In the case of a single Kondo impurity (with one f-orbital) the ground state wavefunction can be written in the form [2.209]:

$$|\psi_0\rangle = A\left[a|FS\rangle + \int \mathrm{d}^3\boldsymbol{k}\; b(\boldsymbol{k})\cdot\left(\sum_\sigma f_\sigma^+ c_{\boldsymbol{k}\sigma}\right)|FS\rangle\right],$$
$$a \ll \int \mathrm{d}^3\boldsymbol{k} b(\boldsymbol{k})\,. \tag{2.96}$$

Here f_σ^+ and $c_{\boldsymbol{k}\sigma}$ create an electron with spin σ in the f-orbital, taken ($c_{\boldsymbol{k}\sigma}$) from the momentum space of the conduction electrons. $|FS\rangle$ denotes the filled Fermi sea, the f-orbital being unoccupied. In the present case of lanthanocenes, the sea of itinerant conduction electrons present in the Kondo or heavy-fermion [2.15] systems stands for the π electrons of the lowest level of the $(COT)^{2-}$ ligand ions. The correlations due to Coulomb repulsion between the f-electrons have to be sufficiently strong in order to give states with unusual numbers of f-electrons at unfavorably high energies. On the other hand, the hybridization between ligand- and f-electrons has to be weak.

The quantum chemical calculations have arrived at the conclusion that the ground state configurations for the lanthanocenes are $4f^1\pi^3$ while for actinocenes they are $5f^{n-1}\pi^4$. In the case of cerocene the unpaired f-electron on the metal atom is hybridized with an unpaired electron in the ligand orbitals to form a singlet ground state with a triplet excited state about 0.5 eV above. The configuration $4f^1\pi^3$ contributes 80% of the ground state of cerocene. The diamagnetism of this ground state has been proven recently by K-edge X-ray absorption spectra [2.210].

Another promising organometallic complex is the tris (cyclopentadienyl)-lanthanide(III) complex Cp_3RE, especially its ester adducts, e.g., the butylacetate adduct $C_{21}H_{27}O_2RE$ [2.211], which can be grown in single crystal with a monoclinic structure. In Fig. 2.46 Raman spectra of $Cp_3Ce(NCCH_3)_2$ and $Cp_3La(NCCH_3)_2$ have been reproduced, which show an ERS transition of $Ce^{3+}, {}^2F_{\frac{5}{2}}$ at 90 K, other transitions have been observed at 2129 cm^{-1} and at 2154 cm^{-1}, ($\rightarrow {}^2F_{\frac{7}{2}}$).

These examples demonstrate that this is a highly challenging field for future work also with optical techniques.

2.4.3 Localized Excitations in Semimagnetic Semiconductors

Diluted magnetic (semimagnetic) semiconductors (DMSs) are compound semiconductors in which a fraction of the cations are magnetic. A typical

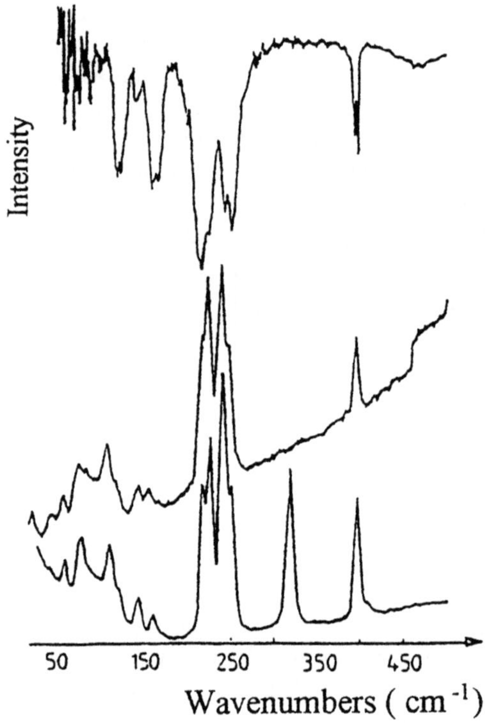

Fig. 2.46. (a): Raman spectrum of $Cp_3Ce(NCCH_3)_2$ at ≈ 30 K. **(b)**: Raman spectrum of $Cp_3La(NCCH_3)_2$ at ≈ 90 K; **(c)**: FIR spectrum of $Cp_3La(NCCH_3)_2$ at ≈ 90 K. An electronic Raman transition occurring at 320 cm^{-1} is displayed in **(a)**. From [2.211]

representative is $Cd_{1-x}Mn_xTe$, containing manganese as magnetic ion. Its unique properties [2.16] arise from the *sp–d* exchange interaction between the localized electrons in the 3*d* shell and *s*-like electrons and the *p*-like holes near the semiconductor band edges. The properties of this DMS are also affected by the antiferromagnetic (af) exchange interaction between nearest-neighbor (nn) magnetic ions. This interaction leads to the formation of af clusters of Mn ions, spin-glass behavior etc. In this section isolated pairs of af coupled Mn^{2+} ions and their energy levels, with a separation determined by the nn exchange constant J_{nn}, are discussed. Raman transitions were observed between the $S = 0$ ground state and the magnetically split $S = 1$ levels of these pairs in $Cd_{1-x}Mn_xTe/Cd_{1-y}Mg_yTe$ quantum well structures comprising semimagnetic quantum wells ($Cd_{1-x}Mn_xTe$) of different widths and Mn concentrations x and diamagnetic $Cd_{1-y}Mg_yTe$ barriers. This example demonstrates how energy levels and Raman intensities of the observed excitations are probes for new boundary conditions or constraints as exerted on unfilled shells, e.g. in a quantum well.

CdTe has the cubic zincblende structure. Mn^{2+}, ($3d^5$, ${}^6S_{\frac{5}{2}}$), on the Cd site is not affected by the cubic CF; it offers an almost ideal example of pure spin magnetism ($g_{\mathrm{Mn}} = 2$). $\Delta M = \pm 1$ transitions within the Zeeman split ground state sextet of isolated ions have been observed in electron-spin resonance (ESR) experiments and in Raman scattering [2.16] analogous to paramagnetic resonance (PR), $E_{\mathrm{PR}} = g_{\mathrm{Mn}}\mu_{\mathrm{B}}B$. The exchange coupling between pairs of spins $\boldsymbol{S}_1, \boldsymbol{S}_2$ in an external field B is described by the Heisenberg Hamiltonian:

$$\mathcal{H}_{\mathrm{pair}} = -2J_{\mathrm{nn}}\boldsymbol{S}_1 \cdot \boldsymbol{S}_2 + g_{\mathrm{Mn}}\mu_{\mathrm{B}}\boldsymbol{S} \cdot \boldsymbol{B}. \tag{2.97}$$

$\boldsymbol{S} = \boldsymbol{S}_1 + \boldsymbol{S}_2$, with energies $E = -J_{\mathrm{nn}}[S(S+1) - \frac{35}{2}] + g_{\mathrm{Mn}}\mu_{\mathrm{B}}M_S B$; $S = 0, 1, \ldots, 5$ and $M_S = S, S-1, \ldots, -S$. Energy differences of the excited pair levels to the ground state are: $2|J_{\mathrm{nn}}|, 6|J_{\mathrm{nn}}|, \ldots, S(S+1)|J_{\mathrm{nn}}|$. Here the $S = 0 \to S = 1, M_S = 0$ transition in a magnetic field is discussed.

Experimental results for Faraday geometry[25] are shown in Fig. 2.47 [2.212]. Besides the intense P^0 line which is independent of B, the PR line due to single Mn^{2+} ions and the weak 2PR line ($\Delta m = 2$) are observed. The $2P^0$ line has been attributed to an $S = 1 \to S = 2, \Delta M = 0$ transition from an excited state due to its intensity increasing with B. The P^0 signal was also observed in resonance with 45, 60, and 100 Å wells with decreasing strength. It was neither found in a 300 Å well nor in bulk samples or epilayers of the zincblende structure. This behavior shows the importance of the confinement of the charge carriers in the quantum well for the scattering intensity of the Mn^{2+} pair signals.

The energy of P^0 depends on x_{Mn} according to $E_{P^0} = (8.49 + 9.17x_{\mathrm{Mn}})\,\mathrm{cm}^{-1}$. It could be shown that this increase is not due to an increase in J_{nn} with x but to interactions of the pair with more distant single Mn^{2+} neighbors. J_{nn} is found as $\lim_{x_{\mathrm{Mn}}\to 0}\{J_{\mathrm{nn}}\} = -4.24 \pm 0.10\,\mathrm{cm}^{-1}$. The exchange interaction is essentially due to superexchange between the d-orbitals of Mn^{2+} via the p orbitals of the intermittent Te^{2-} ions (virtual hopping). From the resonance behavior of the P^0 signals it is derived that localized heavy-hole excitons at the band edge act as intermediate states in the Raman process. Such a scattering mechanism, due to the strong exchange interaction between the carrier spin and the localized moments of the Mn^{2+} ions, is clearly different from the resonance processes in insulators as discussed previously in Sect. 2.2.5.

This scattering mechanism which produces a Raman active $\Delta S = 1$ transition, only occurring in narrow quantum wells, clearly needs some consideration [2.212]. It is related to the exchange interaction of carriers with a coupled Mn^{2+} pair with the total spin $\boldsymbol{S} = \boldsymbol{S}_1 + \boldsymbol{S}_2$ and the spin difference, i.e. the af vector $\boldsymbol{A} = \boldsymbol{S}_1 - \boldsymbol{S}_2$ of the Mn pair. This interaction becomes substantial only under sharp gradients of the carrier wavefunction as encountered for localized

---

[25] In Voigt geometry ($\boldsymbol{B} \perp \boldsymbol{z}$) the Raman spectra were dominated by the intense multiple PR scattering, see [2.16], [2.213].

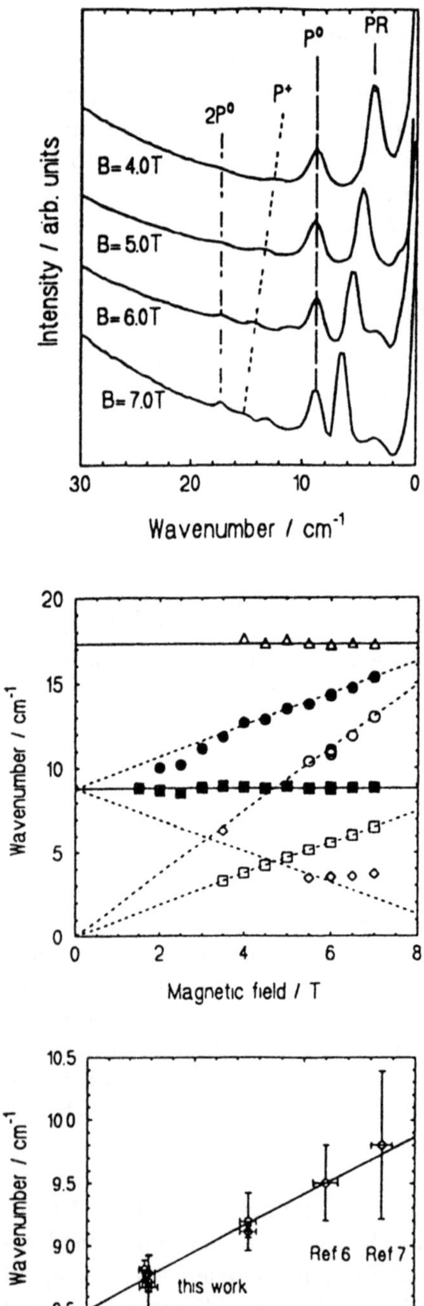

Fig. 2.47. (*Above*) Raman spectra in resonance with the lowest electronic excitation of an 18 Å well of a (Cd,Mn)Te/(Cd,Mg)Te quantum well structure, $x_{\mathrm{Mn}} = 0.027, y_{\mathrm{Mg}} = 0.24$, in Faraday geometry $\bar{z}(\sigma^+, \sigma^+)z$, (see footnote 6), $T = 1.8\,\mathrm{K}$ for various fields B. The increasing background is due to the band-edge luminescence of the σ^+ heavy-hole exciton. PR: paramagnetic resonance of Mn^{2+}; $\lambda_{\mathrm{exc.}} \approx 738\,\mathrm{nm}$; c.f. Fig. 2.4. (*Center*) Observed pair and single ion (PR) transitions in a magnetic field. $\square, P^0 : S = 0 \rightarrow S = 1, M = 0$; $\bullet, P^+ : S = 0 \rightarrow S = 1, M = 1$; $\diamond, P^- : S = 0 \rightarrow S = 1, M = -1$; $\triangle, 2P^0 : S = 1, M = -1 \rightarrow S = 2, M = -1$; $\square$, PR: $g = 2.00$: $\circ$, 2 PR. The dashed lines are guides for the eye, calculated for $g = 2$. (*Below*) Observed Raman shifts of the $S = 0 \rightarrow S = 1, M = 0$ transition P^0 as a function of x_{Mn}. The full line is the best linear fit. From [2.212]

(exciton) states in sufficiently narrow quantum wells. The intensity is determined by an exchange Hamiltonian of the Heisenberg-type comprising besides $\boldsymbol{B}_S \cdot \boldsymbol{S}$ the term: $\boldsymbol{B}_{\mathrm{A}} \cdot \boldsymbol{A}$ with the exchange field $\boldsymbol{B}_{\mathrm{A}} \sim (|\Psi^e(\boldsymbol{R}_1)|^2 - |\Psi^e(\boldsymbol{R}_2)|^2)$ and an analogous term for the holes where $\Psi^{e,h}(\boldsymbol{R}_i)$ are the envelope functions of the electron and hole at the position $i = 1, 2$ of the Mn pair ions. It can be shown by straightforward calculations [2.214] that only the antisymmetric term $B^z_{\mathrm{A}} A^z$ in the exchange Hamiltonian $\mathcal{H}_{exch.} = \boldsymbol{B}_S \cdot \boldsymbol{S} + \boldsymbol{B}_{\mathrm{A}} \cdot \boldsymbol{A}$ produces non-diagonal matrix elements connecting states with $\Delta S = \pm 1$ and $\Delta M = 0$ which cause the P^0 and the $2P^0$ transitions observed in Fig. 2.47.

Thus the confinement of the charge carriers strongly increases the Raman intensity of the scattering from Mn pairs, emphasizing their role as local probes sensitive to specific interactions.

2.5 Conclusions

Electronic Raman scattering (ERS) with metal ions of unfilled shells in solids is, in its 35th year [2.10], a mature and well established method. Is it also an obsolete one? Browsing through the references of this review, it becomes apparent that the major part of the entries has appeared in the seventies and early eighties, but there is also a steady stream of relevant contributions in the mid-nineties. With the foundations of the method firmly laid[26], it now faces new applications. The entry in new areas will be facilitated for the experimentalist both by new instrumental developments and by an extension of sophisticated (numerical) methods for treating quantum chemical problems to a hitherto unknown precision with the help of large computer facilities and elaborate software. The last ten years have shown, however, that the solution of fundamental problems cannot be achieved with a single method alone but only the combination of several carefully selected techniques will pave the way for future progress. ERS will have to cooperate and compete with all one-photon techniques, with inelastic neutron scattering, photoelectron and X-ray spectral (EXAFS) methods using modern synchrotron sources.

Future areas of spectroscopic activities with localized excited electronic levels which can be foreseen are:

1. *Stoichiometry, reactivity, and catalytic behavior of d- and f-metallo-complexes.* Sect. 2.4.2 has given an impressive example of what can be achieved. Other examples of complexes have also been studied [2.215]: A particularly interesting subject are the endofullerenes [2.216]. All the elements of the periodic table can be fitted into a cage of C_{60} or one of the higher fullerenes; the fullerenolanthanides, especially La@C_{82}, were among the first to be produced in sufficient quantities for experimentation by laser ablation. Results are available for La@C_{82}, Ce@C_{82}, Gd@C_{82}, Y@C_{82}.

[26] The very complicated field of Raman intensities requires considerably more attention.

2. *Raman Scattering in RE intermetallic compounds.* The intermetallic compounds of Ce and U, which demonstrate intermediate valence and heavy-fermion behavior due to strong f-electron correlation, continue to remain in the focus of very active and successful basic research in solid state physics [2.15].
3. *Rare earths in semiconductors.* Semiconductors doped with RE ions have been studied for decades from the viewpoint of applications because of their efficient luminescence. The transfer of energy from hot carriers or excitons to the $4f$ shell has been well documented. However, systematic investigations of the f-states on the various sites and in various defect complexes, and CF effects in semiconductors, are still lacking. RE ions can be incorporated in both III–V [2.217], in II–VI and IV–VI compounds [2.218], [2.219]. The techniques of nanostructuring provide new openings also in this well established area (Sect. 2.4.3).
   Besides (Cd,Mn)Te, (Cd,Mn)Se and other Mn compounds there are many other interesting II–VI semimagnetics: (Hg,Fe)Se, (Cd,Fe)Se, (Cd,Co)Se [2.16]. The IV–VI compounds (Pb,Mn)Te and (Pb,Eu)Se have been extensively investigated by CARS spectroscopy [2.219].
4. *Ion–solid interaction mechanisms.* The extremely narrow spectral features of special $4f^n \to 4f^n$ transitions in solids, in combination with the use of nonlinear spectroscopies (photon echoes, free-induction decay), have been investigated for years [2.131]. Recently this field has received new momentum due to the development of techniques for data storage and reprocessing in a dynamical optical memory in materials with an inhomogeneously broadened absorption line and a narrow homogeneous linewidth. The memory capacity is given by the ratio N of the inhomogeneous-to-homogeneous linewidths. In Eu^{3+}:Y_2SiO_5 ($^7F_0 \to {}^5D_0$, 579.88 nm, 4 GHz inhom. width) and in Pr^{3+}:Y_2SiO_5, for example, $N \leq 10^7$ can be reached [2.220].[27]
5. *Luminescent materials, phosphors.* In most inorganic phosphors used in a wealth of applications d- or f-ions play the essential role in the sequence of energy absorption, transfer, and radiative re-emission. Until recently, progress in this field was mostly achieved on an empirical basis [2.221]. For the characterization of the low-lying final levels in the emission process, Raman data, besides (time-resolved) luminescence and excitation spectroscopy, should be very useful. The recent development [2.222] of white

[27] Hole-burning memory in the frequency domain and photon-echo memory in the time domain have been developed beyond the usual holographic techniques. Both memories store information with a spatial resolution limited by diffraction. With hole-burning memory, information is stored as the frequency distribution of holes that can be read out by observing the transmission of a frequency-tunable laser. With photon-echo memory, on the other hand, the Fourier transform of N-bit temporal data is impressed into the inhomogeneous spectrum. For read out a laser pulse is applied, to which the medium responds with a sequence of photon echoes that mimics the original input.

LUCOLEDs (luminescence conversion light emitting diodes) based on the well known $Y_3Al_5O_{12}$:Ce ($4f^1$) phosphor (YAG:Ce^{3+}) is an important example. The blue emission of a GaN LED is partially down-converted by the yellow emitting phosphor, rendering a whitish emission which can be fine-tuned by various other substitutions in the garnet lattice. The crystal field acting on the Ce^{3+} ion is extremely strong (Sect. 2.4.2). The lowest CF component of the $4f^0, 5d^1$ configuration is shifted down to ≈ 2.7 eV into the blue–green region. The overall CF splitting of the $^2F_{\frac{7}{2}}$ component of $4f^1$ is also extreme: 1777 cm^{-1}. The CF splitting of the $^2F_{\frac{5}{2}}$ component is unknown, probably the strong $4f$ electron–phonon interaction suppresses the purely electronic transitions (strong coupling case, Sect. 2.3.3.2). Here again Raman spectroscopy is ideally suited to provide valuable experimental information.

Acknowledgements: The author is indebted to H.-D. Amberger, M. Cardona, M. Dahl, J. Geurts, G. Güntherodt, J. Heber, H.G. Kahle, T. Ruf, J. Schneider, and G. Schütz. They have supported different aspects of the preparation of the manuscript. The Physikalisches Institut of the University at Würzburg and all colleagues there have shown friendly hospitality and have provided all necessary facilities.

2.A Appendix

2.A.1 Representations of the Scattering Tensor

Experimental results involving tensor quantities are conveniently expressed in terms of Cartesian coordinate systems (linear or cylindrical), while theoretical discussions often take advantage of the concept of spherical tensors, well adapted to quantum mechanical calculations. Unitary transformations between the two representations can be found in the literature [2.44], [2.45] and are compiled here for convenience. There exists some ambiguity of the signs used in the literature. Here the phase convention of Fano and Racah is chosen [2.44].

The spherical components of an arbitrary tensor $\boldsymbol{a} = a^{(1,1)}, a^{(1,0)}, a^{(1,-1)}$ of rank 1 (vector) in terms of the Cartesian components $\boldsymbol{a} = a_x, a_y, a_z$ are given by the unitary transformation $\boldsymbol{a}^{(k,q)} = \mathcal{V}^{(1)} \cdot \boldsymbol{a}_i$ [2.44], [2.45]:[28] [29]

(2.98)

$\mathcal{V}^{(1)}$	$a^{(1,1)}$	$a^{(1,0)}$	$a^{(1,-1)}$
(cf)	$2^{-\frac{1}{2}}$	1	$2^{-\frac{1}{2}}$
a_x	$-\mathrm{i}$		i
a_y	1		1
a_z		i	

[28] The common factor (cf) multiplies each entry in the column or line.

[29] The inverse unitary transformation is $\boldsymbol{a}_{ij} = \mathcal{V}^{-1} \cdot \boldsymbol{a}^{(k,q)}$; $\mathcal{V}^{-1} = (\mathcal{V}^T)^*$.

For a tensor of rank 2, $\boldsymbol{a}^{(k,q)} = \mathcal{V}^{(2)} \cdot \boldsymbol{a}_{ij}$, the unitary transformation coefficients are given in the next table [2.44], [2.45].

$\mathcal{V}^{(2)}$	$a^{(0,0)}$	$a^{(2,0)}$	$a^{(2,2)}$	$a^{(2,-2)}$	$a^{(1,0)}$	$a^{(1,1)}$	$a^{(1,-1)}$	$a^{(2,1)}$	$a^{(2,-1)}$
(cf)[27]	$3^{-\frac{1}{2}}$	$6^{-\frac{1}{2}}$	2^{-1}	2^{-1}	$2^{-\frac{1}{2}}$	2^{-1}	2^{-1}	2^{-1}	2^{-1}
a_{xx}	1	1	−1	−1					
a_{yy}	1	1	1	1					
a_{zz}	1	−2							
a_{xy}			−i	i	−i				
a_{yx}			−i	i	i				
a_{xz}						1	1	1	−1
a_{yz}						i	−i	i	i
a_{zx}						−1	−1	1	−1
a_{zy}						−i	i	i	i

(2.99)

For practical reasons the transformations between Cartesian linear ($\boldsymbol{a}_{\mathrm{ij}}$) and cylindrical coordinates ($\boldsymbol{a}_{\mathrm{c,c}}$) for a rank 2 tensor ($\boldsymbol{a}$) and to spherical coordinates ($\boldsymbol{a}^{(\mathrm{k,q})}$) are also given. Transformations between linear ($\boldsymbol{e}_i$, i=x, y, z) and circular polarization $\mathbf{e}_{\mathrm{c}}(\boldsymbol{e}_{+1} = \boldsymbol{e}^{(\mathrm{l})}, \boldsymbol{e}_{-1} = \boldsymbol{e}^{(\mathrm{r})}, \boldsymbol{e}_0 = \boldsymbol{e}_z)$ use

$$\boldsymbol{e}^{(\mathrm{l})} = \pm 2^{-\frac{1}{2}}(\boldsymbol{e}_x + \mathrm{i}\boldsymbol{e}_y); \quad \boldsymbol{e}^{(\mathrm{r})} = 2^{-\frac{1}{2}}(\boldsymbol{e}_x - \mathrm{i}\boldsymbol{e}_y); \tag{2.100}$$

with corresponding signs in (2.100) and (2.102). $\boldsymbol{a}_{\mathrm{c,c}} = \mathcal{P}^{(2)} \cdot \boldsymbol{a}_{ij} = \mathcal{V}_{\mathrm{c}}^{(2)} \cdot \boldsymbol{a}^{(k,q)} = \mathcal{P}^{(2)} \cdot (\mathcal{V}^{(2)})^{-1} \cdot \boldsymbol{a}^{(k,q)}$.

$\mathcal{P}^{(2)}$	$a_{1,1}$	$a_{-1,-1}$	$a_{-1,1}$	$a_{1,-1}$	$a_{1,0}$	$a_{-1,0}$	$a_{0,1}$	$a_{0,-1}$	$a_{0,0}$
(cf)[27]	2^{-1}	2^{-1}	2^{-1}	2^{-1}	$2^{-\frac{1}{2}}$	$2^{-\frac{1}{2}}$	$2^{-\frac{1}{2}}$	$2^{-\frac{1}{2}}$	1
a_{xx}	1	1	±1	±1					
a_{yy}	−1	−1	±1	±1					
a_{xy}	i	−i	±i	∓i					
a_{yx}	i	$-i$	∓i	±i					
a_{xz}					1	±1			
a_{yz}					i	∓i			
a_{zx}							1	±1	
a_{zy}							i	∓i	
a_{zz}									1

(2.101)

$\mathcal{V}_c^{(2)}$	$a_{1,1}$	$a_{-1,-1}$	$a_{-1,1}$	$a_{1,-1}$	$a_{1,0}$	$a_{0,1}$	$a_{-1,0}$	$a_{0,-1}$	$a_{0,0}$	$(\mathrm{cf})^{28}$
$a^{(2,2)}$	1									1
$a^{(2,-2)}$		1								1
$a^{(0,0)}$			± 1	± 1					-1	$3^{-\frac{1}{2}}$
$a^{(2,0)}$			± 1	± 1					2	$6^{-\frac{1}{2}}$
$a^{(1,0)}$			∓ 1	± 1						$2^{-\frac{1}{2}}$
$a^{(1,1)}$					1	-1				$2^{-\frac{1}{2}}$
$a^{(2,1)}$					1	1				$2^{-\frac{1}{2}}$
$a^{(1,-1)}$							∓ 1	± 1		$2^{-\frac{1}{2}}$
$a^{(2,-1)}$							± 1	± 1		$2^{-\frac{1}{2}}$

(2.102)

Antisymmetric scattering due to spherical tensor components $a^{(1,0)}$, $a^{(1,1)}$, $a^{(1,-1)}$ can be observed preferentially in circular polarization: $a_{-1,1}$, $a_{1,-1}$, $a_{-1,0}$, $a_{0,-1}$.

2.A.2 Selection Rules

This section presents, in the form of a series of tables, selection rules for crystal quantum numbers and conversion tables.

Table 2.A.1. One photon and Raman selection rules for intra configurational (parity conserving) electronic transitions in ions with unfilled shells in site symmetries of the 32 point groups $\boldsymbol{G}$. In the case of odd numbered electron configurations the selection rules based on the double group irreducible representations (reps) ($\bar{\Gamma}_i$) have to be used. Active components: x, y, z: electric dipole transitions. M_i: magnetic dipole transitions or antisymmetric pseudovector scattering; $xx, yy, ...yz$: quadrupolar (symmetric scattering) transitions. $M_x \equiv yz - zy$, and cyclic permutations. The symbols of the reps follow [50]. $\rightarrow$: to be replaced by ... with identical result. The main axes are always oriented along the z-direction, $C_s \perp z.\Gamma_i^*$: complex conjugate to Γ_i. $[\Gamma_i, \Gamma_j]$: degenerate rep. $xx+yy, (xx-yy)$: diagonal tensor elements of equal size and the same, (opposite) sign. (ij, ik): off-diagonal symmetric tensor elements (or vector components) of equal size; ia: optically inactive. The reps are also given in the Mulliken notation (A, B, E, F, $\bar{\text{E}}$ [53]). $\bar{\text{E}}^{\frac{1}{2},\frac{3}{2}}$ indicates a fourfold degenerate double group rep in cubic symmetry. The reps are also classified according to Hellwege's crystal quantum numbers μ, μ_I, ν, S, I [56, 57] in the sequence given in the headline of each group. For the different groups $\boldsymbol{G}$, we have given the abelian subgroups $\boldsymbol{G_B}$ (except C_1), which arise by application of a homogeneous magnetic field B in the sequence of directions: $B\|z; \perp z(\|y; \|x)$; in cubic groups: $\|z, [001]; \|C_3, [111]; \|[110]$. The halving subgroups $\boldsymbol{H}$ [see (2.52)] have also been given. From [2.9, 49–57]

$G, G_B \mid H$	REPs	Transitions	Active Components
$\mathrm{C_1}$	$\Gamma_1; A; \mu = 0$	$(\Gamma_1, \Gamma_1), (\bar{\Gamma}_2, \bar{\Gamma}_2)$	All polariz. allowed
1	$\bar{\Gamma}_2; \bar{E}; \frac{1}{2}$		
$\mathrm{C_i}$	$\Gamma_1^{+,-}; A_{g,u}; (\mu_I = 0, \frac{1}{2})$	$(\Gamma_1^+, \Gamma_1^+), (\bar{\Gamma}_2^+, \bar{\Gamma}_2^+), (\Gamma^+ \rightarrow \Gamma^-)$	M_x, M_y, M_z;
$\bar{1}$	$\bar{\Gamma}_2^{+,-}; \bar{E}_{g,u}^{\frac{1}{2}}; (\frac{1}{2}, 0)$		all $ij, (i, j, = x, y, z)$
$C_i \mid C_1$			
C_2	$\Gamma_{1,2};\ A, B; (\mu = 0, 1)$	$(\Gamma_1, \Gamma_1), (\Gamma_2, \Gamma_2), (\bar{\Gamma}_3, \bar{\Gamma}_3^*)$,	$z; M_z; xx, yy$,
2		$(\bar{\Gamma}_4, \bar{\Gamma}_4^*)$	zz, xy
$C_2 \mid C_1$	$[\bar{\Gamma}_3, \bar{\Gamma}_4]; \bar{E}^{\frac{1}{2}}; \pm\frac{1}{2}$	$(\Gamma_1, \Gamma_2), (\bar{\Gamma}_3, \bar{\Gamma}_4^*), (\bar{\Gamma}_4, \bar{\Gamma}_3^*)$	$x, y; M_x, M_y; yz, zx$
$\mathrm{C_s}$	$\Gamma_{1,2}; A', A''; (\mu_I = 0, 1)$	$(\Gamma_1, \Gamma_1), (\Gamma_2, \Gamma_2), (\bar{\Gamma}_3, \bar{\Gamma}_3^*)$,	$x, y; M_z; xx, yy$,
m		$(\bar{\Gamma}_4, \bar{\Gamma}_4^*)$	zz, xy
$C_s \mid C_1$	$[\bar{\Gamma}_3, \bar{\Gamma}_4]; \bar{E}^{\frac{1}{2}}; \pm\frac{1}{2}$	$(\Gamma_1, \Gamma_2), (\bar{\Gamma}_3, \bar{\Gamma}_4^*), (\bar{\Gamma}_4, \bar{\Gamma}_3^*)$	$z; M_x, M_y; yz, zx$

Table 2.A.1. (Continued)

$G, G_B \vert H$	REPs	Transitions	Active Components
$\mathbf{C_{2h}}$	$\Gamma_1^{+,-}; A_{g,u}; (I = \pm 1, \mu = 0)$	$(\Gamma_i^+, \Gamma_i^{+,(*)}), i = 1, \dots 4;$	$M_z; xx, yy, zz, xy$
$2/m$	$\Gamma_2^{+,-}; B_{g,u}; (1,1), (-1,1)$	$(\Gamma_i^+ \to \Gamma_i^-)$	
C_{2h}, C_i	$[\bar{\Gamma}_3^{+,-}, \bar{\Gamma}_4^{+,-}]; \bar{E}_{g,u}^{\frac{1}{2}}; (\pm 1, \pm \frac{1}{2})$	$(\Gamma_1^+, \Gamma_2^+), (\bar{\Gamma}_3^+, \bar{\Gamma}_4^{+*}), (\bar{\Gamma}_4^+, \bar{\Gamma}_3^{+*})$	$M_x, M_y; yz, zx$
C_i, C_2, C_s			
$\mathbf{D_2}$	$\Gamma_1; A; (\nu = 0, \mu = 0)$	$(\Gamma_i, \Gamma_i), i = 1, \dots 4$	xx, yy, zz
222	$\Gamma_2; B_2; (0,1)$	$(\Gamma_1, \Gamma_2), (\Gamma_3, \Gamma_4)$	$y; M_y; zx$
$C_2, C_2,$	$\Gamma_3; B_1; (1,0)$	$(\Gamma_1, \Gamma_3), (\Gamma_2, \Gamma_4)$	$z; M_z; xy$
C_2	$\Gamma_4; B_3; (1,1)$	$(\Gamma_1, \Gamma_4), (\Gamma_2, \Gamma_3)$	$x; M_x; yz$
C_2	$\bar{\Gamma}_5; \bar{E}^{\frac{1}{2}}; (\pm \frac{1}{2})$	$(\bar{\Gamma}_5, \bar{\Gamma}_5)$	All polariz. allowed
$\mathbf{C_{2v}}$	$\Gamma_1; A_1; (S = 1, \mu = 0)$	$(\Gamma_i, \Gamma_i), i = 1 \dots, 4$	$z; xx, yy, zz$
$mm2$	$\Gamma_2; B_1; (1,1)$	$(\Gamma_1, \Gamma_2), (\Gamma_3, \Gamma_4)$	$x; M_y; zx$
$C_2, C_s,$	$\Gamma_3; A_2; (-1,0)$	$(\Gamma_1, \Gamma_3), (\Gamma_2, \Gamma_4)$	$M_z; xy$
C_s	$\Gamma_4; B_2; (-1,1)$	$(\Gamma_1, \Gamma_4), (\Gamma_2, \Gamma_3)$	$y; M_x; yz$
C_s, C_2	$\bar{\Gamma}_5; \bar{E}^{\frac{1}{2}}; \pm \frac{1}{2}$	$(\bar{\Gamma}_5, \bar{\Gamma}_5)$	All polariz. allowed
$\mathbf{D_{2h}}$	$\Gamma_1^{+,-}; A_{g,u};$	$(\Gamma_i^+, \Gamma_i^+), i = 1, \dots, 4;$	xx, yy, zz
mmm	$(I = \pm 1, \nu = \mu = 0)$	$(\Gamma_i^+ \to \Gamma_i^-)$	
	$\Gamma_2^{+,-}; B_{2g,2u}; (\pm 1, 0, 1)$	$(\Gamma_1^+, \Gamma_2^+), (\Gamma_3^+, \Gamma_4^+)$	$M_y; zx$
$C_{2h}, C_{2h},$	$\Gamma_3^{+,-}; B_{1g,1u}; (\pm 1, 1, 0)$	$(\Gamma_1^+, \Gamma_3^+), (\Gamma_2^+, \Gamma_4^+)$	$M_z; xy$
C_{2h}	$\Gamma_4^{+,-}; B_{3g,3u}; (\pm 1, 1, 1)$	$(\Gamma_1^+, \Gamma_4^+), (\Gamma_2^+, \Gamma_3^+)$	$M_x; yz$
$C_{2h}, D_2,$	$\bar{\Gamma}_5^{+,-}; \bar{E}_{g,u}^{\frac{1}{2}}; (\pm 1, -, \pm \frac{1}{2})$	$(\bar{\Gamma}_5^+, \bar{\Gamma}_5^+)$	All M_i and ij
C_{2v}		$(\Gamma^+ \to \Gamma^-)$	$(i, j = x, y, z)$ allowed

Table 2.A.1. (Continued)

$G, G_B \| H$	REPs	Transitions	Active Components
$\mathbf{C_4}$	$\Gamma_1; A; \mu = 0$	$(\Gamma_1, \Gamma_1), (\Gamma_2, \Gamma_2)$	$z; M_z; xx + yy, zz$
4	$\Gamma_2; B; 2$	(Γ_1, Γ_2)	$xx - yy, xy$
C_4	$[\Gamma_3, \Gamma_4]; E; \pm 1$	$(\Gamma_{1,2}, [\Gamma_3^*, \Gamma_4^*])$	$(x, y); (M_x, M_y); (yz, zx)$
		$([\Gamma_3, \Gamma_4], [\Gamma_3^*, \Gamma_4^*])$	$z; M_z; xx, yy, zz, xy$
C_2	$[\bar{\Gamma}_5, \bar{\Gamma}_6]; \bar{E}^{\frac{1}{2}}; \pm\frac{1}{2}$	$([\bar{\Gamma}_5, \bar{\Gamma}_6], [\bar{\Gamma}_5^*, \bar{\Gamma}_6^*])$, $(\bar{\Gamma}_{5,6} \to \bar{\Gamma}_{7,8})$	$z, (x, y); M_z, (M_x, M_y);$ $xx + yy, zz, (yz, zx)$
	$[\bar{\Gamma}_7, \bar{\Gamma}_8]; \bar{E}^{\frac{3}{2}}; \pm\frac{3}{2}$	$([\bar{\Gamma}_5, \bar{\Gamma}_6], [\bar{\Gamma}_7^*, \bar{\Gamma}_8^*])$	$(x, y); (M_x, M_y);$ $(xx - yy, xy), (yz, zx)$
$\mathbf{S_4}$	$\Gamma_1; A; \mu_I = 0$	$(\Gamma_1, \Gamma_1), (\Gamma_2, \Gamma_2)$	$M_z; xx + yy, zz$
$\bar{4}$	$\Gamma_2; B; 2$	(Γ_1, Γ_2)	$z; xx - yy, xy$
S_4	$[\Gamma_3, \Gamma_4]; E, \pm 1$	$(\Gamma_{1,2}, [\Gamma_3^*, \Gamma_4^*])$	$(x, y); (M_x, M_y); (yz, zx)$
		$([\Gamma_3, \Gamma_4], [\Gamma_3^*, \Gamma_4^*])$	$z; M_z; xx, yy, zz, xy$
C_2	$[\bar{\Gamma}_5, \bar{\Gamma}_6]; \bar{E}^{\frac{1}{2}}; \pm\frac{1}{2}$	$([\bar{\Gamma}_5, \bar{\Gamma}_6], [\bar{\Gamma}_5^*, \bar{\Gamma}_6^*])$, $(\bar{\Gamma}_{5,6} \to \bar{\Gamma}_{7,8})$	$(x, y); M_z, (M_x, M_y);$ $xx + yy, zz, (yz, zx)$
	$[\bar{\Gamma}_7, \bar{\Gamma}_8]; \bar{E}^{\frac{3}{2}}; \pm\frac{3}{2}$	$([\bar{\Gamma}_5, \bar{\Gamma}_6], [\bar{\Gamma}_7^*, \bar{\Gamma}_8^*])$	$z, (x, y); (M_x, M_y);$ $(xx - yy, xy), (yz, zx)$

Table 2.A.1. (Continued)

$G, G_B \| H$	REPs	Transitions	Active Components
$\mathbf{C_{4h}}$	$\Gamma_1^{+,-}; A_{g,u}; (I = \pm 1, \mu = 0)$	$(\Gamma_1^+, \Gamma_1^+), (\Gamma_2^+, \Gamma_2^+), (\Gamma_{1,2}^+ \rightarrow \Gamma_{1,2}^-)$	$M_z; xx + yy, zz$
$4/m$	$\Gamma_2^{+,-}; B_{g,u}; (\pm 1, 2)$	(Γ_1^+, Γ_2^+), $(\Gamma_i^+ \rightarrow \Gamma_i^-)$	$xx - yy, xy$
C_{4h},	$[\Gamma_3^{+,-}, \Gamma_4^{+,-}]; E_{g,u}; (\pm 1, \pm 1)$	$(\Gamma_{1,2}^+, [\Gamma_3^{+*}, \Gamma_4^{+*}])$	$(M_x, M_y); (yz, zx)$
C_i, C_i		$([\Gamma_3^+, \Gamma_4^+], [\Gamma_3^{+*}, \Gamma_4^{+*}])$	$M_z; xx, yy, zz, xy$
	$[\bar{\Gamma}_5^{+,-}, \bar{\Gamma}_6^{+,-}]; \bar{E}_{g,u}^{\frac{1}{2}}; (\pm 1, \pm \frac{1}{2})$	$([\bar{\Gamma}_5^+, \bar{\Gamma}_6^+], [\bar{\Gamma}_5^{+*}, \bar{\Gamma}_6^{+*}])$, }	{ $M_z, (M_x, M_y)$;
C_{2h}, C_4,		$(\bar{\Gamma}_{5,6} \rightarrow \bar{\Gamma}_{7,8}), (\Gamma_i^+ \rightarrow \Gamma_i^-)$ }	$xx + yy, zz, (yz, zx)$
S_4	$[\bar{\Gamma}_7^{+,-}, \bar{\Gamma}_8^{+,-}]; \bar{E}_{g,u}^{\frac{3}{2}}; (\pm 1, \pm \frac{3}{2})$	$([\bar{\Gamma}_5^+, \bar{\Gamma}_6^+], [\bar{\Gamma}_7^{+*}, \bar{\Gamma}_8^{+*}])$, }	{ (M_x, M_y);
		$(\Gamma_i^+ \rightarrow \Gamma_i^-)$ }	$(xx - yy, xy), (yz, zx)$
$\mathbf{D_4}$	$\Gamma_1; A_1; (\nu = 0, \mu = 0)$	$(\Gamma_i, \Gamma_i), i = 1 \ldots 4$	$xx + yy, zz$
422	$\Gamma_2; A_2; (1, 0)$	$(\Gamma_1, \Gamma_2), (\Gamma_3, \Gamma_4)$	$z; M_z$
C_4,	$\Gamma_3; B_1; (0, 2)$	$(\Gamma_1, \Gamma_3), (\Gamma_2, \Gamma_4)$	$xx - yy$
C_2, C_2	$\Gamma_4; B_2; (1, 2)$	$(\Gamma_1, \Gamma_4), (\Gamma_2, \Gamma_3)$	xy
	$\Gamma_5; E; (-, \pm 1)$	$(\Gamma_i, \Gamma_5), i = 1, \ldots 4$	$(x, y); (M_x, M_y); (yz, zx)$
		(Γ_5, Γ_5)	$z; M_z; xx, yy, zz, xy$
			{ $(x, y), z; (M_x, M_y), M_z$;
C_4, D_2	$\bar{\Gamma}_6; \bar{E}'^{\frac{1}{2}}; (-, \pm \frac{1}{2})$	$(\bar{\Gamma}_6, \bar{\Gamma}_6), (\bar{\Gamma}_7, \bar{\Gamma}_7)$	$(xx + yy), zz, (yz, zx)$
			{ $(x, y); (M_x, M_y)$;
	$\bar{\Gamma}_7; \bar{E}''^{\frac{1}{2}}; (-, \pm \frac{3}{2})$	$(\bar{\Gamma}_6, \bar{\Gamma}_7)$	$(xx - yy), xy, (yz, zx)$

Table 2.A.1. (Continued)

$G, G_B \| H$	REPs	Transitions	Active Components
$\mathbf{C_{4v}}$	$\Gamma_1; A_1; (S=1, \mu=0)$	$(\Gamma_i, \Gamma_i), i=1\ldots4$	$z; xx+yy, zz$
$4mm$	$\Gamma_2; A_2; (-1, 0)$	$(\Gamma_1, \Gamma_2), (\Gamma_3, \Gamma_4)$	M_z
C_4,	$\Gamma_3; B_1; (1, 2)$	$(\Gamma_1, \Gamma_3), (\Gamma_2, \Gamma_4)$	$xx-yy$
C_s, C_s	$\Gamma_4; B_2; (-1, 2)$	$(\Gamma_1, \Gamma_4), (\Gamma_2, \Gamma_3)$	xy
	$\Gamma_5; E; (-, \pm1)$	$(\Gamma_i, \Gamma_5), i=1\ldots4$	$(x,y); (M_x, M_y); (yz, zx)$
		(Γ_5, Γ_5)	$z; M_z; xx, yy, zz, xy$
C_4, C_{2v}	$\bar{\Gamma}_6; \bar{E}_1^{\frac{1}{2}}; (-, \pm\frac{1}{2})$	$(\bar{\Gamma}_6, \bar{\Gamma}_6), (\bar{\Gamma}_7, \bar{\Gamma}_7)$	{ $(x,y), z; (M_x, M_y), M_z;$ $(xx+yy), zz, (yz, zx)$
	$\bar{\Gamma}_7; \bar{E}_2^{\frac{1}{2}}; (-, \pm\frac{3}{2})$	$(\bar{\Gamma}_6, \bar{\Gamma}_7)$	{ $(x,y); (M_x, M_y);$ $(xx-yy), xy, (yz, zx)$
$\mathbf{D_{2d}}$	$\Gamma_1; A_1; (S=1, \mu_I=0)$	$(\Gamma_i, \Gamma_i), i=1\ldots4$	$xx+yy, zz$
$\bar{4}2m$	$\Gamma_2; A_2; (-1, 0)$	$(\Gamma_1, \Gamma_2), (\Gamma_3, \Gamma_4)$	M_z
S_4,	$\Gamma_3; B_1; (-1, 2)$	$(\Gamma_1, \Gamma_3), (\Gamma_2, \Gamma_4)$	$xx-yy$
C_2, C_2	$\Gamma_4; B_2; (1, 2)$	$(\Gamma_1, \Gamma_4), (\Gamma_2, \Gamma_3)$	$z; xy$
	$\Gamma_5; E; (-, \pm1)$	$(\Gamma_i, \Gamma_5), i=1\ldots4$	$(x,y); (M_x, M_y); (yz, zx)$
		(Γ_5, Γ_5)	$z; M_z; xx, yy, zz, xy$
D_2, C_{2v},	$\bar{\Gamma}_6; \bar{E}_1^{\frac{1}{2}}; (-, \pm\frac{1}{2})$	$(\bar{\Gamma}_6, \bar{\Gamma}_6), (\bar{\Gamma}_7, \bar{\Gamma}_7)$	{ $(x,y); (M_x, M_y), M_z;$ $(xx+yy), zz, (yz, zx)$
S_4	$\bar{\Gamma}_7; \bar{E}_2^{\frac{1}{2}}; (-, \pm\frac{3}{2})$	$(\bar{\Gamma}_6, \bar{\Gamma}_7)$	{ $(x,y), z; (M_x, M_y);$ $(xx-yy), xy, (yz, zx)$

Table 2.A.1. (Continued)

G, G_B | H	REPs	Transitions	Active Components
$\mathbf{D_{4h}}$	$\Gamma_1^{+,-}; A_{1g,1u};$	$(\Gamma_i^+, \Gamma_i^+), i = 1 \ldots, 4;$ }	$xx + yy, zz$
	$(I = \pm 1, \nu = \mu = 0)$	$(\Gamma_i^+ \to \Gamma_i^-)$	
$4/mmm$	$\Gamma_2^{+,-}; A_{2g,2u}; (\pm 1, 1, 0)$	$(\Gamma_5^+, \Gamma_5^+), (\Gamma_5^-, \Gamma_5^-)$	$M_z; xx, yy, zz, xy$
$C_{4h},$	$\Gamma_3^{+,-}; B_{1g,1u}; (\pm 1, 0, 2)$	$(\Gamma_1^+, \Gamma_2^+), (\Gamma_3^+, \Gamma_4^+)$	M_z
C_{2h}, C_{2h}	$\Gamma_4^{+,-}; B_{2g,2u}; (\pm 1, 1, 2)$	$(\Gamma_1^+, \Gamma_3^+), (\Gamma_2^+, \Gamma_4^+);$ $(\Gamma_i^+ \to \Gamma_i^-)$	$xx - yy$
	$\Gamma_5^{+,-}; E_{g,u}; (\pm 1, -, \pm 1)$	$(\Gamma_1^+, \Gamma_4^+), (\Gamma_2^+, \Gamma_3^+)$	xy
		$(\Gamma_i^+, \Gamma_5^+), i = 1 \ldots, 4$	$(M_x, M_y); (yz, zx)$
$C_{4h}, D_{2h},$	$\bar{\Gamma}_6^{+,-}; \bar{E}'^{\pm,\frac{1}{2}}; (\pm 1, -, \pm\frac{1}{2})$	$(\bar{\Gamma}_6^+, \bar{\Gamma}_6^+), (\bar{\Gamma}_7^+, \bar{\Gamma}_7^+)$ {	$(M_x, M_y), M_z;$
$D_4,$			$(xx + yy), zz, (yz, zx)$
D_{2d}, C_{4v}	$\bar{\Gamma}_7^{+,-}; \bar{E}''^{\pm,\frac{1}{2}}; (\pm 1, -, \pm\frac{3}{2})$	$(\bar{\Gamma}_6^+, \bar{\Gamma}_7^+);$ $(\bar{\Gamma}_i^+ \to \bar{\Gamma}_i^-)$ {	$(M_x, M_y);$
			$(xx - yy), xy, (yz, zx)$
$\mathbf{C_3}$	$\Gamma_1; A; \mu = 0$	(Γ_1, Γ_1)	$z; M_z; xx + yy, zz$
3	$[\Gamma_2, \Gamma_3]; E; \pm 1$	$(\Gamma_1, [\Gamma_2^*, \Gamma_3^*]$ {	$(x, y); (M_x, M_y);$
			$(xx - yy, xy), (yz, zx)$
C_3		$([\Gamma_2, \Gamma_3], [\Gamma_2^*, \Gamma_3^*])$	All polariz. allowed
—	$[\bar{\Gamma}_4, \bar{\Gamma}_5]; \bar{E}^{\frac{1}{2}}; \pm\frac{1}{2}$	$[\bar{\Gamma}_4, \bar{\Gamma}_5], [\bar{\Gamma}_4^*, \bar{\Gamma}_5^*]$	All polariz. allowed
	$\bar{\Gamma}_6, \bar{E}^{\frac{3}{2}}; \frac{3}{2}$	$(\bar{\Gamma}_6, \bar{\Gamma}_6)$	$z; M_z; xx + yy, zz$
		$([\bar{\Gamma}_4^*, \bar{\Gamma}_5^*], \bar{\Gamma}_6)$ {	$(x, y); (M_x, M_y);$
			$(xx - yy, xy), (yz, zx)$

Table 2.A.1. (Continued)

$G, G_B \| H$	REPs	Transitions		Active Components
$\mathbf{C_{3i}}$	$\Gamma_1^{+,-}; A_{g,u}; \mu_I = 0$	(Γ_1^+, Γ_1^+)		$M_z; xx+yy, zz$
$\bar{3}$				(M_x, M_y);
	$[\Gamma_2^{+,-}, \Gamma_3^{+,-}]; E_{g,u}; \pm 1$	$(\Gamma_1^+, [\Gamma_2^{+*}, \Gamma_3^{+*}]$,	$(\Gamma_i^+ \to \Gamma_i^-)$	
				$(xx-yy, xy), (yz, zx)$
C_{3i},				$(M_x, M_y), M_z$;
	(see footnote 9)	$([\Gamma_2^+, \Gamma_3^+], [\Gamma_2^{+*}, \Gamma_3^{+*}])$		
C_i, C_i				$xx, yy, zz, xy), (yz, zx)$
				$M_z, (M_x, M_y)$;
	$[\bar{\Gamma}_4^{+,-}, \bar{\Gamma}_5^{+,-}]; \bar{E}^{\frac{1}{2}}_{g,u}; \pm\frac{1}{2}$	$([\bar{\Gamma}_4^+, \bar{\Gamma}_5^+], [\bar{\Gamma}_4^{+*}, \bar{\Gamma}_5^{+*}])$		
				$xx, yy, zz, xy, (yz, zx)$
C_3	$\bar{\Gamma}_6^{+,-}, \bar{E}^{\frac{3}{2}}_{g,u}; \frac{3}{2}$	$(\bar{\Gamma}_6^+, \bar{\Gamma}_6^+)$	$(\bar{\Gamma}_i^+ \to \bar{\Gamma}_i^-)$	$M_z; xx+yy, zz$
				(M_x, M_y);
		$([\bar{\Gamma}_4^{+*}, \bar{\Gamma}_5^{+*}], \bar{\Gamma}_6^+)$		
				$(xx-yy, xy), (yz, zx)$
$\mathbf{D_3}$	$\Gamma_1; A_1; (\nu = 0, \mu = 0)$	$(\Gamma_1, \Gamma_1), (\Gamma_2, \Gamma_2)$		$xx+yy, zz$
32	$\Gamma_2; A_2; (1, 0)$	(Γ_1, Γ_2)		$z; M_z$
				$(x, y); (M_x, M_y)$;
	$\Gamma_3; E; (-, \pm 1)$	$(\Gamma_{1,2}, \Gamma_3)$		
C_3, C_2				$(xx-yy, xy), (yz, zx)$
		$(\Gamma_3, \Gamma_3), (\bar{\Gamma}_4, \bar{\Gamma}_4)$		All polariz. allowed
				$(x, y); (M_x, M_y)$;
	$\bar{\Gamma}_4; \bar{E}^{\frac{1}{2}}; (-, \pm\frac{1}{2})$	$([\bar{\Gamma}_5^*, \bar{\Gamma}_6^*], \bar{\Gamma}_4)$		
C_3				$(xx-yy, xy), (yz, zx)$
	$\bar{\Gamma}_{5,6}; \bar{E}^{\frac{3}{2}}; (\pm\frac{1}{2}, \frac{3}{2})$	$([\bar{\Gamma}_5, \bar{\Gamma}_6], [\bar{\Gamma}_5^*, \bar{\Gamma}_6^*]$		$z; M_z; xx+yy, zz$

Table 2.A.1. (Continued)

$G, G_B \vert H$	REPs	Transitions	Active Components
$\mathbf{C_{3v}}$	$\Gamma_1; A_1; (S=1, \mu=0)$	$(\Gamma_1, \Gamma_1), (\Gamma_2, \Gamma_2)$	$z; xx+yy, zz$
$3m$	$\Gamma_2; A_2; (-1, 0)$	(Γ_1, Γ_2)	M_z
C_3, C_s	$\Gamma_3; E; (-, \pm 1)$	$(\Gamma_{1,2}, \Gamma_3)$	$\{$ $(x, y); (M_x, M_y)$; $(xx-yy, xy), (yz, zx)$
		$(\Gamma_3, \Gamma_3), (\bar{\Gamma}_4, \bar{\Gamma}_4)$	All polariz. allowed
C_3	$\bar{\Gamma}_4; \bar{E}^{\frac{1}{2}}; (-, \pm\frac{1}{2})$	$([\bar{\Gamma}_5^*, \bar{\Gamma}_6^*], \bar{\Gamma}_4)$	$\{$ $(x, y); (M_x, M_y)$; $(xx-yy, xy), (yz, zx)$
	$\bar{\Gamma}_{5,6}; \bar{E}^{\frac{3}{2}}; (\pm i, \frac{3}{2})$	$([\bar{\Gamma}_5, \bar{\Gamma}_6], [\bar{\Gamma}_5^*, \bar{\Gamma}_6^*]$	$z; M_z; xx+yy, zz$
$\mathbf{D_{3d}}$	$\Gamma_1^{+,-}; A_{1g,1u}$; $(I=\pm 1, \nu=\mu=0)$	$(\Gamma_1^+, \Gamma_1^+), (\Gamma_2^+, \Gamma_2^+)$,	$xx+yy, zz$
$\bar{3}m$	$\Gamma_2^{+,-}; A_{2g,2u}; (\pm 1, 1, 0)$	(Γ_1^+, Γ_2^+) $(\Gamma_i^+ \to \Gamma_i^-)$	M_z
C_{3i}, C_{2h}	$\Gamma_3^{+,-}; E_{g,u}; (\pm 1, -, \pm 1)$	$((\Gamma_1^+, \Gamma_2^+), \Gamma_3^+)$	$\{$ (M_x, M_y); $(xx-yy, xy), (yz, zx)$
	$\bar{\Gamma}_4^{+,-}; \bar{E}_{g,u}^{\frac{1}{2}}; (\pm 1, -, \pm\frac{1}{2})$	$(\Gamma_3^+, \Gamma_3^+), (\bar{\Gamma}_4^+, \bar{\Gamma}_4^+)$	$\{$ $M_z, (M_x, M_y)$; $xx, yy, zz, xy, (yz, zx)$
$C_{3i}, D_3,$	$[\bar{\Gamma}_5^+, \bar{\Gamma}_6^+]; \bar{E}_g^{\frac{3}{2}}; (+1, \pm\frac{1}{2}, \frac{3}{2})$	$([\bar{\Gamma}_5^+, \bar{\Gamma}_6^+], [\bar{\Gamma}_5^{+*}, \bar{\Gamma}_6^{+*}])$, $(\bar{\Gamma}_i^+ \to \bar{\Gamma}_i^-)$	$M_z; xx+yy, zz$ (M_x, M_y);

Table 2.A.1. (Continued)

$G, G_B \mid H$	REPs	Transitions	Active Components
$\mathbf{C_6}$	$\Gamma_1; A; \mu = 0$	$(\Gamma_1, \Gamma_1), (\Gamma_4, \Gamma_4)$	$z, M_z; xx + yy, zz$
6	$\Gamma_4; B; 3$	$([\Gamma_2, \Gamma_3], [\Gamma_2^*, \Gamma_3^*]); (\Gamma_{2,3} \rightarrow \Gamma_{5,6})$	$z; M_z; xx, yy, zz, xy$
	$[\Gamma_{2,3}]; E_2; \pm 2$	$(\Gamma_1, [\Gamma_2^*, \Gamma_3^*]); (\Gamma_4, [\Gamma_5^*, \Gamma_6^*])$	$(xx - yy, xy)$
C_6	$[\Gamma_{5,6}]; E_1; \pm 1$	(Γ_1, Γ_4)	ia
		$(\Gamma_1, [\Gamma_5^*, \Gamma_6^*]); (\Gamma_4, [\Gamma_2^*, \Gamma_3^*])$	$(x, y); (M_x, M_y); (yz, zx)$
		$([\Gamma_2, \Gamma_3], [\Gamma_5^*, \Gamma_6^*])$	$(x, y); (M_x, M_y); (yz, zx)$
	$[\bar{\Gamma}_{7,8}]; \bar{E}^{\frac{1}{2}}; \pm\frac{1}{2}$	$([\bar{\Gamma}_7, \bar{\Gamma}_8], [\bar{\Gamma}_7^*, \bar{\Gamma}_8^*]);$ }	{ $(x, y), z; (M_x, M_y), M_z;$
		$(\bar{\Gamma}_{7,8} \rightarrow \bar{\Gamma}_{9,10})$ }	{ $xx + yy, zz, (yz, zx)$
C_3	$[\bar{\Gamma}_{9,10}]; \bar{E}^{\frac{5}{2}}; \pm\frac{5}{2}$	$([\bar{\Gamma}_{11}, \bar{\Gamma}_{12}], [\bar{\Gamma}_{11}^*, \bar{\Gamma}_{12}^*])$	$z; M_z; xx + yy, zz$
	$[\bar{\Gamma}_{11,12}]; \bar{E}^{\frac{3}{2}}; \pm\frac{3}{2}$	$([\bar{\Gamma}_7, \bar{\Gamma}_8], [\bar{\Gamma}_9^*, \bar{\Gamma}_{10}^*])$	$(xx - yy, xy)$
		$([\bar{\Gamma}_7, \bar{\Gamma}_8], [\bar{\Gamma}_{11}^*, \bar{\Gamma}_{12}^*]),$ }	{ $(x, y); (M_x, M_y);$
		$(\Gamma_{7,8} \rightarrow \Gamma_{9,10})$ }	{ $(xx - yy, xy), (yz, zx)$
$\mathbf{C_{3h}}$	$\Gamma_1; A'; \mu_I = 0$	$(\Gamma_1, \Gamma_1), (\Gamma_4, \Gamma_4)$	$M_z; xx + yy, zz$
$\bar{6}$	$\Gamma_4; A''; 3$	$([\Gamma_2, \Gamma_3], [\Gamma_2^*, \Gamma_3^*]), (\Gamma_{2,3} \rightarrow \Gamma_{5,6})$	$(x, y); M_z; xx, yy, zz, xy$
	$[\Gamma_{2,3}]; E'; \pm 2$	$(\Gamma_1, [\Gamma_2^*, \Gamma_3^*]); (\Gamma_4, [\Gamma_5^*, \Gamma_6^*])$	$(x, y); (xx - yy, xy)$
C_{3h}	$[\Gamma_{5,6}]; E''; \pm 1$	(Γ_1, Γ_4)	z
		$(\Gamma_1, [\Gamma_5^*, \Gamma_6^*]); (\Gamma_4, [\Gamma_2^*, \Gamma_3^*])$	$(M_x, M_y); (yz, zx)$
		$([\Gamma_2, \Gamma_3], [\Gamma_5^*, \Gamma_6^*])$	$z; (M_x, M_y); (yz, zx)$

Table 2.A.1. (Continued)

$G, G_B \| H$	REPs	Transitions	Active Components
	$[\bar{\Gamma}_{7,8}]; \bar{E}^{\frac{1}{2}}; \pm\frac{1}{2}$	$([\bar{\Gamma}_7, \bar{\Gamma}_8], [\bar{\Gamma}_7^*, \bar{\Gamma}_8^*]),$ }	{ $(M_x, M_y), M_z;$
		$(\bar{\Gamma}_{7,8} \rightarrow \bar{\Gamma}_{9,10})$	$xx + yy, zz, (yz, zx)$
C_3	$[\bar{\Gamma}_{9,10}]; \bar{E}^{\frac{5}{2}}; \pm\frac{5}{2}$	$([\bar{\Gamma}_{11}, \bar{\Gamma}_{12}], [\bar{\Gamma}_{11}^*, \bar{\Gamma}_{12}^*])$	$z; M_z; xx + yy, zz$
	$[\bar{\Gamma}_{11,12}]; \bar{E}^{\frac{3}{2}}; \pm\frac{3}{2}$	$([\bar{\Gamma}_7, \bar{\Gamma}_8], [\bar{\Gamma}_9^*, \bar{\Gamma}_{10}^*])$	$(x, y), z; (xx - yy, xy)$
		$([\bar{\Gamma}_7, \bar{\Gamma}_8], [\bar{\Gamma}_{11}^*, \bar{\Gamma}_{12}^*]),$ }	{ $(x, y); (M_x, M_y);$
		$(\bar{\Gamma}_{7,8} \rightarrow \bar{\Gamma}_{9,10})$	$(xx - yy, xy), (yz, zx)$
C_{6h}	$\Gamma_1^{+,-}; A_{g,u};$	$(\Gamma_1^+, \Gamma_1^+), (\Gamma_4^+, \Gamma_4^+), (\Gamma_i^+ \rightarrow \Gamma_i^-)$	$M_z; xx + yy, zz$
$6/m$	$(I = \pm 1, \mu = 0)$	$([\Gamma_2^+, \Gamma_3^+], [\Gamma_2^{+*}, \Gamma_3^{+*}]),$ }	
			$M_z; xx, yy, zz, xy$
	$\Gamma_4^{+,-}; B_{g,u}; (\pm 1, 3)$	$(\Gamma_{2,3} \rightarrow \Gamma_{5,6})$	
C_{6h}	$[\Gamma_{2,3}^{+,-}]; E_{2g,2u}; (\pm 1, \pm 2)$	$(\Gamma_1^+, [\Gamma_2^{+*}, \Gamma_3^{+*}]); (\Gamma_4^+, [\Gamma_5^{+*}, \Gamma_6^{+*}])$	$(xx - yy, xy)$
	$[\Gamma_{5,6}^{+,-}]; E_{1g,1u}; (\pm 1, \pm 1)$	(Γ_1^+, Γ_4^+)	ia
		$(\Gamma_1^+, [\Gamma_5^{+*}, \Gamma_6^{+*}]); (\Gamma_4^+, [\Gamma_2^{+*}, \Gamma_3^{+*}])$	$(M_x, M_y); (yz, zx)$
		$([\Gamma_2^+, \Gamma_3^+], [\Gamma_5^{+*}, \Gamma_6^{+*}]), (\Gamma_i^+ \rightarrow \Gamma_i^-)$	$(M_x, M_y); (yz, zx)$
		$([\bar{\Gamma}_7^+, \bar{\Gamma}_8^+], [\bar{\Gamma}_7^{+*}, \bar{\Gamma}_8^{+*}]),$ }	{ $(M_x, M_y), M_z;$
		$(\bar{\Gamma}_{7,8} \rightarrow \bar{\Gamma}_{9,10})$	$xx + yy, zz, (yz, zx)$
C_{3h}	$[\bar{\Gamma}_{7,8}^{+,-}]; \bar{E}_{g,u}^{\frac{1}{2}}; (\pm 1, \pm\frac{1}{2})$	$([\bar{\Gamma}_{11}^+, \bar{\Gamma}_{12}^+], [\bar{\Gamma}_{11}^{+*}, \bar{\Gamma}_{12}^{+*}])$	$M_z; xx + yy, zz$
$C_{3i}, C_6,$	$[\bar{\Gamma}_{9,10}^{+,-}]; \bar{E}_{g,u}^{\frac{5}{2}}; (\pm 1, \pm\frac{5}{2})$	$([\bar{\Gamma}_7^+, \bar{\Gamma}_8^+], [\bar{\Gamma}_9^{+*}, \bar{\Gamma}_{10}^{+*}]),\ (\bar{\Gamma}_i^+ \rightarrow \bar{\Gamma}_i^-)$	$(xx - yy, xy)$
	$[\bar{\Gamma}_{11,12}^{+,-}]; \bar{E}_{g,u}^{\frac{3}{2}}; (\pm 1, \pm\frac{3}{2})$	$([\bar{\Gamma}_7^+, \bar{\Gamma}_8^+], [\bar{\Gamma}_{11}^{+*}, \bar{\Gamma}_{12}^{+*}]),$ }	{ $(M_x, M_y);$
		$(\bar{\Gamma}_{7,8} \rightarrow \bar{\Gamma}_{9,10})$	$(xx - yy, xy), (yz, zx)$

Table 2.A.1. (Continued)

$G, G_B \| H$	REPs	Transitions	Active Components
$\mathbf{D_6}$	$\Gamma_1; A_1; (\nu = 0, \mu = 0)$	$(\Gamma_i, \Gamma_i), i = 1 \ldots, 4$	$xx + yy, zz$
622	$\Gamma_2; A_2; (1, 0)$	$(\Gamma_1, \Gamma_2), (\Gamma_3, \Gamma_4)$	z, M_z
C_6,	$\Gamma_3; B_1; (0, 3)$	$(\Gamma_1, \Gamma_3), (\Gamma_2, \Gamma_4)$	ia
C_2, C_2	$\Gamma_4; B_2; (1, 3)$	$(\Gamma_1, \Gamma_4), (\Gamma_2, \Gamma_3)$	ia
	$\Gamma_5; E_1; (-, \pm 1)$	$(\Gamma_{1,2}, \Gamma_5), (\Gamma_{3,4}, \Gamma_6)$	$\{$ $(x, y), (M_x, M_y)$; (yz, zx)
	$\Gamma_6; E_2; (-, \pm 2)$	$(\Gamma_{3,4}, \Gamma_5), (\Gamma_{1,2}, \Gamma_6)$	$xx - yy, xy$
		$(\Gamma_5, \Gamma_5), (\Gamma_6, \Gamma_6)$	$z; M_z; xx, yy, zz, xy$
		(Γ_5, Γ_6)	$\{$ $(x, y); (M_x, M_y)$; (yz, zx)
	$\bar{\Gamma}_7; \bar{E}_1^{\frac{1}{2}}; (-, \pm\frac{1}{2})$	$(\bar{\Gamma}_7, \bar{\Gamma}_7), (\bar{\Gamma}_8, \bar{\Gamma}_8)$	$\{$ $z, (x, y); M_z, (M_x, M_y)$; $xx + yy, zz, (yz, zx)$
C_6, D_3	$\bar{\Gamma}_8; \bar{E}_2^{\frac{1}{2}}; (-, \pm\frac{5}{2})$	$(\bar{\Gamma}_9, \bar{\Gamma}_9)$	$z; M_z; xx + yy, zz$
	$\bar{\Gamma}_9; \bar{E}^{\frac{3}{2}}; (-, \pm\frac{3}{2})$	$(\bar{\Gamma}_7, \bar{\Gamma}_8)$	$xx - yy, xy$
		$(\bar{\Gamma}_7, \bar{\Gamma}_9), (\bar{\Gamma}_8, \bar{\Gamma}_9)$	$\{$ $(x, y); (M_x, M_y)$; $(xx - yy, xy), (yz, zx)$

Table 2.A.1. (Continued)

$G, G_B \mid H$	REPs	Transitions	Active Components
$\mathbf{C_{6v}}$	$\Gamma_1; A_1; (S=1, \mu=0)$	$(\Gamma_i, \Gamma_i), i = 1 \ldots, 4$	$z; xx+yy, zz$
$6mm$	$\Gamma_2; A_2; (-1, 0)$	$(\Gamma_1, \Gamma_2), (\Gamma_3, \Gamma_4)$	M_z
	$\Gamma_3; B_1; (1, 3)$	$(\Gamma_1, \Gamma_3), (\Gamma_2, \Gamma_4)$	ia
$C_6, C_s,$	$\Gamma_4; B_2; (-1, 3)$	$(\Gamma_1, \Gamma_4), (\Gamma_2, \Gamma_3)$	ia
C_s	$\Gamma_5; E_1; (-, \pm 1)$	$(\Gamma_{1,2}, \Gamma_5), (\Gamma_{3,4}, \Gamma_6)$	$(x,y), (M_x, M_y);$ (yz, zx)
	$\Gamma_6; E_2; (-, \pm 2)$	$(\Gamma_{3,4}, \Gamma_5), (\Gamma_{1,2}, \Gamma_6)$	$xx-yy, xy$
		$(\Gamma_5, \Gamma_5), (\Gamma_6, \Gamma_6)$	$z; M_z; xx, yy, zz, xy$
		(Γ_5, Γ_6)	$(x,y); (M_x, M_y);$ (yz, zx)
	$\bar{\Gamma}_7; \bar{E}_1^{\frac{1}{2}}; (-, \pm\frac{1}{2})$	$(\bar{\Gamma}_7, \bar{\Gamma}_7), (\bar{\Gamma}_8, \bar{\Gamma}_8)$	$z, (x,y); M_z, (M_x, M_y);$ $xx+yy, zz, (yz, zx)$
C_6, C_{3v}	$\bar{\Gamma}_8; \bar{E}_2^{\frac{1}{2}}; (-, \pm\frac{5}{2})$	$(\bar{\Gamma}_9, \bar{\Gamma}_9)$	$z; M_z; xx+yy, zz$
	$\bar{\Gamma}_9; \bar{E}^{\frac{3}{2}}; (-, \pm\frac{3}{2})$	$(\bar{\Gamma}_7, \bar{\Gamma}_8)$	$xx-yy, xy$
		$(\bar{\Gamma}_7, \bar{\Gamma}_9), (\bar{\Gamma}_8, \bar{\Gamma}_9)$	$(x,y); (M_x, M_y);$ $(xx-yy, xy), (yz, zx)$

Table 2.A.1. (Continued)

$G, G_B \| H$	REPs	Transitions	Active Components
$\mathbf{D_{3h}}$	$\Gamma_1; A_1'; (S=1, \mu_I=0)$	$(\Gamma_i, \Gamma_i), i=1\ldots,4$	$xx+yy, zz$
$\bar{6}m2$	$\Gamma_2; A_2'; (-1,0)$	$(\Gamma_1,\Gamma_2),(\Gamma_3,\Gamma_4)$	M_z
	$\Gamma_3; A_1''; (-1,3)$	$(\Gamma_1,\Gamma_3),(\Gamma_2,\Gamma_4)$	ia
$C_{3h}, C_2,$	$\Gamma_4; A_2''; (1,3)$	$(\Gamma_1,\Gamma_4),(\Gamma_2,\Gamma_3)$	z
C_s	$\Gamma_5; E''; (-,\pm1)$	$(\Gamma_{1,2},\Gamma_5),(\Gamma_{3,4},\Gamma_6)$	$(M_x,M_y);(yz,zx)$
	$\Gamma_6; E'; (-,\pm2)$	$(\Gamma_{3,4},\Gamma_5),(\Gamma_{1,2},\Gamma_6)$	$(x,y); xx-yy, xy$
		$(\Gamma_5,\Gamma_5),(\Gamma_6,\Gamma_6)$	$\{(x,y); M_z;$ xx, yy, zz, xy
		(Γ_5,Γ_6)	$z; (M_x,M_y); (yz,zx)$
	$\bar{\Gamma}_7; \bar{E}'\frac{1}{2}; (-,\pm\frac{1}{2})$	$(\bar{\Gamma}_7,\bar{\Gamma}_7),(\bar{\Gamma}_8,\bar{\Gamma}_8)$	$\{M_z, (M_x,M_y);$ $xx+yy, zz, (yz,zx)$
$C_{3h}, D_3,$	$\bar{\Gamma}_8; \bar{E}''\frac{1}{2}; (-,\pm\frac{5}{2})$	$(\bar{\Gamma}_9,\bar{\Gamma}_9)$	$z; M_z; xx+yy, zz$
C_{3v}	$\bar{\Gamma}_9; \bar{E}'\frac{3}{2}; (-,\pm\frac{3}{2})$	$(\bar{\Gamma}_7,\bar{\Gamma}_8)$	$z,(x,y); xx-yy, xy$
		$(\bar{\Gamma}_7,\bar{\Gamma}_9),(\bar{\Gamma}_8,\bar{\Gamma}_9)$	$\{(x,y); (M_x,M_y);$ $(xx-yy,xy),(yz,zx)$
$\mathbf{D_{6h}}$	$\Gamma_1^{+,-}; A_{1g,1u};$	$(\Gamma_i^+,\Gamma_i^+), i=1\ldots,4;$	$xx+yy, zz$
	$(I=\pm1, \nu=\mu=0)$	$(\Gamma_1^+,\Gamma_2^+),(\Gamma_3^+,\Gamma_4^+)$	M_z
$6/mmm$	$\Gamma_2^{+,-}; A_{2g,2u}; (\pm1,1,0)$	$(\Gamma_1^+,\Gamma_3^+),(\Gamma_2^+,\Gamma_4^+)$	ia
	$\Gamma_3^{+,-}; B_{1g,1u}; (\pm1,0,3)$	$(\Gamma_1^+,\Gamma_4^+),(\Gamma_2^+,\Gamma_3^+)$	ia
$C_{6h}, C_{2h},$	$\Gamma_4^{+,-}; B_{2g,2u}; (\pm1,1,3)$	$(\Gamma_{1,2}^+,\Gamma_5^+),(\Gamma_{3,4}^+,\Gamma_6^+)$	$(M_x,M_y);(yz,zx)$
C_{2h}	$\Gamma_5^{+,-}; E_{1g,1u}; (\pm1,-,\pm1)$	$(\Gamma_{3,4}^+,\Gamma_5^+),(\Gamma_{1,2}^+,\Gamma_6^+)$	$xx-yy, xy$
	$\Gamma_6^{+,-}; E_{2g,2u}; (\pm1,-,\pm2)$	$(\Gamma_5^+,\Gamma_5^+),(\Gamma_6^+,\Gamma_6^+)$	$M_z; xx, yy, zz, xy$
		$(\Gamma_5^+,\Gamma_6^+),$ $(\Gamma_i^+ \rightarrow \Gamma_i^-)$	$(M_x,M_y);(yz,zx)$

Table 2.A.1. (Continued)

$G, G_B \mid H$	REPs	Transitions	Active Components
	$\bar{\Gamma}_7^{+,-}; \bar{E}_{g,u}^{\frac{1}{2}}; (\pm 1, -, \pm\frac{1}{2})$	$(\bar{\Gamma}_7^+, \bar{\Gamma}_7^+), (\bar{\Gamma}_8^+, \bar{\Gamma}_8^+)$	$\{$ $M_z, (M_x, M_y)$; $xx+yy, zz, (yz, zx)$
C_{6h}, D_{3d}, C_{6v}, D_6, D_{3h}	$\bar{\Gamma}_8^{+,-}; \bar{E}_{g,u}^{\frac{5}{2}}; (\pm 1, -, \pm\frac{5}{2})$	$(\bar{\Gamma}_9^+, \bar{\Gamma}_9^+)$, $(\bar{\Gamma}_i^+ \to \bar{\Gamma}_i^-)$	$M_z; xx+yy, zz$
	$\bar{\Gamma}_9^{+,-}; \bar{E}_{g,u}^{\frac{3}{2}}; (\pm 1, -, \pm\frac{3}{2})$	$(\bar{\Gamma}_7^+, \bar{\Gamma}_8^+)$	$xx-yy, xy$
		$(\bar{\Gamma}_7^+, \bar{\Gamma}_9^+), (\bar{\Gamma}_8^+, \bar{\Gamma}_9^+)$	$\{$ (M_x, M_y); $(xx-yy, xy), (yz, zx)$
T	$\Gamma_1; A; (\kappa=0, \mu=0)$	(Γ_1, Γ_1)	$xx+yy+zz$
23	$[\Gamma_2, \Gamma_3]; E; (0, \pm 1)$	$(\Gamma_1, [\Gamma_2^*, \Gamma_3^*])$	$(xx+yy-2zz, xx-yy)$
	$\Gamma_4; F; (-, [0, \pm 1])$	$([\Gamma_2, \Gamma_3], [\Gamma_2^*, \Gamma_3^*])$	xx, yy, zz
C_2, C_3		$(\Gamma_1, \Gamma_4), ([\Gamma_2, \Gamma_3], \Gamma_4)$	$\{$ $(x, y, z); (M_x, M_y, M_z)$; (xy, yz, zx)
		(Γ_4, Γ_4)	All polariz. allowed
	$\bar{\Gamma}_5; \bar{E}^{\frac{1}{2}}; (-, \pm\frac{1}{2})$	$(\bar{\Gamma}_5, \bar{\Gamma}_5)$	$\{$ $(x, y, z); (xx+yy+zz)$, $(M_x, M_y, M_z); (xy, yz, zx)$
—	$[\bar{\Gamma}_6, \bar{\Gamma}_7]; \bar{E}^{\frac{1}{2}, \frac{3}{2}}$; $(-, [\pm\frac{1}{2}, \mp\frac{3}{2}])$	$(\bar{\Gamma}_5, [\bar{\Gamma}_6^*, \bar{\Gamma}_7^*])$	$\{$ $(x, y, z), (M_x, M_y, M_z)$, $(xx+yy-2zz, xx-yy)$, (xy, yz, zx)
		$([\bar{\Gamma}_6, \bar{\Gamma}_7], [\bar{\Gamma}_6^*, \bar{\Gamma}_7^*]$	All polariz. allowed

Table 2.A.1. (Continued)

$G, G_B \mid H$	REPs	Transitions	Active Components
$\mathbf{T_h}$	$\Gamma_1^{+,-}; A_{g,u};$	(Γ_1^+, Γ_1^+), $\quad (\Gamma_i^+ \to \Gamma_i^-)$	$xx + yy + zz$
$m3$	$(I = \pm 1, \kappa = \mu = 0)$.	$(\Gamma_1^+, [\Gamma_2^{+*}, \Gamma_3^{+*}])$	$(xx + yy - 2zz, xx - yy)$
$C_{2h}, C_{3i},$	$[\Gamma_2^{+,-}, \Gamma_3^{+,-}]; E_{g,u};$	$([\Gamma_2^+, \Gamma_3^+], [\Gamma_2^{+*}, \Gamma_3^{+*}])$	xx, yy, zz
C_i	$(\pm 1, 0, \pm 1)$.	$(\Gamma_1^+, \Gamma_4^+), ([\Gamma_2^{+*}, \Gamma_3^{+*}], \Gamma_4^+)$	$(M_x, M_y, M_z); (xy, yz, zx)$
	$\Gamma_4^{+,-}; F_{g,u}; (\pm 1, -, [0, \pm 1])$	(Γ_4^+, Γ_4^+)	$(M_x, M_y, M_z); xx, yy, ...yz$
	$\bar{\Gamma}_5^{+,-}; \bar{E}_{g,u}^{\frac{1}{2}}; (\pm 1, -, \pm\frac{1}{2})$		$(xx + yy + zz)$,
		$(\bar{\Gamma}_5^+, \bar{\Gamma}_5^+)$	
T	$[\bar{\Gamma}_6^{+,-}, \bar{\Gamma}_7^{+,-}]; \bar{E}_{g,u}^{\frac{1}{2},\frac{3}{2}};$		$(xy, yz, zx), (M_x, M_y, M_z)$
	$(\pm 1, -, [\pm\frac{1}{2}, \mp\frac{3}{2}])$		(M_x, M_y, M_z),
		$(\bar{\Gamma}_5^+, [\bar{\Gamma}_6^{+*}, \bar{\Gamma}_7^{+*}]);$ $\quad (\bar{\Gamma}_i^+ \to \bar{\Gamma}_i^-)$	$(xx + yy - 2zz, xx - yy)$,
			(xy, yz, zx)
		$([\bar{\Gamma}_6^+, \bar{\Gamma}_7^+], [\bar{\Gamma}_6^{+*}, \bar{\Gamma}_7^{+*}]$	$(M_x, M_y, M_z); xx, ...yz$
$\mathbf{O}$	$\Gamma_1; A_1; (\kappa = 0, \mu = 0)$	$(\Gamma_1, \Gamma_1), (\Gamma_2, \Gamma_2)$	$xx + yy + zz$
432	$\Gamma_2; A_2; (0, 2)$	(Γ_1, Γ_2)	ia
	$\Gamma_3; E; (-, [0, 2])$	$(\Gamma_{1,2}, \Gamma_3)$	$(xx + yy - 2zz, xx - yy)$
$C_4, C_3,$	$\Gamma_4; F_1; (-, [0, \pm 1])$	$(\Gamma_1, \Gamma_4), (\Gamma_2, \Gamma_5)$	$(M_x, M_y, M_z), (x, y, z)$
C_2	$\Gamma_5; F_2; (-, [2, \pm 1])$	$(\Gamma_2, \Gamma_4), (\Gamma_1, \Gamma_5)$	(xy, yz, zx)
		(Γ_3, Γ_3)	xx, yy, zz
		$(\Gamma_4, \Gamma_4), (\Gamma_5, \Gamma_5)$	All polariz. allowed
			$(x, y, z); (M_x, M_y, M_z)$
		$(\Gamma_3, \Gamma_{4,5})$	
			(xy, yz, zx)
			$(x, y, z); (M_x, M_y, M_z);$
		(Γ_4, Γ_5)	$(xx + yy - 2zz, xx - yy)$,
			(xy, yz, zx)

Table 2.A.1. (Continued)

G, G_B	H	REPs	Transitions	Active Components
		$\bar{\Gamma}_6; \bar{E}_1^{\frac{1}{2}}; (-, \pm\frac{1}{2})$	$(\bar{\Gamma}_6, \bar{\Gamma}_6), (\bar{\Gamma}_7, \bar{\Gamma}_7)$	$(xx + yy + zz)$; $(x, y, z); (M_x, M_y, M_z)$
	T	$\bar{\Gamma}_7; \bar{E}_2^{\frac{1}{2}}; (-, \pm\frac{3}{2})$	$(\bar{\Gamma}_6, \bar{\Gamma}_7)$	(xy, yz, zx)
		$\bar{\Gamma}_8; \bar{E}^{\frac{1}{2}, \frac{3}{2}}; (-, [\pm\frac{1}{2}, \pm\frac{3}{2}])$	$(\bar{\Gamma}_{6,7}, \bar{\Gamma}_8)$	$(x, y, z); (M_x, M_y, M_z)$ $(xx + yy - 2zz, xx - yy)$, (xy, yz, zx)
			$(\bar{\Gamma}_8, \bar{\Gamma}_8)$	All polariz. allowed
$\mathbf{T_d}$ $\bar{4}3m$		$\Gamma_1; A_1; (\kappa = 0, \mu_I = 0)$	$(\Gamma_1, \Gamma_1), (\Gamma_2, \Gamma_2)$	$xx + yy + zz$
		$\Gamma_2; A_2; (0, 2)$	(Γ_1, Γ_2)	ia
	$S_4, C_3,$ C_s	$\Gamma_3; E; (-, [0, 2])$	$(\Gamma_{1,2}, \Gamma_3)$	$(xx + yy - 2zz, xx - yy)$
		$\Gamma_4; F_1; (-, [0, \pm 1])$	$(\Gamma_1, \Gamma_4), (\Gamma_2, \Gamma_5)$	(M_x, M_y, M_z)
		$\Gamma_5; F_2; (-, [2, \pm 1])$	$(\Gamma_2, \Gamma_4), (\Gamma_1, \Gamma_5)$	$(x, y, z); (xy, yz, zx)$
			(Γ_3, Γ_3)	xx, yy, zz
			$(\Gamma_4, \Gamma_4), (\Gamma_5, \Gamma_5)$	All polariz. allowed
			$(\Gamma_3, \Gamma_{4,5})$	$(x, y, z); (M_x, M_y, M_z)$ (xy, yz, zx)
			(Γ_4, Γ_5)	$(x, y, z); (M_x, M_y, M_z)$; $(xx + yy - 2zz, xx - yy)$, (xy, yz, zx)
		$\bar{\Gamma}_6; \bar{E}_1^{\frac{1}{2}}; (-, \pm\frac{1}{2})$	$(\bar{\Gamma}_6, \bar{\Gamma}_6), (\bar{\Gamma}_7, \bar{\Gamma}_7)$	$(xx + yy + zz)$; (M_x, M_y, M_z)
	T	$\bar{\Gamma}_7; \bar{E}_2^{\frac{1}{2}}; (-, \pm\frac{3}{2})$	$(\bar{\Gamma}_6, \bar{\Gamma}_7)$	$(x, y, z); (xy, yz, zx)$
		$\bar{\Gamma}_8; \bar{E}^{\frac{1}{2}, \frac{3}{2}}; (-, [\pm\frac{1}{2}, \pm\frac{3}{2}])$	$(\bar{\Gamma}_{6,7}, \bar{\Gamma}_8)$	$(x, y, z); (M_x, M_y, M_z)$ $(xx + yy - 2zz, xx - yy)$, (xy, yz, zx)
			$(\bar{\Gamma}_8, \bar{\Gamma}_8)$	All polariz. allowed

Table 2.A.1. (Continued)

$G, G_B \vert H$	REPs	Transitions		Active Components
$\mathrm{O_h}$	$\Gamma_1^+; A_{1g};$	$(\Gamma_1^+, \Gamma_1^+), (\Gamma_2^+, \Gamma_2^+)$		$xx + yy + zz$
$m3m$	$(I = 1, \kappa = \mu = 0)$	(Γ_1^+, Γ_2^+)		ia
	$\Gamma_2^+; A_{2g}; (1, 0, 2)$	$(\Gamma_{1,2}^+, \Gamma_3^+)$		$(xx + yy - 2zz, xx - yy)$
$C_{4h}, C_{3i},$	$\Gamma_3^+; E_g; (1, -, [0, 2])$	$(\Gamma_1^+, \Gamma_4^+), (\Gamma_2^+, \Gamma_5^+)$		(M_x, M_y, M_z)
C_{2h}	$\Gamma_4^+; F_{1g}; (1, -, [0, \pm 1])$	$(\Gamma_2^+, \Gamma_4^+), (\Gamma_1^+, \Gamma_5^+)$		(xy, yz, zx)
	$\Gamma_5^+; F_{2g}; (1, -, [2, \pm 1])$	(Γ_3^+, Γ_3^+)		xx, yy, zz
		$(\Gamma_4^+, \Gamma_4^+), (\Gamma_5^+, \Gamma_5^+)$		All polariz. allowed *)
		$(\Gamma_3^+, \Gamma_{4,5}^+)$		$\left\{ \begin{array}{l} (M_x, M_y, M_z) \\ (xy, yz, zx) \end{array} \right.$
	$I = +1, \Gamma_i^+, (g) \rightarrow$ $I = -1, \Gamma_i^-, (u)$	$(\Gamma_4^+, \Gamma_5^+),$	$(\Gamma_i^+ \rightarrow \Gamma_i^-)$	$\left\{ \begin{array}{l} (M_x, M_y, M_z); \\ (xx + yy - 2zz, xx - yy), \\ (xy, yz, zx) \end{array} \right.$
	$\bar{\Gamma}_6^+; \bar{E}_{1g}^{\frac{1}{2}}; (1, -, \pm\frac{1}{2})$	$(\bar{\Gamma}_6^+, \bar{\Gamma}_6^+), (\bar{\Gamma}_7^+, \bar{\Gamma}_7^+)$		$\left\{ \begin{array}{l} (xx + yy + zz); \\ (M_x, M_y, M_z) \end{array} \right.$
T_d, O, T_h	$\bar{\Gamma}_7^+; \bar{E}_{2g}^{\frac{1}{2}}; (1, -, \pm\frac{3}{2})$	$(\bar{\Gamma}_6^+, \bar{\Gamma}_7^+)$		(xy, yz, zx)
	$\bar{\Gamma}_8^+; \bar{E}_g^{\frac{1}{2},\frac{3}{2}}; (1, -, [\pm\frac{1}{2}, \pm\frac{3}{2}])$	$(\bar{\Gamma}_{6,7}^+, \bar{\Gamma}_8^+)$		$\left\{ \begin{array}{l} (M_x, M_y, M_z) \\ (xx + yy - 2zz, xx - yy), \\ (xy, yz, zx) \end{array} \right.$
	$I = +1, \bar{\Gamma}_i^+, (g) \rightarrow$ $I = -1, \bar{\Gamma}_i^-, (u)$	$(\bar{\Gamma}_8^+, \bar{\Gamma}_8^+),$	$(\bar{\Gamma}_i^+ \rightarrow \bar{\Gamma}_i^-)$	All polariz. allowed *)
				* : except (x, y, z)

Table 2.A.2. Relations between the Γ_i symbols of the reps [2.50], the crystal quantum numbers μ, [2.56, 2.57] and the quantum numbers M of the free-ion for the cyclic groups C_p. $M \equiv \mu \pmod p$. Pairs of reps in square brackets are Kramers degenerate

p	n even, (M, μ integer)		n odd, (M, μ half-integer)	
1	Γ_1	$\mu = 0$	$\bar{\Gamma}_2$	$\mu = [\frac{1}{2}]$
2	$\Gamma_{1,2}$	$\mu = 0, 1$	$[\bar{\Gamma}_{3,4}]$	$\mu = [\pm\frac{1}{2}]$
3	$\Gamma_1, [\Gamma_{2,3}]$	$\mu = 0, [\pm 1]$	$[\bar{\Gamma}_{4,5}], [\bar{\Gamma}_6]$	$\mu = [\pm\frac{1}{2}], [\pm\frac{3}{2}]$
4	$\Gamma_1, \Gamma_2, [\Gamma_{3,4}]$	$\mu = 0, 2, [\pm 1]$	$[\bar{\Gamma}_{5,6}], [\bar{\Gamma}_{7,8}]$	$\mu = [\pm\frac{1}{2}], [\pm\frac{3}{2}]$
6	$\Gamma_1, \Gamma_4, [\Gamma_{5,6}], [\Gamma_{2,3}]$	$\mu = 0, 3, [\pm 1], [\pm 2]$	$[\bar{\Gamma}_{7,8}], [\bar{\Gamma}_{9,10}], [\bar{\Gamma}_{11,12}]$	$\mu = [\pm\frac{1}{2}], [\pm\frac{5}{2}], [\pm\frac{3}{2}]$

Table 2.A.3. Selection rules based on crystal quantum numbers for one-photon electric dipole transitions in non-cubic groups. The various quantum numbers apply for different classes of point groups, see Table 2.A.1, [2.56, 2.57]. If I is defined, intra configurational electric dipole transitions are forbidden

Pol.	Matrix Element	$\Delta\mu$	$\Delta\mu_I$	$\Delta\nu$	I	S
π	$\langle i \vert z \vert k\rangle$	$\equiv 0$ (mod$p$)	$\equiv \frac{p}{2}$ (mod$p$)	± 1	$I_i \neq I_k$	$S_i = S_k$
σ	$\langle i \vert x \pm \mathrm{i}y \vert k\rangle$	$\equiv \pm 1$ (mod$p$)	$\equiv \pm(1 + \frac{p}{2})$ (mod$p$)	–	$I_i \neq I_k$	–
σ	$\langle i \vert x \vert k\rangle$	$\equiv \pm 1$ (mod$p$)	$\equiv \pm(1 + \frac{p}{2})$ (mod$p$)	± 1	$I_i \neq I_k$	$S_i = S_k$
σ	$\langle i \vert y \vert k\rangle$	$\equiv \pm 1$ (mod$p$)	$\equiv \pm(1 + \frac{p}{2})$ (mod$p$)	0	$I_i \neq I_k$	$S_i = -S_k$

Table 2.A.4. Selection rules based on crystal quantum numbers for one-photon magnetic dipole transitions or antisymmetric (pseudovector) Raman scattering in non-cubic groups. $M_z \to (xy - yx)$, $M_x \to (yz - zy)$, $M_y \to (zx - xz)$; $M_x \pm \mathrm{i}M_y \to (y \mp \mathrm{i}x)z - z(y \mp \mathrm{i}x) = \binom{\mathrm{L}}{\mathrm{R}}z - z\binom{\mathrm{L}}{\mathrm{R}}$

Matrix Element	$\Delta\mu$	$\Delta\mu_I$	$\Delta\nu$	I	S
$\langle i \vert M_z \vert k\rangle$	$\equiv 0$ (mod$p$)	$\equiv 0$ (mod$p$)	± 1	$I_i = I_k$	$S_i = -S_k$
$\langle i \vert M_x \pm iM_y \vert k\rangle$	$\equiv \pm 1$ (mod$p$)	$\equiv \pm 1$ (mod$p$)	–	$I_i = I_k$	–
$\langle i \vert M_x \vert k\rangle$	$\equiv \pm 1$ (mod$p$)	$\equiv \pm 1$ (mod$p$)	± 1	$I_i = I_k$	$S_i = -S_k$
$\langle i \vert M_y \vert k\rangle$	$\equiv \pm 1$ (mod$p$)	$\equiv \pm 1$ (mod$p$)	0	$I_i = I_k$	$S_i = S_k$

Table 2.A.5. Selection rules based on crystal quantum numbers for quadrupolar scattering in non-cubic groups, $I_i = I_k$. Trace (polarized) scattering only for $\Delta\mu_{(I)} \equiv 0\,(\mathrm{mod}\,p)$; [2.56]. The lower signs in column S have to be used in combination with $\Delta\mu_I$. $z^2 \to zz, x^2+y^2 \to xx+yy$, etc

Matrix Element	$\Delta\mu$	$\Delta\mu_I$	$\Delta\nu$	S
$\langle i\|z^2\|k\rangle$	$\equiv 0$ (mod$p$)	$\equiv 0(mod p)$	0	$S_i = S_k$
$\langle i\|x^2+y^2\|k\rangle$	$\equiv 0$ (mod$p$)	$\equiv 0(mod p)$	0	$S_i = S_k$
$\langle i\|z(x \pm \mathrm{i}y\|k\rangle$	$\equiv \pm 1$ (mod$p$)	$\equiv \pm 1$ (mod$p$)	–	–
$\langle i\|zx\|k\rangle$	$\equiv \pm 1$ (mod$p$)	$\equiv \pm 1$ (mod$p$)	0	$S_i = S_k$
$\langle i\|yz\|k\rangle$	$\equiv \pm 1$ (mod$p$)	$\equiv \pm 1$ (mod$p$)	± 1	$S_i = -S_k$
$\langle i\|(x \pm \mathrm{i}y)^2\|k\rangle$	$\equiv \pm 2$ (mod$p$)	$\equiv \pm 2$ (mod$p$)	–	–
$\langle i\|(x^2-y^2)\|k\rangle$	$\equiv \pm 2$ (mod$p$)	$\equiv \pm 2$ (mod$p$)	0	$S_i = \pm S_k$
$\langle i\|(xy)\|k\rangle$	$\equiv \pm 2$ (mod$p$)	$\equiv \pm 2$ (mod$p$)	± 1	$S_i = \mp S_k$

References

2.1 P. Fulde, M. Löwenhaupt: Magnetic Excitations in Crystal-Field Split 4*f* Systems, in *Adv. Phys.* **34**, 589–661 (1986)
2.2 M. Löwenhaupt K.H. Fischer: Neutron Scattering on Heavy Fermion and Valence Fluctuation 4*f*-Systems, in *Handbook of Magnetic Materials,* Vol. 7, ed. by K.H.J. Buschow (Elsevier, Amsterdam 1993) pp. 503–608
2.3 G. Placzek: Rayleigh-Streuung und Raman-Effekt, in *Handbuch der Radiologie*, Vol. 6/II, ed. by E. Marx, (Akademische Verlagsgesellschaft, Leipzig, 1934) pp. 205–374
2.4 P. Myslinski, J.A. Koningstein: Chem. Phys. **114**, 137–147 (1987)
2.5 J.Y. Zhou, P.A. Tanner, W.J. Peng, P.Q. Yang: *Proc. XIV. Int. Conf. on Raman Spectroscopy* , Hongkong 1994, ed. by N.-T. Yu, X.-Y. Li, (Wiley, Chichester 1994) pp. 1098–1099
2.6 K. Shimoda (ed.): High-Resolution Laser Spectroscopy, *(Topics Appl. Phys.,* Vol. 13) (Springer, Berlin, Heidelberg 1976)
2.7 S.A. Asher, Z. Chi, J.S.W. Holtz, I.K. Lednev, A.S. Karnoup, M.C. Sparrow: *Proc. XVI. Int. Conf. on Raman Spectroscopy* , Cape Town 1998, ed. by A.M. Heyns (Wiley, Chichester 1998) pp. 11–14;
S.A. Asher, C.H. Munro, Z. Chi, *Laser Focus World* **33**, No. 7, 99–109 (1997)
2.8 J. Sawatzki, C. Lehner, N.T. Kawai: in [2.5], pp. 1120–1121
2.9 A. Kiel, S.P.S. Porto: J. Mol. Spectrosc. **32**, 458–468 (1969)
2.10 J. Hougen, S. Singh: Phys. Rev. Lett. **10**, 406–407 (1963)
2.11 B. Halperin, J.A. Koningstein: J. Chem. Phys. **69**, 3302–3310 (1978)
2.12 J.A. Koningstein, O.Sonnich Mortensen: Electronic Raman Transitions, in *The Raman Effect*, Vol. 2: *Applications*, ed. by A. Anderson (Dekker, New York, 1973) pp. 519–542
2.13 R.J.H. Clark, T.J. Dines: Electronic Raman Spectroscopy, in *Advances in Infrared and Raman Spectroscopy*, Vol. 9, ed. by R.J.H. Clark, R.E. Hester (Heyden, London, 1982) pp. 282–360

2.14 M.V. Klein: Electronic Raman Scattering, in *Light Scattering in Solids I (Topics Appl. Phys.)* Vol. **8**, ed. by M. Cardona (Springer, Berlin 1975) pp. 147–207
2.15 E. Zirngiebel, G. Güntherodt: Light Scattering in Rare Earth and Actinide Intermetallic Compounds, in *Light Scattering in Solids VI, (Topics Appl. Phys.) Vol.* **68** ed. by M. Cardona, G. Güntherodt (Springer, Berlin, Heidelberg 1991) pp. 207–284
2.16 A.K. Ramdas, S. Rodriguez: Raman Scattering in Diluted Magnetic Semiconductors, in [2.15], pp. 137–206
2.17 C. Thomsen: Light Scattering in High-T_c Superconductors, in [2.15],/break pp. 285–360
2.18 J.A. Sanjurjo, C. Rettori, S. Oseroff: Z. Fisk. Phys. Rev. B **49**, 4391–4394 (1994)
2.19 R. Loudon: The Quantum Theory of Light, 2nd edition (Clarendon Press, Oxford 1983)
2.20 M.G. Cottam, D.L. Lockwood: Light Scattering in Magnetic Solids, (Wiley, New York, Chichester 1986)
2.21 N. Bloembergen: Nonlinear Optics (W.A. Benjamin, Inc., New York 1965)
2.22 Y.R. Shen: The Principles of Nonlinear Optics (Wiley, New York 1984)
2.23 W. Hayes, R. Loudon: Scattering of Light by Crystals (Wiley, New York 1978)
2.24 M. Cardona: Resonance Phenomena, in *Light Scattering in Solids II (Topics Appl. Phys., Vol.* **50***)* ed. by M. Cardona, G. Güntherodt (Springer, Berlin, Heidelberg 1982) pp. 19–178
2.25 L.D. Landau, E.M. Lifshitz: Lehrbuch der Theoretischen Physik V,1; Statistische Physik, part I, § 124 (Akademie-Verlag Berlin 1976)
2.26 C.J. Ballhausen: Introduction to Ligand Field Theory, (McGraw-Hill, New York 1962)
2.27 S. Sugano, Y. Tanabe, H. Kamimura: Multiplets of Transition-Metal Ions in Crystals (Academic Press, New York 1970)
2.28 H.M. Crosswhite, H.W. Moos (eds.): Optical Properties of Ions in Crystals (Interscience Publishers, New York 1967)
2.29 S. Hüfner: Optical Spectra of Transparent Rare Earth Compounds, (Academic Press, New York 1978)
2.30 C.A. Morrison, R.P. Leavitt: Spectroscopic Properties of Triply Ionized Lanthanides in Transparent Host Crystals, in *Handbook on the Physics and Chemistry of Rare Earths*, ed. by K.A. Gschneidner, Jr. and L. Eyring, Vol.5, ch. 46; (North-Holland, Amsterdam 1982)
2.31 G.M. Williams, P.C. Becker, J.G. Conway, N. Edelstein, L.A. Boatner, M.M. Abraham: Phys. Rev. B **40**, 4132–4142 (1989)
2.32 P.C. Becker, N. Edelstein, G.M. Williams, J.J. Bucher, R.E. Russo, J.A. Koningstein, L.A. Boatner, M.M. Abraham: Phys. Rev. B **31**, 8102–8110 (1985)
2.33 J. Kraus: Spektroskopische Untersuchungen zur Elektron–Phonon-Wechselwirkung in den Selten-Erd-Verbindungen $TbVO_4$, TbF_3 und HoF_3; Dissertation, Univ. Würzburg (1987) unpublished
2.34 K.M. Häussler, A. Lehmeyer, L. Merten: phys. stat. sol. (b) **111**, 513–518 (1982)
2.35 J.D. Jackson: Classical Electrodynamics, Wiley, New York (1962)

2.36 T. Kurosawa: J. Phys. Soc. Japan **16**, 1298 (1961); A.S. Barker, Jr. Phys. Rev. **136**, A 1290–1295 (1964)
2.37 O. Hittmair: Lehrbuch der Quantentheorie (Thiemig, München (1972)
2.38 A. Abragam, B. Bleaney: Electron Paramagnetic Resonance of Transition Ions (Clarendon Press, Oxford 1970)
2.39 L.D. Barron, E. Nørby Svendsen: Antisymmetric Light Scattering and Time Reversal, in *Advances in Infrared and Raman Spectroscopy,* ed. by R.J.H. Clark, R.E. Hester (Wiley-Heyden, Chichester 1980) pp. 322–339; R. Loudon: J. Raman Spectrosc. **7**, 10–14 (1978)
2.40 A.R. Edmonds: Angular Momentum in Quantum Mechanics, (Princeton University Press, Princeton 1960)
2.41 B.R. Judd: Operator Techniques in Atomic Spectroscopy, (McGraw-Hill, New York 1963)
2.42 B.G. Wybourne: Spectroscopic Properties of Rare Earths (Interscience Publishers, New York 1965)
2.43 R.N. Zare: Angular Momentum (Wiley, New York 1988)
2.44 A.J. Stone: Mol. Phys. **29**, 1461–1471 (1975)
2.45 E.H. van Kleef: Am. J. Phys. **63**, 626–633 (1995)
2.46 T.C. Damen, S.P.S. Porto, B. Tell: Phys. Rev. **142**, 570–574 (1966)
2.47 B.H. Bransden, C.J. Joachain: *Physics of Atoms and Molecules* (Longman Scientific & Technical, Harlow 1983); O. Laporte, in *Handbuch der Astrophysik III* (part 2) (Springer, Heidelberg, 1930) p. 684
2.48 A. Furrer, J. Mesot, W. Henggeler, G. Böttger: J. of Superconductivity **10**, 273–277 (1997)
2.49 H.A. Bethe, Ann. Phys. (V) **3**, 133–208 (1929)
2.50 G.F. Koster, J.O. Dimmock, R.G. Wheeler, H. Statz: Properties of the Thirty-Two Point Groups (M.I.T. Press, Cambridge (MA) 1963)
2.51 M. Lax: Symmetry Principles in Solid State and Molecular Physics, (Wiley, New York 1974)
2.52 D.L. Rousseau, R.P. Bauman, S.P.S. Porto: J. Raman Spectrosc. **10**, 253–290 (1981)
2.53 E. Bright Wilson, Jr., J.C. Decius, P.C. Cross: Molecular Vibrations, The Theory of Infrared and Raman Spectra (McGraw-Hill, New York 1955)
2.54 R. Loudon: Adv. Phys. **13**, 423–482 (1964)
2.55 G. Turrell: Infrared and Raman Spectra of Crystals, (Academic Press, London 1972)
2.56 K.H. Hellwege, Ann. Phys. (Leipzig) (6) **4**, 95, 127, 136, 143, 150 (1948).
2.57 E. Fick, G. Joos: Kristallspektren, *Handbuch der Physik / Encyclopdia of Physics* **XXVIII** (Spectroscopy II) (Springer, Berlin, Heidelberg 1957) pp. 205–295
2.58 W. Beeckman, J. Goffart: Theoretica Chimica Acta **74**, 219–228 (1988)
2.59 K.H. Hellwege: Z. Phys. **127**, 513–521 (1950)
2.60 B.R. Judd: Phys. Rev. **127**, 750–761 (1962)
2.61 G.S. Ofelt: J. Chem. Phys. **37**, 511–520 (1962)
2.62 J.D. Axe, Jr: Phys. Rev. **136**, A 42–A 45 (1964)
2.63 O. Sonnich Mortensen, J.A. Koningstein: J. Chem. Phys. **48**, 3971–3977 (1968); Phys. Rev. **168**, 75–78 (1968)
2.64 J.D. Axe: J. Chem. Phys. **39**, 1154–1160 (1963)
2.65 M.C. Downer, A. Bivas, N. Bloembergen: Opt. Commun. **41**, 335–340 (1982)

2.66 M.C. Downer, A. Bivas: Phys. Rev. B **28**, 3677–3696 (1983).
2.67 L. Smentek-Mielczarek: J. Chem. Phys. **94**, 5369 (1991); Phys. Rev. B **40**, 6499–6504 (1989); Phys. Rev. B **40**, 2019–2023 (1989); Phys. Rev. B **46**, 14 453–14 459 (1992); Phys. Rev. B **46**, 14 460–14 466 (1992); Phys. Rev. B **46**, 14 467–14 477 (1992)
2.68 A. Ceulemans, G.M. Vandenberghe: J. Chem. Phys. **98**, 9372–9378 (1993)
2.69 M.F. Reid, F.S. Richardson: Phys. Rev. B **29**, 2830–2832 (1984)
2.70 B.R. Judd, D.R. Pooler: J. Phys. C **15**, 591–598 (1982)
2.71 B.R. Judd: Rep. Prog. Phys. **48**, 907–954 (1985)
2.72 M.C. Downer: Ph. D. Thesis, Harvard University, Cambridge (MA) (1983)
2.73 P.C. Becker: Electronic Raman Scattering in Rare Earth Phosphate Crystals; Ph. D. Thesis, Lawrence Berkeley Laboratory, Univers. of California (1986)
2.74 K.P. Traar: Phys. Rev. B **35**, 3111–3115 (1987)
2.75 C.W. Nielson, G.F. Koster: Spectroscopic Coefficients of the p^n, d^n, and f^n Configurations (M.I.T. Press, Cambridge (MA) 1964)
2.76 G.M. Williams, N. Edelstein, L.A. Boatner, M.M. Abraham: Phys. Rev. B **40**, 4143–4152 (1989)
2.77 P.A. Tanner, S. Xia, Y.-L. Liu, Y. Ma: Phys. Rev. B **55**, 12 182–12 195 (1997); P.A. Tanner, M.H.M. Chua, in *Proc. Sixteenth Int. Conf. on Raman Spectroscopy,* (Cape Town, 1998) ed. by A.M. Heyns, (Wiley, Chichester) pp. 600–601 (1998)
2.78 S. Xia, G.M. Williams, N. Edelstein: Chem. Phys. **138**, 255–261 (1989)
2.79 A.D. Nguyen, K. Murdoch, N. Edelstein, L.A. Boatner, M.M. Abraham: Phys. Rev. B **56**, 7974–7987 (1997)
2.80 A.D. Nguyen: Phys. Rev. B **55**, 5786–5798 (1997)
2.81 G.M. Williams, P.C. Becker, N. Edelstein, L.A. Boatner, and M.M. Abraham: Phys. Rev. B **40**, 1288–1296 (1989)
2.82 R.L. Wadsack, R.K. Chang: Solid State Commun. **10**, 45–48 (1972)
2.83 D. Nicollin, J.A. Koningstein: Chem. Phys. **49**, 377 (1980)
2.84 G.M. Williams: Resonance Electronic Raman Scattering in Rare Earth Crystals; Ph. D. Thesis, Lawrence Berkeley Laboratory, Univers. of California (1988)
2.85 D. Piehler, N. Edelstein: Phys. Rev. A **41**, 6406–6414 (1990)
2.86 M.L. Shand: J. Appl. Phys. **52**, 1470–1472 (1981)
2.87 H. Lotem, R.T. Lynch, Jr., N. Bloembergen: Phys. Rev. A **14**, 1748–1754 (1976)
2.88 J.F. Nye: Physical Properties of Crystals (Clarendon Press, Oxford 1957)
2.89 D. Piehler: J. Opt. Soc. Am. B **8**, 1889–1898 (1991)
2.90 R.C. Powell, St.A. Payne, L.L. Chase, G.D. Wilke: Phys. Rev. B **41**, 8593–8602 (1990)
2.91 D.A. Ender, M.S. Otteson, R.L. Cone, M.R. Ritter, H.J. Guggenheim: Opt. Lett. **7**, 611–613 (1982)
2.92 R.L. Cone, D.A. Ender, M.S. Otteson, P.L. Fisher, J.M. Friedman, H.J. Guggenheim: Appl. Phys. B **28**, 143 ff. (1982)
2.93 J. Huang, G.K. Liu, R.L. Cone: Phys. Rev. B **39**, 6348–6354 (1989)
2.94 F. Moshary, R. Kichinski, R.J. Beach, S.R. Hartmann: Phys. Rev. A **40**, 4426–4434 (1989)
2.95 A.P. Cracknell: Prog. Theor. Phys. **35**, 196 (1966); J. Phys. C (Solid State Phys.) **2**, 500–511 (1969)

2.96 C.J. Bradley, B.L. Davies: Rev. Mod. Phys. **40**, 359–379 (1968)
2.97 J.O. Dimmock, R.G. Wheeler: in *The Mathematics of Physics and Chemistry,* Vol. II, ed. by H. Margenau, G.M. Murphy, D. Van Nostrand Co. Inc., New York (1964)
2.98 E. Anastassakis, E. Burstein: J. Phys. C: Solid State Phys. **5**, 2468–2481 (1972)
2.99 E. Anastassakis: Morphic Effects in Lattice Dynamics, in *Dynamical Properties of Solids,* Vol. 4, ed. by G.K. Horton, A.A. Maradudin (North-Holland, Amsterdam 1980) pp. 157–375
2.100 E.F. Bertaut: Spin Configurations of Ionic Structures: Theory and Practice, in *Magnetism* , Vol. III, ed by G.T. Rado, H. Suhl, (Academic Press, New York, 1963) pp. 149–209.
2.101 W. Schneider, H. Weitzel: Solid State Commun. **18**, 995–997 (1976); H. Weitzel, J. Hirte: Phys. Rev. B **37**, 5414–5422 (1988)
2.102 L. Graf: phys. stat. sol. (b) **88**, 429–437 (1978)
2.103 L. Graf: J. Magn. Magn. Mater. **6**, 124–128 (1977)
2.104 L.Graf, G. Schaack: J. Magn. Magn. Mater. **9**, 86–90 (1978)
2.105 R.L. Cone, R.S. Meltzer: Ion–Ion Interactions and Exciton Effects in Rare Earth Insulators, in *Spectroscopy of Solids Containing Rare Earth Ions* , Vol. 21 of *Modern Problems in Condensed Matter Sciences* , ed. by A.A. Kaplyanskii, R.M. Macfarlane (North-Holland, Amsterdam 1987)
2.106 T. Fujiwara, Y. Tanabe: J. Phys. Soc. Jpn. **39**, 7–17 (1975)
2.107 A.T. Abdalian, J. Cibert, P. Moch: J. Phys. C: Solid State Phys. **13**, 5587–5602 (1980); J. Cibert, Y. Merle d' Aubigné, J. Ferré, M. Regis: (ibidem) **13**, 2781–2789 (1980)
2.108 J.A. Koningstein: Chem. Phys. Letters **146**, 576–581 (1988)
2.109 D.L. Rousseau, P.F. Williams: J. Chem. Phys. **64**, 3519–3537 (1976); R.M. Hochstrasser, C.A. Nyi: J. Chem. Phys. **70**, 1112–1128 (1979);
J.M. Friedman: J. Chem. Phys. **71**, 3147–3157 (1979);
F.A. Novak, J.M. Friedman, R.M. Hochstrasser: in *Laser and Coherence Spectroscopy* , ed. by J.I. Steinfeld (Plenum Press, New York, 1978)
2.110 M.P. Hehlen, A. Kuditcher, S. C. Rand, M.A. Tischler: J. Chem. Phys. **107**, 4886–4892 (1997)
2.111 F.S. Ham: Jahn–Teller Effects in Electron Paramagnetic Resonance, in *Electron Paramagnetic Resonance* , ed. by S. Geschwind (Plenum Press, New York, London 1972) pp. 1–119
2.112 W. Moffit, A.D. Liehr: Phys. Rev. **106**, 1195–1200 (1957)
2.113 M.S. Child, H.C. Longuet-Higgins: Phil. Transactions of the Roy. Soc. (London) Ser. **A**, Math. and Phys. Sciences **254**, 259–294 (1961)
2.114 E. Mulazzi, N. Terzi: Solid State Commun. **18**, 721–724 (1976); Phys. Rev. **19**, 2332–2342 (1979)
2.115 C.D. Churcher, G.E. Stedman: J. Phys. C: Solid State Phys. **14**, 2237–2264 (1981)
2.116 R.P. Bauman, S.P.S. Porto: Phys. Rev. **161**, 842–847 (1967)
2.117 M. Dahl, G. Schaack: Z. Phys. B **56**, 279–288 (1984)
2.118 G.A. Gehring, K. Gehring: Cooperative Jahn–Teller Effects, in Rep. Prog. Phys. **38**, 1–89 (1975)
2.119 R.J. Elliott, R.T. Harley, W. Hayes, S.R.P. Smith: Proc. Roy. Soc. Lond. **A, 328**, 217–266 (1972)

2.120 M.D. Sturge: The Jahn–Teller Effects in Solids, in *Solid State Physics* ed. by F. Seitz, D. Turnbull, H. Ehrenreich, **20** (New York, Academic Press 1967) pp. 91–213

2.121 S. Guha, L.L. Chase: Phys. Rev. Lett. **32**, 869–872 (1974); Phys. Rev. B **12**, 1658–1675 (1975)

2.122 H.C. Longuet-Higgins, U. Öpik, M.H.L. Pryce, R.A. Sack, Proc. Roy. Soc. (London) Ser. **A**, **244**, 1 (1958); M.C.M. O'Brien, (ibidem) **A**, **281**, 323 (1964)

2.123 L.L. Chase, C.H. Hao: Phys. Rev. B **12**, 5990–5994 (1975)

2.124 R.J. Elliott, Physica **86–88B**, 1118–1121 (1977)

2.125 K. Sugihara: J. Phys. Soc. Jpn. **14**, 1231 (1959); R. Orbach and M. Tachiki: Phys. Rev. **158**, 524–529 (1967)

2.126 R.T. Harley, K.B. Lyons, P.A. Fleury: J. Phys. C: Solid State Phys. **13**, L447–L453 (1980); R.T. Harley, K.B. Lyons, P.A. Fleury, S.R.P. Smith, J. Phys. C: Solid State Phys. **16**, 1407–1422 (1983)

2.127 G.E. Devlin, J.L. Davis, L. Chase, S. Geschwind, Appl. Phys. Letters, **19**, 138–141 (1971);
P.A. Fleury, K.B. Lyons: Central Peaks near Structural Phase Transitions, in *Light Scattering near Phase Transitions* , Vol. 5 of *Modern Problems in Condensed Matter Sciences* , ed. by H.Z. Cummins, A.P. Levanyuk (North-Holland, Amsterdam 1983)

2.128 R.T. Harley, W. Hayes, A.M. Perry, S.R.P. Smith, R.J. Elliott, I.D. Saville: J. Phys. C: Solid State Phys. **7**, 3145–3160 (1974)

2.129 R.T. Harley, R.M. Macfarlane: J. Phys. C: Solid State Phys. **8**, L451–L455 (1975)

2.130 P.J. Becker, M.J.M. Leask, R.N. Tyte: J. Phys. C: Solid State Phys. **5**, 2027–2036 (1972)

2.131 R.M. Macfarlane, R.M. Shelby: Coherent Transient and Holeburning Spectroscopy of Rare Earth Ions in Solids, in [2.105], pp. 51–184

2.132 A.P. Young: J. Phys. C: Solid State Phys. **8**, 3158–3170 (1975)

2.133 P. Thalmeier, P. Fulde: Z. Phys. B **26**, 323–328 (1977)

2.134 P. Thalmeier, B. Lüthi: Electron–Phonon Interaction in Intermetallic Rare Earth Compounds, in *Handbook on the Physics and Chemistry of Rare Earths*, ed. by K.A. Gschneidner,Jr. and L. Eyring, Vol.14, ch. 96; (Elsevier, Amsterdam 1991)

2.135 K. Ahrens, G. Schaack: Proc. Int. Conf. Lattice Dynamics, Paris 1977; ed. by Balkanski, M. (Flammarion, Paris 1977) pp. 257–261; K. Ahrens, G. Schaack: Phys. Rev. Lett. **42**, 1488–1491 (1979); K. Ahrens, Z. Phys. B **40**, 45–54 (1980)

2.136 C.Y. She, T.W. Broberg, L.S. Wall, D.F. Edwards: Phys. Rev. B **6**, 1847–1850 (1972)

2.137 M. Dahl, G. Schaack: Phys. Rev. Lett. **56**, 232–235 (1986); M. Dahl: Z. Phys. B **72**, 87–99 (1988)

2.138 M. Dahl, G. Schaack, B. Schwark: Europhys. Lett. **4**, 929–934 (1987)

2.139 K. Elk, W. Gasser: Die Methode der Greenschen Funktionen in der Festkörperphysik (Akademie-Verlag, Berlin 1979).

2.140 G. Schaack: Z. Phys. B **26**, 49–58 (1977); Physica **89 B**, 195–200 (1977)

2.141 G. Schaack: Solid State Commun. **17**, 505–509 (1975)

2.142 H. Gerlinger, G. Schaack: Phys. Rev. B **33**, 7438–7450 (1986)

2.143 K. Ahrens, H. Gerlinger, H. Lichtblau, G. Schaack, G. Abstreiter, S. Mroczkowski: J. Phys. C: Solid State Phys. **13**, 4545–4564 (1980)
2.144 G. Schaack: J. Phys. C: Solid State Phys. **9**, L297–L301 (1976); K. Ahrens, G. Schaack, Indian J. Pure Appl. Phys. **16**, 311–318 (1978)
2.145 M. Dahl, G. Janotta, G. Schaack, S. Shi: Z. Phys. B **76**, 327–334 (1989)
2.146 W. Dörfler, H.D. Hochheimer, G. Schaack: Z. Phys. B **51**, 153–163 (1983)
2.147 Y. Yamada: J. Phys. Soc. Jpn. **50**, 2996–3004 (1981)
2.148 J.Kraus, W. Görlitz, M. Hirsch, R. Roth, G. Schaack: Z. Phys. B **74**, 247–258 (1989)
2.149 W. Dörfler, G. Schaack: Z. Phys. B **59**, 283–291 (1985)
2.150 A.K. Kupchikov, B.Z. Malkin, D.A. Rzaev, A.I. Ryskin: Fiz. Tverd. Tela **24**, 2373–2380 (1982); Sov. Phys. Solid State **24**, 1348–1352 (1982)
2.151 S.A. Al'tshuler, B.Z. Malkin, M.A. Teplov, D.N. Terpilovskiĭ: Magnetoelastic Interactions in Rare-Earth Paramagnetics $LiLnF_4$, in Sov. Sci. Rev. A. Phys. **6**, 61–160 (1985)
2.152 G. Schaack, K. Leiteritz: unpublished
2.153 P. Thalmeier, P. Fulde: Phys. Rev. Lett. **49**, 1588–1591 (1982)
2.154 G. Güntherodt, A. Jayaraman, G. Batlogg, M. Croft, E. Melczer: Phys. Rev. Lett. **51**, 2330–2332 (1983)
2.155 M. Löwenhaupt, B.D. Rainford, F. Steglich: Phys. Rev. Lett. **42**, 1709–1712 (1979)
2.156 P.C. Becker, G.M. Williams, N.M. Edelstein, J.A. Koningstein, L.A. Boatner, M.M. Abraham: Phys. Rev. B **45**, 5027–5030 (1992)
2.157 K. Leiteritz, G. Schaack: J. Raman Spectrosc. **10**, 36–43 (1981)
2.158 A. Kiel, T. Damen, S.P.S. Porto, S. Singh, F. Varsanyi: Phys. Rev. **178**, 1518–1524 (1969)
2.159 D.C. Johnston: Normal-State Magnetic Properties of Single-Layer Cuprate High-Temperature Superconductors and Related Materials, in *Handbook of Magnetic Materials* , Vol. 10, ed. by K.H.J. Buschow (Elsevier, Amsterdam 1997) pp. 1–237; M.A. Kastner, R.J. Birgeneau, G. Shirane, and Y. Endoh, Rev. Mod. Phys. **70**, 897–928 (1998)
2.160 M. Cardona: Raman Scattering in high-T_c Superconductors, in *Proc. Sixteenth Int. Conf. on Raman Spectroscopy,* (Cape Town, 1998) ed. by A.M. Heyns, Wiley, Chichester, pp. 21–24 (1998)
2.161 W. Henggeler, A. Furrer: J. Phys.: Condens. Matter **10**, 2579–2596 (1998)
2.162 J.A. Sanjurjo, G.B. Martins, P.G. Pagliuso, E. Granado, I. Torriani, C. Rettori, S. Oseroff: Z. Fisk: Phys. Rev. B **51**, 1185–1189 (1995)
2.163 S. Jandl, P. Dufour, T. Strach, T. Ruf, M. Cardona, V. Nekvasil, C. Chen, B. M. Wanklyn: Phys. Rev. B **52**, 15 558–15 564 (1995)
2.164 T. Strach, T. Ruf, M. Cardona, C.T. Lin, S. Jandl, V. Nekvasil, D.I. Zhigunov, S.N. Barilo, S.V. Shiryaev: Phys. Rev. B **54**, 4276–4282 (1996)
2.165 T. Strach, T. Ruf, M. Cardona, S. Jandl, V. Nekvasil, C. Chen, B.M. Wanklyn, D.I. Zhigunov, S.N. Barilo, S.V. Shiryaev: Phys. Rev. B **56**, 5578–5582 (1997)
2.166 S. Jandl, T. Strach, T. Ruf, M. Cardona, V. Nekvasil, M. Iliev, D.I. Zhigunov, S.N. Barilo, S.V. Shiryaev: Phys. Rev. B **56**, 5049–5052 (1997)
2.167 T. Strach, T. Ruf, M. Cardona, S. Jandl, V. Nekvasil: Physica B **234–236**, 810–811 (1997)
2.168 A.T. Boothroyd, S.M. Doyle, D.M^{c}K. Paul, R. Osborn: Phys. Rev. B **45**, 10 075–10 086 (1992)

2.169 S. Jandl, P. Dufour, T. Strach, T. Ruf, M. Cardona, V. Nekvasil, C. Chen, B.M. Wanklyn, S. Piñol: Phys. Rev. B **53**, 8632–8637 (1996)
2.170 U. Staub, R. Osborn, E. Balcar, L. Soderholm, V. Trounov: Europhys. Lett. **31**, 175–180 (1995); Phys. Rev. B **55**, 11 629–11 636 (1997)
2.171 J. Pérez-Mato, M. Aroyo, J. Hlinka, M. Quilichini, R. Curat: Phys. Rev. Lett. **81**, 2462–2465 (1998).
2.172 S.L.J. Billinge, T. Egami: Phys. Rev. B **47**, 14 386–14 405 (1993)
2.173 S. Jandl, M. Iliev, C. Thomsen, T. Ruf, M. Cardona, B.M. Wanklyn, C. Chen: Solid State Commun. **87**, 609–612 (1993)
2.174 M.L. Sanjuán, M.A. Laguna: Phys. Rev. B **52**, 13 000–13 005 (1995); M.A. Laguna, M.L. Sanjuán, S. Piñol: Solid State Commun. **102**, 413–418 (1997)
2.175 A. Furrer, P. Brüesch, P. Unternährer: Phys. Rev. B **38**, 4616–4623 (1988)
2.176 L. Soderholm, C.-K. Loong, G.L. Goodman, B.D. Dabrowski: Phys. Rev. B **43**, 7923–7935 (1991)
2.177 G. L. Goodman, C.-K. Loong, L. Soderholm: J. Phys.: Condens. Matter **3**, 49–67 (1991)
2.178 P. Allenspach, J. Mesot, U. Straub, M. Guillaume, A. Furrer, S.-I. Yoo, M.J. Kramer, R.W. McCallum, I. Maletta, H. Blank, H. Mutka, R. Osborn, M. Arai, Z. Bowden, A.D. Taylor: Z. Phys. B **95**, 301–310 (1994)
2.179 U. Staub, J. Mesot, M. Guillaume, P. Allenspach, A. Furrer, H. Mutka, Z. Bowden, A. Taylor: Phys. Rev. B **50**, 4068–4074 (1994)
2.180 J. Mesot, P. Allenspach, U. Staub, A. Furrer, H. Mutka, R. Osborn, A. Taylor: Phys. Rev. B **47**, 6027–6036 (1993)
2.181 G.A. Stewart, S.J. Harker, D.M. Pooke: J. Phys. Condens. Matter **10**, 8269–8278 (1998)
2.182 J.H. van Vleck: The Theory of Electric and Magnetic Susceptibilities (Oxford University Press, London 1932)
2.183 R.M. White: Quantum Theory of Magnetism (McGraw-Hill, New York 1970)
2.184 E.T. Heyen, R. Wegerer, M. Cardona: Phys. Rev. Lett. **67**, 144–147 (1991); E.T. Heyen, R. Wegerer, E. Schönherr, M. Cardona: Phys. Rev. **44**, 10 195–10 205 (1991)
a. T. Strach, T. Ruf, A.M. Niraimathi, A.A. Martin, M. Cardona: Physica C **301**, 9–16 (1998)
b. A.A. Martin, T. Ruf, T. Strach, M. Cardona, T. Wolf: Phys. Rev. B **58**, 14 349–14 355 (1998)
c. A.A. Martin, V. Hadjiev, T. Ruf, M. Cardona, T. Wolf: Phys. Rev. B **58**, 14 211–14 214 (1998)
2.185 T. Ruf, E.T. Heyen, M. Cardona, J. Mesot, A. Furrer: Phys. Rev. **46**, 11 792–11 797 (1992)
2.186 P. Allenspach, A. Furrer, P. Brüesch, P. Unternährer: Physica **156–157 B**, 864 (1989)
2.187 R. Wegerer, C. Thomsen, T. Ruf, E. Schönherr, M. Cardona, M. Reedyk, J.S. Xue, J.E. Greedan, A. Furrer: Phys. Rev. B **48**, 6413–6419 (1993)
2.188 T. Ruf, R. Wegerer, E.T. Heyen, M. Cardona, A. Furrer: Solid State Commun. **85**, 297–300 (1993)
2.189 A.F. Goncharov, V.V. Struzhkin, T. Ruf, K. Syassen: Phys. Rev. **50**, 13 841–13 844 (1994)
2.190 J. Mesot, P. Allenspach, U. Staub, A. Furrer, H. Mutka: Phys. Rev. Lett. **70**, 865–868 (1993); W. Henggeler, G. Cuntze, J. Mesot, M. Klauda, G. Saemann–Ischenko, A. Furrer: Europhys. Lett. **29**, 233–238 (1995)

2.191 M. Gutmann, P. Allenspach, S. Rosenkranz, A. Furrer: J. Phys. Condens. Matter **10**, 7369–7382 (1998)
2.192 D. Boal, P. Grünberg, J.A. Koningstein: Phys. Rev. B **7**, 4757–4763 (1973)
2.193 I. Dabrowski, P. Grünberg, J.A. Koningstein: J. Chem. Phys. **56**, 1264–1268 (1972)
2.194 J.A. Koningstein, G. Schaack: Phys. Rev. **B 2**, 1242–1250 (1970)
2.195 J.A. Koningstein, G. Schaack: J. Opt. Soc. Am. **60**, 755–758 (1970)
2.196 P. Grünberg, S. Hüfner, E. Orlich, J. Schmitt: Phys. Rev. **184**, 285–293 (1969)
2.197 M.T. Hutchings: Point Charge Calculations of Energy Levels of Magnetic Ions in Crystalline Electric Fields, in *Solid State Physics* , Vol. **16**, ed. by F. Seitz, D. Turnbull (Academic Press, New York 1964) pp. 227–273
2.198 D.J. Newman: Theory of Lanthanide Crystal Fields, in Adv. Phys. **20**, 197–256 (1971)
2.199 R. Bayerer, J. Heber, D. Mateika: Z. Phys. B - Condensed Matter **64**, 201–210 (1986); H. Gross, J. Neukum, J. Heber, D. Mateika, and T. Xiao: Phys. Rev. B **48**, 9264–9272 (1993)
2.200 H.-D. Amberger, G.G. Rosenbauer, R.D. Fischer: Mol. Phys. **32**, 1291–1298 (1976);
B. Kanellakopoulos, H.-D. Amberger, G.G. Rosenbauer, R.D. Fischer: J. Inorg. Nucl. Chem. **39**, 607–611 (1977)
2.201 H.-D. Amberger, R.D. Fischer, G.G. Rosenbauer: Transit. met. Chem. **1**, 242–246 (1976);
H.-D. Amberger, G.G. Rosenbauer, R.D. Fischer: J. Phys. Chem. Solids, **38**, 379–385 (1977);
R. Nevald, F.W. Voss, O.V. Nielsen, H.-D. Amberger, R.D. Fischer: Solid State Commun. **32**, 1223–1225 (1979)
2.202 P.A. Tanner, Y.-L. Liu, N. Edelstein, K. Murdoch, N. Khaidukov: J. Phys.: Condens. Matter **9**, 7817–7836 (1997)
2.203 E. Finkman, E. Cohen, L.G. van Uitert: Phys. Rev. B **7**, 2899–2909 (1973)
2.204 T.J. Marks, R.D. Fischer (eds.): *Organometallics of the f-Elements* (Reidel, Dordrecht 1979)
2.205 H.-D. Amberger, F.T. Edelmann: J. Organomet. Chem. **508**, 275–279 (1996)
2.206 A. Zalkin, K.N. Raymond: J. Am. Chem. Soc. **91**, 5667 (1969)
2.207 R.F. Dallinger, P. Stein, T.G. Spiro: J. Am. Chem. Soc. **100**, 7865 (1978)
2.208 F. Auzel, O.L. Malta: J. Phys. (Paris) **44**, 201–206 (1983)
2.209 C.-S. Neumann, P. Fulde: Z. Phys. B **74**, 277–278 (1989);
M. Dolg, P. Fulde, W. Küchle, C.-S. Neumann, H. Stoll: J. Chem. Phys. **94**, 3011–3017 (1991);
M. Dolg, P. Fulde, H. Stoll, H. Preuss, A. Chang, R.M. Pitzer: Chem. Phys. **195**, 71–82 (1995);
W. Liu, M. Dolg, P. Fulde: J. Chem. Phys. **107**, 3584–3591 (1997)
2.210 N.M. Edelstein, P.G. Allen, J.J. Bucher, D.K. Shuh, C.D. Sofield: J. Am. Chem. Soc. **118**, 13 115–13 116 (1996)
2.211 H. Schulz, H.-D. Amberger: J. Organomet. Chem. **443**, 71–78 (1993);
S. Jank: Diploma Thesis (Univ. Hamburg 1995)
2.212 J. Stühler, G. Schaack, M. Dahl, A. Waag, G. Landwehr, K.V. Kavokin, I.A. Merkulov: J. Raman Spectrosc. **27**, 281–287 (1996)

2.213 J. Stühler, G. Schaack, M. Dahl, A. Waag, G. Landwehr, K.V. Kavokin, I.A. Merkulov: Phys. Rev. Lett. **74**, 2567–2570 (1995); erratum **74**, 4966 (1995)

2.214 J. Stühler: Resonante Ramanstreuung an Anregungen der Mn^{2+}-Ionen in $Cd_{1-x}Mn_xTe$ und $Cd_{1-x}Mn_xTe/Cd_{1-y}Mg_yTe$-Quantentrogstrukturen, Dissertation, Univ. Würzburg, 1995 unpublished

2.215 J.-C. Bünzli, G.A. Leonard, D. Plancherel, G. Chapuis: Helv. Chim. Acta **69**, 288–297 (1986)

2.216 J.H. Weaver, D.M. Poirier: Solid State Properties of Fullerenes and Fullerene-Based Materials, in *Solid State Physics* , Vol. **48**, ed. by H. Ehrenreich, F. Spaepen, (Academic Press, Boston 1994) pp. 1–108; S. Hino, H. Takahashi, K. Iwasaki, K. Matsumoto, T. Miyazaki, S. Hasegawa, K. Kikuchi, Y. Achiba: Phys. Rev. Lett. **71**, 4261–4263 (1993);
T. Suzuki, K. Kikuchi, F. Oguri, Y. Nakao, S. Suzuki, Y. Achiba, K. Yamamoto, H. Funasaka, T. Takahashi: Tetrahedron **52**, 4973–4982 (1996)

2.217 L.B. Chang, C.C. Liu, Y.C. Cheng: in *Compound Semiconductors* ed. by M.S. Shur, R.A. Suris: Inst. Phys. Conf. Ser. No. 155 (1997) p. 291–294; N.N. Loyko, V.M. Konnov, T.V. Larikova, V.A Dravin, V.V. Ushakov, A.A. Gippius: ibid., p. 569–572

2.218 M. Godlewski, K. Świątek, D. Hommel: Radiative Recombination Processes in Rare Earth Doped II–VI Materials: in *II–VI Semiconductor Compounds* ed. by M. Jain (World Scientific Publishing, Singapore 1993) pp. 131–152

2.219 G. Bauer, H. Pascher: Diluted Magnetic IV–VI Compounds, in *Diluted Magnetic Semiconductors,* ed. by M. Jain, World Scientific Publishing, Singapore (1991) pp. 339–407

2.220 T.W. Mossberg: Opt. Lett. **7**, 77–79 (1982); M. Mitsunaga, R. Yano, N. Uesugi: Opt. Lett. **16**, 1890–1892 (1991); F.R. Graf, B.H. Plagemann, E.S. Maniloff, S.B. Altner, A. Renn, U.P. Wild: Opt. Lett. **21**, 284–286 (1996); X.A. Chen, A. Nguyen, J. Perry, D. Huestis, R. Kachru: Science **278**, 96–100 (1996).

2.221 G.Blasse, B.C. Grabmaier: *Luminescent Materials,* (Springer, Berlin, Heidelberg 1994)

2.222 J. Baur, P. Schlotter, J. Schneider: White Light Emitting Diodes in *Festkörperprobleme / Advances in Solid State Physics* **37**, ed. by R. Helbig (Vieweg, Braunschweig / Wiesbaden, 1997) pp. 67–78; J. Baur: Dissertation (Univ. Freiburg 1998)

3 Brillouin Light Scattering from Layered Magnetic Structures

Burkard Hillebrands

3.1 Introduction

One of the most prospering areas of modern solid state physics is the field of magnetism in artificially layered structures. The magnetic properties of these structures, which comprise single magnetic films, magnetic double layers, multilayers consisting of a few magnetic and nonmagnetic layers or periodically stacked superlattice structures, differ from those of the constituent bulk materials in many characteristic ways. Sometimes surprisingly novel phenomena are found, which are not known from the bulk material [3.1].

Layered magnetic structures are characterized by a substantially large number of magnetic atoms located at lattice sides near surfaces or interfaces. The magnetic environments of these atoms are different from those in the bulk. The coordination number, and therefore the hybridization of electronic states with states of neighboring atoms, is reduced and, at the interfaces, the Fermi level is adjusted between the electronic states of the interfacing materials. It is therefore not very surprising that these systems exhibit novel magnetic properties, in particular in some cases surprisingly large magnetic anisotropy contributions, oscillating interlayer exchange coupling, giant magnetoresistance, spin polarized quantum well states, as well as spin wave excitations, which depend on the layering geometry.

Magnetic anisotropies reflect, to a large degree, the symmetries of the system. On the one hand, they are closely connected to the crystallographic symmetry since the latter is related via spin–orbit coupling to the free energy of the magnetization. On the other hand any symmetry-breaking mechanism, like the broken translational symmetry perpendicular to the layers at the interfaces, as well as e.g. magnetoelastic interaction due to layer stresses, is associated with corresponding anisotropy contributions. In particular, magnetic properties of ultrathin pseudomorphic films are inherently connected to magnetic anisotropies, which in turn are dominated by interface and magnetoelastic contributions.

Beside these static phenomena the properties of dynamic excitations in layered structures are found to be very distinct from their counterparts

Topics in Applied Physics, Vol. 75
Light Scattering in Solids VII
Eds.: M. Cardona, G. Güntherodt

in bulk materials. The spectrum of dynamic excitations in ultrathin magnetic structures is largely determined by the presence of interfaces which break the translational symmetry along the layer normal, and often confine the wavevector direction. In ultrathin magnetic films thermally activated spin wave modes propagate only along the film planes, and thus they exhibit truly two-dimensional behavior.

Magnetic superlattices, constructed from a large number of alternating magnetic and nonmagnetic layers, may not only be considered as an extension of ultrathin films, in the sense that for superlattices many thin film properties, like, e.g., interface anisotropies, are just multiplied by the number of magnetic layers. A new interaction is introduced by the magnetic coupling between the individual magnetic layers. This interaction may be of dipolar type as well as in many cases of exchange type. The magnetic properties of superlattices will therefore depend on the properties of the individual layers as well as on the details of the interlayer coupling.

Perhaps the most important consequence of these interlayer coupling mechanisms in superlattices is the formation of so-called collective spin wave excitations due to the stacking periodicity (modulation wavelength), which are coherent throughout all layers of the superlattice structure. These excitations are unique to magnetic superlattice structures and not known for bulk materials. Of great importance is also the fact that the type of interlayer coupling, as well as its sign and strength, can be taylored by appropriate choice of the spacer material and its thickness. For instance the discovery of the existence of antiferromagnetic interlayer coupling between ferromagnetic layers has boosted the discovery and investigation of a large variety of new magnetization ground states in superlattices. Coupling phenomena like oscillatory, bilinear and biquadratic coupling have been widely found, and associated transport phenomena, like the giant magnetoresistance effect, have been already employed for technological applications.

Brillouin light scattering (BLS) from spin waves has grown during the last decade into a powerful method for investigating magnetic properties in magnetic films and multilayers. The current status is characterized by ongoing increasing use as a standard research tool in thin film magnetism for determining magnetic parameters, as well as by new advances in instrumentation and by applications to new fields.

Light is inelastically scattered from propagating spin waves with wavevectors in the range of 10^3–$10^5\,\mathrm{cm}^{-1}$. In this range dipolar type modes (Damon–Eshbach modes) and exchange modes exist. From measurements of the spin wave frequencies as a function of the direction and magnitude of the in-plane wavevector, $q_{\|}$, and the direction and strength of the externally applied field, magnetic anisotropies as well as intra- and interlayer coupling constants are determined. The specific advantages of the method are characterized by high sensitivity of, e.g., 10^{-9} emu for Co, nondestructive measurements in saturation, a small laser focus of $\approx$ 30–50 µm in diameter enabling local measurements and therefore the possibility of using samples with controlled lateral variation of parameters like the film thickness, as well as compatibility with ultrahigh vacuum. The sensing length is characterized by the spin wave wavelength, which is typically 300 nm. Thus magnetic phenomena varying on this length scale, like spatially varying internal fields, or patterned structures can be investigated easily.

This Chapter is primarily aimed as a review of experimental BLS work from magnetic films and multilayers. It contains an introduction into the physics of spin waves in artificially layered magnetic structures using a continuum model, which by now is widely used for the calculation of spin wave dispersion in these structures, including intra- and interlayer exchange interaction and magnetic anisotropies (Sect. 3.2). Magnetic anisotropies are discussed in detail, since they are the most easily and most widely investigated properties. The various spin wave modes in magnetic films, bilayers and multilayers are discussed. Another field is the investigation of nonlinear spin-wave excitations by BLS. The basic theory, as far as it is needed for the discussion of the experimental work, is also presented in Sect. 3.2. Section 3.3 discusses the light scattering cross section and Sect. 3.4 the instrumentation. In Sect. 3.5 some key applications are discussed in detail to illustrate the most promising applications of the method. In Sect. 3.6 a summary is given. A compilation of published experimental BLS work is contained in the Appendix.

The first observation of spin waves was made by *Griffiths* in 1946 [3.2] by performing microwave resonance experiments in 25 μm thick electrodeposited Ni, Fe and Co films. The experiments were interpreted by *Kittel* [3.3, 3.4] as a resonance absorption of microwave energy by spin waves. Subsequent improvements in the model calculations were performed by *Ament* and *Rado* [3.5] and *Rado* and *Weertman* [3.6] by inclusion of exchange interaction at the film surfaces (pinning effects). In addition to these $q = 0$ dipolar modes, $q \neq 0$ standing spin wave resonances were predicted by *Kittel* in 1958 [3.7]. These modes exist and can be excited if the surface spins are pinned, as was shown experimentally by *Seavey* and *Tannenwald* in the same year [3.8]. The roots of BLS go back to 1922, where first experiments on light scattering from acoustic phonons were predicted by *Brillouin* [3.9] and independently by *Mandelstam* in 1926 [3.10, 3.11]. First experimental evidence was found by *Gross* [3.12] in transparent liquids in 1930. Due to their low light scattering cross section, spin waves were not detected in the next four decades. The first light scattering experiment involving spin waves was reported by *Fleury* et al. for antiferromagnetic FeF_2 [3.13] and MnF_2 [3.14] alloys. Success in observing dipolar-type spin waves in ferromagnetic systems was achieved following the invention of the multipass Fabry-Perot interferometer by *Sandercock* in 1970 [3.15, 3.16, 3.17, 3.18]. Thermally excited spin waves were first reported by *Sandercock* and *Wettling* in yttrium-iron-garnet (YIG) [3.19]. The first surface spin wave spectrum has been reported in the ferromagnetic semiconductor EuO by *Grünberg* and *Metawe* in 1977 [3.20]. The first surface and bulk spin wave spectra in the transition metals Fe and Ni were reported by *Sandercock* and *Wettling* [3.21, 3.22] and *Malozemoff* et al. [3.23]. The next milestones were set by the first observation of the surface spin wave mode and the standing spin waves in ultrathin Fe-films of thicknesses as small as 30 Å deposited onto sapphire, and the observed crossover between these modes as a function of the film thickness [3.24], the coupled spin wave modes in exchange coupled bilayer systems [3.25, 3.26] as well as in superlattices [3.27, 3.28, 3.29]. These results are discussed in detail in the subsequent sections.

A number of excellent reviews about the field of BLS from spin wave excitations have been published [3.25, 3.30, 3.31, 3.32, 3.33, 3.34, 3.35, 3.36, 3.37, 3.38, 3.39, 3.40, 3.41, 3.42, 3.43, 3.44, 3.45]. The most recent are the reviews by *Cochran* about light scattering from ultrathin magnetic layers and bilayers [3.39] and by *Hillebrands* and *Güntherodt* about BLS from magnetic superlattices [3.40] in the two-volume review book about ultrathin magnetic structures [3.1], the article by *Dutcher* about BLS and microwave resonance work in metallic films and multilayers [3.41], and reviews mostly about experimental work by *Demokritov* and *Tsymbal* [3.43] and about theory by *Cottam* et al. [3.44]. Overviews about the theory of spin waves in films and multilayers are given in the "classic" review of *Wolfram* and *De Wames* [3.30], as well as by *Mills* [3.31], *Borovic-Romanov* and *Kreines* [3.33], *Patton* [3.34], *Mills* [3.35], *Cochran* [3.39] and *Cottam* et al. [3.44]; the latter two references discuss, in particular, the light scattering cross section problem. Recent introductions in the field of magnetism in ultrathin films are found in the two-volume book of *Bland* and *Heinrich* [3.46] as well as in the reviews of *Gradmann* [3.47] and of *Heinrich* and *Cochran* [3.38]. The latter reference contains a discussion of BLS from these structures.

The intention of the present chapter is to give an overview about the current status and potential of experimental BLS work with the main emphasis on the discussion of available experimental work. We shall not be able to discuss in full all aspects of spin wave propagation and BLS in layered structures because this would fill more than the available space. Therefore discussions of the theoretical work going beyond the potential applicability in BLS experiments, in particular a full discussion of the BLS cross section, is not included, the latter due to its still rather limited content of information to be extracted from a BLS experiment (see Sect. 3.3 and the note in the concluding remarks). The reader is referred to existing review literature for these topics [3.39, 3.44].

3.2 Theoretical Background

In the following an introduction into spin wave properties in films and layered structures is given. For the discussion we use a model which contains most salient features of spin waves in layered structures, including exchange contributions and anisotropy.

3.2.1 Continuum Theory of Spin-Wave Excitations

We use a continuum model approach introduced for single layers by *Rado* and *Hicken* [3.48], *Cochran* and *Dutcher* [3.49], and applied to superlattices by *Hillebrands* [3.50,3.51] and *Stamps* and *Hillebrands* [3.52,3.53,3.54,3.55]. The full equivalence to a microscopic model starting from the spin Hamiltonian has been demonstrated by *Stamps* and *Hillebrands* [3.55]. We discuss directly the case of multilayered structures. The application to single films is then straightforward.

The coordinate system used is shown in Fig. 3.1. The $\hat{x}$-axis is perpendicular to the layers. For a N-layer system the positions of the interfaces are defined by d_n, $n = 1...N$, such that for the n-th layer the interfaces lie at $x = d_{n-1}$ and $x = d_n$. We use the index "n" to indicate parameters of the n-th layer, but when appropriate this index is omitted for better clarity of the formulas. Without loss of generality, we choose the $\hat{z}$-axis as the direction of the saturation magnetization, $\boldsymbol{M}_s$, which is the time averaged direction of the precessing magnetization, $\boldsymbol{M}$. θ and ϕ are the polar and azimuthal angles with θ measured against the surface normal. ϕ is measured against a crystallographic reference direction of lowest indices within the film plane, which is usually [3.100] (dashed horizontal line in Fig. 3.1). The direction

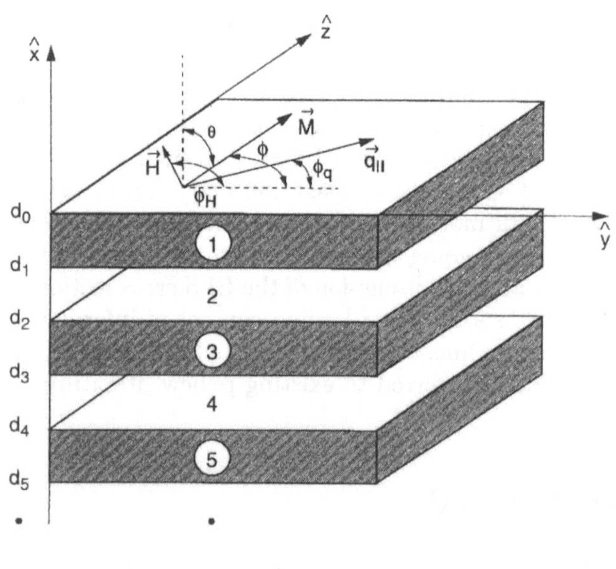

Fig. 3.1. Coordinate system used for calculating spin wave frequencies. The dashed line is a lowest-indexed crystallographic reference direction within the film plane

of the spin wave propagation, defined by the wavevector component of the mode parallel to the film plane, $q_{\parallel}$, is within the (y, z) plane. Its angle with the crystallographic reference direction is ϕ_q.

3.2.1.1 Equation of motion. We begin with the full Landau-Lifshitz torque equation of motion:

$$\frac{1}{\gamma}\frac{\partial \boldsymbol{M}}{\partial t} = \boldsymbol{M} \times \boldsymbol{H}_{\text{eff}} \,, \tag{3.1}$$

where $\gamma = \gamma_e \cdot g/2$ is the gyromagnetic ratio, γ_e = -1.759 $\cdot$ 10^7Hz/Oe is the value of γ for the free electron and g is the spectroscopic splitting factor. The effective magnetic field acting on the magnetization, $\boldsymbol{H}_{\text{eff}}$, is given by

$$\boldsymbol{H}_{\text{eff}} = \boldsymbol{H} - \frac{1}{M_s}\boldsymbol{\nabla}_{\boldsymbol{\alpha}} E_{\text{ani}} + \frac{2A}{M_s^2}\nabla^2 \boldsymbol{M} \,. \tag{3.2}$$

We have omitted the layer index, n, for clarity. E_{ani} is the volume anisotropy energy density and A is the exchange stiffness constant. The first term in the right hand side of (3.2) is a field which includes the external applied field and the fluctuating fields generated by the precessing spins. The second term is an effective field due to magnetic anisotropies with $\boldsymbol{\nabla}_{\boldsymbol{\alpha}}$ the gradient operator for which the differentiation variables are the components of the unit vector $\boldsymbol{\alpha}$ pointing into the direction of $\boldsymbol{M}_s$. The various contributions to the anisotropy will be discussed in Sect. 3.2.2. The last term is the exchange field due to volume exchange interaction.

Together with the equation of motion the magnetostatic Maxwell equations need to be fulfilled [3.56]:

$$\boldsymbol{\nabla} \times \boldsymbol{H} = 0 \tag{3.3}$$

$$\boldsymbol{\nabla} \cdot (\boldsymbol{H} + 4\pi \boldsymbol{M}_s) = 0 \,. \tag{3.4}$$

We will assume that the fluctuations in $\boldsymbol{M}$ and $\boldsymbol{H}$ associated with the spin waves are small compared to the static values. This condition is almost always fulfilled for thermally driven spin waves at temperatures considerably less than T_c [3.57]. We split $\boldsymbol{M}(t)$ and $\boldsymbol{H}(t)$ into time-independent static parts $\boldsymbol{M}_s$ and $\boldsymbol{H}$ and dynamic parts $\boldsymbol{m}(t)$ and $\boldsymbol{h}(t)$:

$$\boldsymbol{M}(t) = \boldsymbol{M}_s + \boldsymbol{m}(t), \qquad |\, \boldsymbol{m}(t) \,| \ll |\, \boldsymbol{M}_s \,| \,, \tag{3.5}$$

$$\boldsymbol{H}(t) = \boldsymbol{H} + \boldsymbol{h}(t), \qquad |\, \boldsymbol{h}(t) \,| \ll |\, \boldsymbol{H} \,| \,. \tag{3.6}$$

Let us define the spin-wave wavevector $\boldsymbol{q} = (q_x, q_y, q_z)$ and $q_{\parallel}(q_{\perp})$ the component of $\boldsymbol{q}$ parallel (perpendicular) to the interfaces. Thus $q^2 = q_x^2 + q_y^2 + q_z^2$, $q_{\parallel}^2 = q_y^2 + q_z^2$, $q_{\perp} = q_x$. In a BLS experiment $q_{\parallel}$, and thus $q_y = q_{\parallel} \cos\phi_q$ and $q_z = q_{\parallel} \sin\phi_q$ are defined by the scattering geometry as a consequence of wavevector conservation in the scattering process. We assume that inside the ferromagnetic layers $\boldsymbol{m}$ and $\boldsymbol{h}$ are proportional to

$$\exp[\mathrm{i}(\omega t - q_x x - q_y y - q_z z)] \;. \tag{3.7}$$

Outside the magnetic layers $\boldsymbol{h}$ is proportional to

$$\exp[\mathrm{i}(\omega t - q_y y - q_z z) - q_x^e x] \;, \tag{3.8}$$

with q_x^e a real number and $\boldsymbol{m}$ equal to zero. Using (3.5) and (3.6) we linearize (3.1–4) by dropping all terms that are of quadratic or higher order in components of $\boldsymbol{m}$ and $\boldsymbol{h}$.

Thus the equation of motion (3.1) becomes a system of linear equations in the components of $\boldsymbol{m}$ and $\boldsymbol{h}$, which is in general inhomogeneous. The inhomogeneous part, which is essentially a static torque density acting on the magnetization, contains the equilibrium direction of the magnetization. It becomes zero for $\boldsymbol{M}_\mathrm{s}$ along the equilibrium direction. Recall that we have chosen the coordinate system such that $\boldsymbol{M}_\mathrm{s}$ points into the $\hat{z}$-direction. We obtain assuming $q_y \neq 0$ and with $q^2 = q_x^2 + q_y^2 + q_z^2$:

$$-q_z h_y + q_y h_z = 0 \;, \tag{3.9}$$

$$-q_y h_x + q_x h_y = 0 \;, \tag{3.10}$$

$$q_x h_x + q_y h_y + q_z h_z + 4\pi q_x m_x + 4\pi q_y m_y + 4\pi q_z m_z = 0 \;, \tag{3.11}$$

$$M_\mathrm{s} h_y + \left(\frac{\mathrm{i}\omega}{\gamma} - H_\gamma\right) m_x - \left(H \cos(\phi_H - \phi) + H_\beta + \frac{2A}{M_\mathrm{s}} q^2\right) m_y = 0 \;, \tag{3.12}$$

$$\begin{aligned} &-M_\mathrm{s} h_x + \left(H \cos(\phi_H - \phi) + H_\alpha + \frac{2A}{M_\mathrm{s}} q^2\right) m_x \\ &+ \left(\frac{i\omega}{\gamma} + H_\gamma\right) m_y = 0 \;, \end{aligned} \tag{3.13}$$

$$\frac{i\omega}{\gamma} m_z = 0 \;, \tag{3.14}$$

with

$$\frac{\mathrm{d}E_\mathrm{ani}}{\mathrm{d}\alpha_x} = H_\alpha m_x + H_\gamma m_y \;, \tag{3.15}$$

$$\frac{\mathrm{d}E_\mathrm{ani}}{\mathrm{d}\alpha_y} = H_\gamma m_x + H_\beta m_y \;, \tag{3.16}$$

where H_α, H_β and H_γ are the volume anisotropy fields (see Sect. 3.2.2) and α_i, $i = x, y, z$ are the direction cosines of $\boldsymbol{M}_\mathrm{s}$. The angle ϕ_H of the external field, $\boldsymbol{H}$, is measured against the in-plane reference direction. Thus $\phi_H - \phi$ is the angle between $\boldsymbol{H}$ and $\boldsymbol{M}$. For $\boldsymbol{M}_s$ lying in-plane and E_{ani} expressed in polar coordinates, with θ the polar and ϕ the azimuthal angle, the anisotropy fields can alternatively be calculated with

$$H_\alpha = \frac{\partial^2}{M_s \partial \theta^2} E_{ani} , \tag{3.17}$$

$$H_\beta = \frac{\partial^2}{M_s \partial \phi^2} E_{ani} , \tag{3.18}$$

$$H_\gamma = \frac{\partial^2}{M_s \partial \theta \partial \phi} E_{ani} , \tag{3.19}$$

evaluated at the equilibrium direction of the magnetization. The second order derivatives of E_{ani} with respect to θ and ϕ are often used in formalisms using a suitably defined susceptibility tensor [3.58, 3.59, 3.60, 3.61].

In order that (3.9–14)) have a nontrivial solution, the secular determinant of their coefficients must vanish. Neglecting the unphysical solution $q_y = 0$, the solutions are given by the zeros of the relation:

$$\left(\frac{\omega^2}{\gamma^2} + H_\gamma^2 - H_a H_b \right) q^2 - 4\pi M_s \left(H_b q_x^2 + H_a q_y^2 - 2 H_\gamma q_x q_y \right) , \tag{3.20}$$

with

$$H_a = H \cos(\phi_H - \phi) + H_\alpha + \frac{2A}{M_s} q^2 , \tag{3.21}$$

$$H_b = H \cos(\phi_H - \phi) + H_\beta + \frac{2A}{M_s} q^2 , \tag{3.22}$$

and $q^2 = q_x^2 + q_y^2 + q_z^2$. The parameter $2A/M_s$ is also called the spin wave stiffness constant. Since the in-plane components q_y and q_z are constants defined by the scattering geometry and the wavevector of the incident light, equation (3.20) is the dispersion relation between the spin wave frequency ω and the wavevector component perpendicular to the layers, $q_\perp = q_x$. We label the six roots of $q_\perp$ by $q_{x1} ... q_{x6}$. The calculation of these quantities is performed numerically. The corresponding dynamic fields and magnetizations are labeled h_{xi}, h_{yi} and m_{xi}, m_{yi}, respectively, for $i = 1...6$.

For an easier analysis of the boundary conditions it is convenient to express h_{yi}, m_{xi} and m_{yi} in terms of h_{xi}. From (3.10) we find

$$h_{yi} = \frac{q_y}{q_{xi}} h_{xi} . \tag{3.23}$$

For m_{xi} and m_{yi} we define the quantities u_i and v_i:

$$m_{xi} = u_i h_{xi} \tag{3.24}$$

$$m_{yi} = v_i h_{xi} \tag{3.25}$$

From (3.12) and (3.13) we are able to calculate u_i and v_i:

$$u_i = \frac{M_s}{D}\left(H_b - Q\left(\frac{\mathrm{i}\omega}{\gamma} + H_\gamma\right)\right), \tag{3.26}$$

$$v_i = \frac{M_s}{D}\left(H_a Q + \left(\frac{\mathrm{i}\omega}{\gamma} - H_\gamma\right)\right), \tag{3.27}$$

with

$$D = H_a H_b - H_\gamma^2 - \left(\frac{\omega}{\gamma}\right)^2 \tag{3.28}$$

and

$$Q = \frac{q_y}{q_{xi}} = \frac{q_\|}{q_{xi}} \sin(\phi - \phi_q) . \tag{3.29}$$

In the nonmagnetic layers we obtain from (3.3) and (3.4):

$$\mathrm{i} q_\| h_x^{\mathrm{e}} - q_x^{\mathrm{e}} h_\|^{\mathrm{e}} = 0 , \tag{3.30}$$

$$q_x^{\mathrm{e}} h_x^{\mathrm{e}} + \mathrm{i} q_\| h_\|^{\mathrm{e}} = 0 , \tag{3.31}$$

where $h_\|$ is the component of $\boldsymbol{h}$ in the direction of $q_\|$. A non-vanishing solution of (3.30) and (3.31) requires

$$q_x^{\mathrm{e}} = \pm q_\| , \tag{3.32}$$

where we label the two solutions of (3.32) by q_{x1}^{e} and q_{x2}^{e}. We then obtain:

$$h_\|^{\mathrm{e}+} = i h_{x1}^{\mathrm{e}} , \tag{3.33}$$

$$h_\|^{\mathrm{e}-} = -i h_{x2}^{\mathrm{e}} . \tag{3.34}$$

Outside the layers we must require $h_{x2}^{\mathrm{e}} = 0$ for the vacuum above the layers and $h_{x1}^{\mathrm{e}} = 0$ for the vacuum below the layers, respectively, since h_x^{e} must approach zero for $x \to \pm\infty$.

3.2.1.2 . The solutions within each layer, h_{xi}, $i = 1...6$, and outside the layers, h_{xi}^{e}, $i = 1, 2$, are interrelated by boundary conditions at each interface. From the equation of motion (3.1) and the Maxwell equations (3.3,3.4), boundary conditions can be derived. At each interface the parallel component of $\boldsymbol{H}(t)$ and the perpendicular component of $\boldsymbol{H}(t) + 4\pi \boldsymbol{M}(t)$ have to be continuous. Continuity of $h_\|$ and of $h_x + 4\pi m_x$ requires by use of (3.23–25):

$$\sum_{i=1}^{6} \frac{q_\|}{q_{xi}} h_{xi} \mathrm{e}^{-\mathrm{i} q_{xi} d_n} - i h_{x1}^{\mathrm{e}} \mathrm{e}^{-q_\| d_n} + \mathrm{i} h_{x2}^{\mathrm{e}} \mathrm{e}^{q_\| d_n} = 0 \tag{3.35}$$

$$\sum_{i=1}^{6}(1+4\pi u_i)h_{xi}\mathrm{e}^{-\mathrm{i}q_{xi}d_n} - h^{\mathrm{e}}_{x1}\mathrm{e}^{-q_{\parallel}d_n} - h^{\mathrm{e}}_{x2}\mathrm{e}^{q_{\parallel}d_n} = 0 \tag{3.36}$$

for magnetic/non-magnetic interfaces, and

$$\sum_{i=1}^{6}\frac{q_{\parallel}}{q_{xi,n}}h_{xi,n}\mathrm{e}^{-\mathrm{i}q_{xi,n}d_n} - \sum_{i=1}^{6}\frac{q_{\parallel}}{q_{xi,n+1}}h_{xi,n+1}\mathrm{e}^{-\mathrm{i}q_{xi,n+1}d_n} = 0 \tag{3.37}$$

$$\sum_{i=1}^{6}(1+4\pi u_{i,n})h_{xi,n}\mathrm{e}^{-\mathrm{i}q_{xi,n}d_n}$$
$$-\sum_{i=1}^{6}(1+4\pi u_{i,n+1})h_{xi,n+1}\mathrm{e}^{-\mathrm{i}q_{xi,n+1}d_n} = 0 \tag{3.38}$$

for interfaces between two magnetic layers.

From the equation of motion (3.1) we obtain the condition that the sum of the interface torques must be zero for each interface. The general boundary conditions at the ($x = d_n$)-interface are given by the so-called Hoffmann boundary conditions, which include exchange coupling to the interface of the next magnetic layer at $x = d_{n'}$ [3.62, 3.63, 3.64]:

$$\boldsymbol{M}_n \times \left[\frac{1}{M_{s,n}}\boldsymbol{\nabla}_{\alpha_n}\sigma_{\mathrm{inter},n} - \frac{2A_n}{M_{s,n}^2}\frac{\partial \boldsymbol{M}_n}{\partial n_n}\right]\Bigg|_{x=d_n}$$
$$-\boldsymbol{M}_n \times \frac{2A_{nn'}}{M_{s,n}M_{s,n'}}\left[\boldsymbol{M}_{n'} + a_{n'}\frac{\partial \boldsymbol{M}_{n'}}{\partial n_{n'}}\right]\Bigg|_{x=d_{n'}} = 0 \tag{3.39}$$

where σ_{inter} is the interface anisotropy energy (Sect. 3.2.2) and ∂/∂_n is the partial derivative with respect to the surface normal unit vector, $\boldsymbol{n}$. The vector $\boldsymbol{n}$ points from the interface into the corresponding magnetic layer. The interlayer exchange constant between layers n and n' is $A_{nn'}$. Without loss of generality we call this parameter A_{12} in the following. The lattice constant is denoted by a_n. The first term in (3.39) is the so-called Rado–Weertman boundary condition, describing the surface torque of a single magnetic film due to interface anisotropies and volume exchange interaction [3.6], whereas the second term describes the exchange coupling between the two layers [3.64]. We need to discuss two limiting cases of the interlayer exchange constant, A_{12}. For $A_{12} = 0$, equation (3.39) resembles the so-called Rado–Weertman boundary conditions, i.e., the interface torques must be separately zero for $x = d_n$ and $x = d_{n'}$. For large absolute values of A_{12} comparable to A_n/a_n we obtain $\boldsymbol{M}_n \times \boldsymbol{M}_{n'} = 0$, i.e., $\boldsymbol{M}_n$ and $\boldsymbol{M}_{n'}$ are aligned either parallel or antiparallel, depending on the sign of A_{12}.

In terms of the components h_{xi} the Hoffmann boundary conditions then read:

$$\sum_{i=1}^{6}(-\mathrm{i}A_n q_{xi,n} - k_{s,n} - k_{p,n}\cos^2\phi + A_{nn'})u_{i,n}h_{xi,n}\mathrm{e}^{-\mathrm{i}q_{xi,n}d_n}$$

$$-\sum_{i=1}^{6}\frac{M_{s,n'}}{M_{s,n}}A_{nn'}u_{i,n'}h_{xi,n'}(1-\mathrm{i}a_{n'}q_{xi,n'})\mathrm{e}^{-\mathrm{i}q_{xi,n'}d_{n'}}=0 \tag{3.40}$$

$$\sum_{i=1}^{6}(-\mathrm{i}A_nq_{xi,n}+k_{p,n}(1-2\cos^2\phi)+A_{nn'})v_{i,n}h_{xi,n}\mathrm{e}^{-\mathrm{i}q_{xi,n}d_n}$$
$$-\sum_{i=1}^{6}\frac{M_{s,n'}}{M_{s,n}}A_{nn'}v_{i,n'}h_{xi,n'}(1-\mathrm{i}a_{n'}q_{xi,n'})\mathrm{e}^{-\mathrm{i}q_{xi,n'}d_{n'}}=0 \tag{3.41}$$

$$\sum_{i=1}^{6}(iA_nq_{xi,n}-k_{s,n}-k_{p,n}\cos^2\phi+A_{nn'})u_{i,n}h_{xi,n}\mathrm{e}^{-\mathrm{i}q_{xi,n}d_n}$$
$$-\sum_{i=1}^{6}\frac{M_{s,n}}{M_{s,n'}}A_{nn'},u_{i,n'}h_{xi,n'}(1+ia_{n'}q_{xi,n'})\mathrm{e}^{-\mathrm{i}q_{xi,n'}d_{n'}}=0 \tag{3.42}$$

$$\sum_{i=1}^{6}(iA_nq_{xi,n}+k_{p,n}(1-2\cos^2\phi)+A_{nn'})v_{i,n}h_{xi,n}\mathrm{e}^{-\mathrm{i}q_{xi,n}d_n}$$
$$-\sum_{i=1}^{6}\frac{M_{s,n}}{M_{s,n'}}A_{nn'}v_{i,n'}h_{xi,n'}(1+ia_{n'}q_{xi,n'})\mathrm{e}^{-\mathrm{i}q_{xi,n'}d_{n'}}=0\,, \tag{3.43}$$

where $A_{nn'}$ is the interlayer exchange constant between layers n and n'. For reasons of clarity only the lowest order anisotropy contributions are considered here. The generalization of these equations by inclusion of higher order terms is straightforward [see, e.g., (3.53, 3.54) below].

Equations (3.35–38,3.40–43) form a set of linear equations in $h_{xi,n}$ and $h^{\mathrm{e}}_{xi,n}$. Since there are six solutions to h_{xi} in each magnetic layer, two solutions h^{e}_{xi} in each nonmagnetic layer, and two solutions outside the multilayer, the dimension of the system of linear equations is given by $2 + 6\ N_{\mathrm{mag}} + 2N_{\mathrm{nonmag}}$, where $N_{\mathrm{mag}}(N_{\mathrm{nonmag}})$ is the total number of magnetic (nonmagnetic) layers. In order to fulfill the boundary conditions simultaneously for all interfaces, the determinant of the system of linear equations in $h_{xi,n}$ must equal zero. For finding the spin wave frequencies, the numerical procedure is as follows: For a given frequency ω, equations (3.20–29) are used to calculate $q_{xi,n}$, $u_{i,n}$ and $v_{i,n}$. Then the value of the boundary condition determinant is evaluated. In a root finding routine ω is varied and the boundary condition determinant is calculated until the value of the boundary condition determinant fulfills a convergence criterion. The calculations are performed by means of appropriate numerical tools.

3.2.2 Magnetic Anisotropies

Magnetic anisotropies reflect the symmetry of the system as well as acting symmetry breaking mechanisms. They are caused either by the dipolar interaction between magnetic moments (e.g. shape anisotropy) or by spin orbit coupling described by the Hamiltonian operator

$$\mathcal{H}_{\mathrm{so}} = \xi \boldsymbol{\ell} \cdot \boldsymbol{s} \,, \tag{3.44}$$

with ξ the spin–orbit coupling constant, $\boldsymbol{\ell}$ the orbital momentum and $\boldsymbol{s}$ the spin momentum. Here we will only discuss the spin–orbit terms. Since the electron orbits are coupled via the electrostatic potential to the crystal lattice, the spin–orbit interaction results in anisotropy contributions which are the magneto-crystalline anisotropy contributions with the same symmetry as the crystal lattice, surface anisotropies caused by the reduced symmetry at interface lattice sites (broken translational symmetry along the interface normal) and strain induced anisotropies caused by a deformation of the lattice unit cell. If the exchange splitting between spin-up and spin-down bands is much larger than the band width (i.e., the crystal field splitting energy), spin–orbit coupling between spin-up and spin-down electrons can be neglected, and the spin–orbit coupling can be expressed by an effective field, $\boldsymbol{H}_{\mathrm{orb}}$, acting on the orbital moment $\boldsymbol{m}_\ell$ as

$$\mathcal{H} = -\boldsymbol{m}_\ell \cdot \boldsymbol{H}_{\mathrm{orb}} = -\boldsymbol{m}_\ell \cdot \frac{\pm\xi\boldsymbol{\alpha}}{2\mu_{\mathrm{B}}} \,, \tag{3.45}$$

with $\boldsymbol{\alpha}$ the unit vector pointing into the direction of magnetization and the $+/-$ sign for bands which are more/less than half filled. The spin–orbit coupling energy, E_{so}, can then be expressed in a power series of $\boldsymbol{H}_{\mathrm{orb}}$ with the two lowest-order terms as

$$\begin{aligned} E_{\mathrm{so}} = & -\frac{1}{2}\sum_{ij} \chi_{ij}^{(2)} H_{\mathrm{orb},i} H_{\mathrm{orb},j} \\ & -\frac{1}{2}\sum_{ijk\ell} \chi_{ijk\ell}^{(4)} H_{\mathrm{orb},i} H_{\mathrm{orb},j} H_{\mathrm{orb},k} H_{\mathrm{orb},\ell} \,, \end{aligned} \tag{3.46}$$

with $\chi^{(2)}$ and $\chi^{(4)}$ the second and fourth order susceptibility tensors. These tensors fulfill the local lattice symmetries.

E_{so} can thus be further expressed as

$$E_{\mathrm{so}} = -\frac{1}{2}\left(\frac{\xi}{2\mu_{\mathrm{B}}}\right)^2 f_2(\boldsymbol{\alpha}) - \frac{1}{2}\left(\frac{\xi}{2\mu_{\mathrm{B}}}\right)^4 f_4(\boldsymbol{\alpha}) \,. \tag{3.47}$$

$f_2(\boldsymbol{\alpha})$ and $f_4(\boldsymbol{\alpha})$ are functions, which contain products of components of the magnetization unit vector $\boldsymbol{\alpha}$ in second and fourth order, respectively. The components α_i are $\alpha_x = \cos\theta$, $\alpha_y = \sin\theta\,\cos\phi$, and $\alpha_z = \sin\theta\sin\phi$ with θ the polar angle measured with respect to the $\hat{x}$-axis and ϕ the azimuthal angle (see Fig. 3.1). In the following we write the free energy density, E_{ani}, in terms of these direction cosines.

It should be noted, that the shape anisotropy does not explicitly enter the free anisotropy energy density, E_{ani}, as used in the equation of motion (3.2), since it is implicitly contained in the boundary conditions.

In the following we denote volume anisotropies by upper case letters and interface anisotropies by lower case letters. We use the general term "interface anisotropy" instead of "surface anisotropy" in the case of surfaces of single films. If appropriate the rotational manifold of an anisotropy contribution is denoted by a superscript in brackets to the corresponding anisotropy constant.

3.2.2.1 Magnetocrystalline Anisotropies. Due to spin–orbit coupling the free energy of the magnetization, E_{ani}, a function of the direction of $\boldsymbol{M}$, is coupled to the crystalline symmetry. Expanding this free energy with respect to functions belonging to irreducible representations of the crystallographic symmetry groups yields the so-called anisotropy constants as expansion coefficients, which depend on the choice of representations, in particular on the choice of the coordinate system. One of the familiar anisotropy constants is the volume anisotropy constant, K_1, of a cubic crystal. In cartesian coordinates the free energy density, E_{ani}, is

$$E_{\text{ani}} = K_1(\alpha_x^2\alpha_y^2 + \alpha_y^2\alpha_z^2 + \alpha_z^2\alpha_x^2) , \tag{3.48}$$

with α_i the direction cosines of the magnetization, here defined with respect to the crystallographic $\langle 100 \rangle$ reference axes. E_{ani} is of fourth order in α_i as first non-vanishing order. For hexagonal structures E_{ani} is given by

$$E_{\text{ani}} = K_1 \sin^2\theta + K_2 \sin^4\theta . \tag{3.49}$$

In Table 3.1 the anisotropy fields H_α, H_β and H_γ derived from E_{ani}, as they enter the spin wave calculations, are listed for cubic and hexagonal anisotropy in polar coordinates with the symmetry axes pointing into different crystallographic directions as applicable for different film orientations.

Often a uniaxial in-plane anisotropy, $K_{\text{p}}^{(2)}$, is found which can be of magnetocrystalline or magnetoelastic (see below) origin. The free energy density of this contribution is, assuming that the $\hat{y}$-axis is the symmetry axis

$$E_{\text{ani}} = K_{\text{p}}^{(2)}\alpha_y^2 \tag{3.50}$$

For $K_{\text{p}}^{(2)} < 0$ the $\hat{y}$-axis is the easy axis with respect to this anisotropy.

3.2.2.2 Interface Anisotropies. Interface anisotropies arise from the broken translational symmetry along the film or stack normal. On the basis of the pair bonding model of *Néel* [3.65, 3.66] they can be calculated by performing the dipolar sums of the magnetic moments located at the interfaces although the obtained values do not agree well, even in sign, with experimental results. To lowest order, the free interface energy density, σ_{inter}, is given by [3.67]

$$\sigma_{\text{inter}} = -k_{\text{s}}\alpha_x^2 + k_{\text{p}}^{(2)}\alpha_y^2 + k_{\text{p}}^{(4)}\alpha_y^2\alpha_z^2 , \tag{3.51}$$

Table 3.1. Anisotropy fields for cubic (100)-, (110)- and (111)-oriented magnetic layers and layers of hexagonal symmetry. ϕ is the angle between the magnetization, $\boldsymbol{M}$, and an in-plane reference axis. The rows with index "iso" show the anisotropy fields for textured layers consisting of a mosaic spread with preferred crystallographic orientation normal to the layer plane and in-plane isotropy. The last row contains the contribution of an in-plane uniaxial anisotropy with K_p the anisotropy constant and $\phi - \phi_0$ the angle between the magnetization and the symmetry axis of this anisotropy. For cubic symmetry, K_1 is the volume anisotropy constant as defined in (3.48). For hexagonal symmetry, K_1 and K_2 are the lowest-order anisotropy constants as defined in (3.49)

	H_α	H_β	H_γ
(100)	$\frac{K_1}{M_s}(2 - 4\sin^2\phi + 4\sin^4\phi)$	$\frac{K_1}{M_s}(2 - 16\sin^2\phi + 16\sin^4\phi)$	-
(110)	$\frac{K_1}{M_s}(2 - 7\sin^2\phi + 3\sin^4\phi)$	$\frac{K_1}{M_s}(2 - 13\sin^2\phi + 12\sin^4\phi)$	-
(111)	$-\frac{K_1}{M}$	-	$-\frac{K_1}{M_s}\sqrt{2}\sin 3\phi$
$(100)_{\mathrm{iso}}$	$\frac{3}{2}\frac{K_1}{M}$	-	-
$(110)_{\mathrm{iso}}$	$-\frac{3}{8}\frac{K_1}{M}$	-	-
$(111)_{\mathrm{iso}}$	$-\frac{K_1}{M}$	-	-
(0001)	$-2\frac{K_1+2K_2}{M}$	-	-
uniaxial	$-\frac{2K_p^{(2)}}{M_s}\cos^2(\phi - \phi_0)$	$-\frac{2K_p^{(2)}}{M_s}(1 - 2\sin^2(\phi - \phi_0))$	-

with k_s the out-of-plane anisotropy constant. The constants $k_p^{(2)}$ and $k_p^{(4)}$ describe the first two nonvanishing orders of the in-plane interface anisotropy. They are of second and fourth order in α_i and correspond to two- and four-fold symmetry in the film plane, respectively. The signs are chosen in the usual way, i.e., a positive sign of k_s indicates a corresponding easy axis perpendicular to the film and a positive sign of $k_p^{(2)}$ and $k_p^{(4)}$ indicates that the [100] axis is hard. Since the second term in (3.51) is of second order in α_i, it only exists for interfaces of symmetry not higher than two-fold.

If the film thickness is smaller than the static exchange correlation length $\sqrt{A/2\pi M_s^2}$ (≈ 30 Å for Fe), the magnetization is homogeneous across the film and the interface torques from interface anisotropies acting on the interface moments can be converted to volume torques acting on the entire film magnetization. This corresponds to replacing the interface energy density, σ_{inter}, by a volume energy density

$$E_{\mathrm{inter}} = \frac{2}{d}\sigma_{\mathrm{inter}} \tag{3.52}$$

with the factor of two counting the two interfaces of the film. *Rado* [3.68] and *Gradmann* et al. [3.69] give estimates for the range of validity for (3.52). From (3.51) and assuming the validity of (3.52), effective anisotropy fields H'_α and

H'_β can be derived for $\theta = \pi/2$ (in-plane orientation of the magnetization) in analogy to (3.15,3.16):

$$H'_\alpha = -\frac{4}{M_s d}\left(k_s + k_p^{(2)}\cos^2\phi + 2k_p^{(4)}\sin^2\phi\cos^2\phi)\right) , \tag{3.53}$$

$$H'_\beta = \frac{4}{M_s d}\left(k_p^{(2)}(2\sin^2\phi - 1) + k_p^{(4)}\left(1 - 8\sin^2\phi\cos^2\phi\right)\right) , \tag{3.54}$$

and $H'_\gamma = 0$. These effective anisotropy fields will be used below.

3.2.2.3 Magnetoelastic Anisotropies. In ultrathin epitaxial films and superlattices large in-plane strains arise from the lattice mismatch at the interfaces. We first consider the case that the elastic strains ϵ_i are not released due to, e.g., formation of misfit dislocations. We first calculate all relevant elastic strain components before we calculate the corresponding anisotropy contributions. Following the approach of *Hillebrands* and *Dutcher* we do this directly for the superlattice case [3.70].

We start with the free energy density, E, of the superlattice system averaged over one bilayer period. The coordinate system is chosen as usual for elastic problems, with the $\hat{x}_3$-axis perpendicular to the layers and the $\hat{x}_1$ axis along the lowest-indexed in-plane axis. Since the layers are homogeneously stressed along the x_1x_2-plane, only the strains ϵ_1, ϵ_2 and ϵ_3 are nonzero. Let f_1 and f_2 be the fractional contributions of the magnetic layer thickness, d_1, and the nonmagnetic layer thickness, d_2, respectively, i.e. $f_i = d_i/(d_1 + d_2)$. With $c_{ij,1}$ being the elastic stiffness constants of the magnetic material, $c_{ij,2}$ those of the nonmagnetic spacer material, $\epsilon_{i,1}$ and $\epsilon_{i,2}$ the corresponding strains, b_{ij} the magneto-elastic tensor of the magnetic material and α_i the magnetization direction cosines for axes $\hat{x}_i$, the free energy density averaged over one bilayer period is

$$E = f_1 \sum_{i,j} c_{ij,1}\epsilon_{i,1}\epsilon_{j,1} + f_2 \sum_{i,j} c_{ij,2}\epsilon_{i,2}\epsilon_{j,2} + f_1 \sum_{i,j} b_{ij}\alpha_i^2\epsilon_{j,1} . \tag{3.55}$$

The first two terms on the right hand side are the elastic energies of the two layers of the bilayer period and the third term is the magnetoelastic energy in the magnetic layer. Here we use the (6×6) matrix notation for c_{ij} and b_{ij}. The tensor components c_{ij} and b_{ij} are rotated from the crystallographic reference frame into the oriented layer frame. Since $\epsilon_i = 0$ for $i > 3$ the sums in (3.55) run over i,j = 1...3. At the interfaces the strains must accommodate the in-plane mismatch:

$$a_1(1 + \epsilon_{i,1}) = a_2(1 + \epsilon_{i,2}), \quad i = 1, 2 \tag{3.56}$$

with a_i the lattice parameters of the two constituent materials. The calculation of the magnetic anisotropy is now performed in two steps. First the equilibrium conditions for (3.55) are derived. Since the magnetostriction constants (see below) are several orders of magnitude smaller than the strains involved, we can neglect the magneto-elastic term in (3.55) for this step. We obtain a system of six linear equations for the six unknowns $\epsilon_{i,1}$ and $\epsilon_{i,2}$, i = 1...3, which is solved numerically.

As a general result we find that in a multilayer or superlattice structure the elastically softer material accommodates the larger fraction of interface strains. In particular in Co/Pt superlattices the Co layers contain the larger part of strains. This in turn (see below) increases the corresponding magnetic anisotropy contributions.

For pseudomorphic films, with lattice parameter a_f on a substrate with lattice parameter a_s, we can directly determine the in-plane strains from

$$a_\mathrm{f}(1+\epsilon_i) = a_\mathrm{s}, \quad i = 1,2 \tag{3.57}$$

The out-of-plane strain ϵ_3 is obtained by evaluating (3.55). For (001)-oriented films of tetragonal symmetry ϵ_3 is given by $\epsilon_3 = -2\epsilon_1 \cdot c_{13}/c_{33}$ [1.20].

The strain-induced anisotropy contributions are obtained from the third term in (3.55) by comparison to a general anisotropy energy expression for each magnetic layer of the form [3.71]:

$$E_\mathrm{ani} = -K_\mathrm{s}\alpha_3^2 + K_\mathrm{p}\alpha_1^2 \tag{3.58}$$

with K_p (K_s) the strain-induced uniaxial in-plane (out-of-plane) anisotropy constant, and $\alpha_1^2+\alpha_2^2+\alpha_3^2 = 1$. The signs in (3.58) are chosen following the usual convention that a positive sign for K_s and a negative sign for K_p denote an easy axis for the corresponding anisotropy contribution. Note that (3.58) is formally close to (3.51), which describes interface anisotropies, since in both cases the lowest order non-vanishing terms are of second order. The results for different types of film orientations are listed in Table 3.2; they are obtained by performing the appropriate tensor rotations on b_{ij}. The magnetoelastic constants $b_{11} - b_{12}$ and b_{44} in Table 3.2 are connected to the magnetostriction constants λ_{100} and λ_{111} via [3.72]

$$b_{11} - b_{12} = -\frac{3}{2}\lambda_{100}(c_{11} - c_{12}) \tag{3.59}$$

$$b_{44} = -3\lambda_{111}c_{44} \tag{3.60}$$

We now consider the formation of misfit dislocations [3.73,3.74,3.75,3.76]. In this case the strains are partly released and the resulting anisotropy constants are smaller. We follow the approach of *Chappert* and *Bruno* [3.77], and *Bruno* and *Seiden* [3.78]. The elastic free energy density E_elast of a magnetic layer can be expressed by

$$E_\mathrm{elast} = C\epsilon_1^2\,, \tag{3.61}$$

with C an expression containing elastic constants. For (001)-oriented films of tetragonal symmetry C is given by $C = 2(c_{11}+c_{12}) - 4c_{13}^2/c_{33}$. We assume that

Table 3.2. Magnetoelastic out-of-plane volume anisotropy constants, K_s, and in-plane volume anisotropy constants, K_p, for different orientations of films with cubic crystallographic symmetry. ϵ_i are the strain components and b_{ij} are the components of the magnetoelastic tensor, as described in the text

	K_s	K_p
(100)	$(b_{11} - b_{12})(\epsilon_1 - \epsilon_3)$	-
(110)	$2\,b_{44}(\epsilon_2 - \epsilon_3)$	$(b_{11} - b_{12})(\epsilon_1 - \epsilon_2/2 - \epsilon_3/2) - b_{44}(\epsilon_2 - \epsilon_3)$
(111)	$2\,b_{44}(4\epsilon_1/3 - \epsilon_3)$	-

the dislocations are formed at the interface. The dislocation energy density, σ_{dis}, which is therefore an interface energy density, is given by

$$\sigma_{\text{dis}} = \alpha\mu\rho_{\text{dis}} , \tag{3.62}$$

with $\alpha \approx 1$ a numerical factor depending on the dislocation geometry, μ the dislocation energy per unit length and ρ_{dis} the dislocation density:

$$\rho_{\text{dis}} = \left| \frac{1}{a_f(1+\epsilon_1)} - \frac{1}{a_s} \right| . \tag{3.63}$$

Interactions between dislocations are neglected. The equilibrium state is obtained by minimizing $E_{\text{elast}} + d\sigma_{\text{dis}}$ with respect to ϵ_1 with d the layer thickness. With ϵ_1 small compared to unity one obtains an analytical expression for the critical thickness, d_{crit}:

$$d_{\text{crit}} \approx \frac{\alpha\mu}{2a_f C \mid \eta \mid} , \tag{3.64}$$

with $\eta = (a_f - a_s)/a_s$ the misfit at the interface. For $d < d_{\text{crit}}$ pseudomorphic film growth is energetically favored. For $d > d_{\text{crit}}$ misfit dislocations are formed and the strain is estimated by

$$\epsilon_1 \approx -\eta \frac{d_{\text{crit}}}{d} . \tag{3.65}$$

Calculating ϵ_2 and ϵ_3 from ϵ_1 as described above the corresponding contributions are obtained. However, since the ϵ_i are proportional to the inverse layer thickness as obtained from (3.65), the derived anisotropy values also are. They can therefore be viewed as effective interface anisotropies. A quantitative analysis of experiments on interface anisotropies with consideration of misfits has so far not been performed due, to a large degree, to the fact that the dislocation energy μ and the dislocation distribution are largely not known for magnetic thin film materials like Co and Fe.

3.2.2.4 Perpendicular Anisotropies. In order to force the magnetization out of the layer plane, the associated gain in magnetocrystalline, magnetoelastic and interface anisotropy energy must be larger than the shape anisotropy energy, $2\pi M_s^2$. We define an effective out-of-plane anisotropy, K_{eff}, by the total change in free energy between perpendicular and in-plane directions of the magnetization, $\boldsymbol{M}$ (including the contribution $2\pi M_s^2$ from the shape anisotropy):

$$K_{\text{eff}} = E_{\text{ani}}(\theta = \pi/2) - E_{\text{ani}}(\theta = 0) . \tag{3.66}$$

For $K_{\text{eff}} > 0$ the system is perpendicularly magnetized at zero field. For a film of thickness d with hexagonal symmetry K_{eff} is given by

$$K_{\text{eff}} = K_1 + K_2 + \frac{2k_s}{d} - 2\pi M_s^2 . \tag{3.67}$$

For layers with in-plane anisotropy contributions, K_p, equation (3.67) reads for $K_p < 0$:

$$K_{\text{eff}} = K_1 + K_2 + K_p + \frac{2k_s}{d} - 2\pi M_s^2 , \tag{3.68}$$

since for $\theta = \pi/2$ (in-plane magnetization) an additional free energy contribution is gained by rotating the magnetization into the corresponding easy in-plane direction.

3.2.3 Spin Waves in Single Magnetic Layers

For single thin magnetic Fe or Co layers of thicknesses d typically smaller than ≈ 30 Å the only spin wave mode accessible in a Brillouin light scattering experiment ($\omega/2\pi \lesssim 100$ GHz) is the dipolar type, so-called Damon–Eshbach mode [3.79]. This mode exists if an external field is applied parallel to the film. The Damon–Eshbach mode is a surface mode, i.e., the mode energy is localized near the film surface and the precession amplitude decays perpendicular to the film with a decay length of the order of $2\pi/q_{\|}$, which is in the range of ≈ 3000 Å in a Brillouin light scattering experiment. Neglecting anisotropies and the weak exchange contribution, the mode frequency for propagation perpendicular to the applied field is

$$\left(\frac{\omega}{\gamma}\right)^2 = H(H + 4\pi M_s) + (2\pi M_s)^2(1 - \mathrm{e}^{-2q_{\|}d}) . \tag{3.69}$$

The Damon–Eshbach mode travels parallel to the layers in an angular range close to perpendicular to the applied field and with a defined sense of revolution about the film.

For larger film thicknesses so-called standing spin waves are accessible; they are of exchange type and consist of two counterpropagating modes travelling almost perpendicular to the film with a wavevector $q \approx n\pi/d$, where n is a positive integer.

We would like to comment on the proper treatment of bulk and interface anisotropies, since the former enter the equation of motion and the latter the boundary conditions. As shown by *Hillebrands* [3.51], different values of the interface anisotropies at both sides of the film result in different spin wave frequencies for the Damon–Eshbach mode and the exchange modes for $q_{\|}$ and $-q_{\|}$. Therefore in a BLS experiment the spin wave frequencies obtained from the Stokes and the anti-Stokes part of the spectrum differ in their absolute values. This can be utilized for the separate determination of the interface anisotropy constants of each interface. However, the experimental resolution of ≈0.3GHz corresponding to a typical light scattering experiment is not high enough compared to a typical frequency difference obtained from this effect ($\lesssim$ 0.3GHz). Apart from this effect, interface anisotropies can be converted into effective volume anisotropies for film thicknesses smaller than the static exchange length (see preceding section).

For ultrathin single magnetic layers of thickness d which fulfill $q_{\|}d \ll 1$, an analytic expression for the spin wave frequencies ω can be derived [3.55]. Under the assumption that the magnetization lies in the film plane ω is given by [3.55]

$$\begin{aligned}\left(\frac{\omega}{\gamma}\right)^2 &= \left(H_a + H'_\alpha + 4\pi M_s f\left(1 - \frac{1}{2}q_{\|}d\right)\right)\\ &\times \left(H_b + H'_\beta + 2\pi M_s f q_{\|} d \sin^2(\phi - \phi_q)\right) - (H_\gamma + H'_\gamma)^2\,, \qquad (3.70)\end{aligned}$$

with H_a and H_b defined in (3.21,3.22). H'_α, H'_β and H'_γ are the anisotropy fields obtained from the interface anisotropy energy, σ_{inter}, as defined in (3.53,3.54), ϕ and ϕ_q are the angles of $\boldsymbol{M}_s$ and $\boldsymbol{q}_{\|}$ with the in-plane reference direction, respectively. The constant f is the demagnetization factor of ultrathin films which may deviate from the thick-film limit of $f = 1$ [3.80, 3.81]: For a magnetic monolayer f is equal to 0.5392. From (3.70) it follows immediately that the spin wave frequency depends (i) on the angle between the direction of magnetization and the wavevector (determined by the scattering geometry), $\phi - \phi_q$, and (ii) on the angle ϕ between the direction of magnetization and the in-plane reference direction due to the anisotropy terms. In the case of zero anisotropies the directional dependence of ω disappears as $q_{\|}d$ approaches zero.

Of particular importance is the situation when the external magnetic field H is applied in a hard magnetic direction. We first consider the case in which the magnetization undergoes an in-plane rotation into the direction of the in-plane applied field. Evaluating (3.70) we find that with increasing external field the spin wave frequency first decreases until the external field forces the magnetization into the direction of the applied field for fields larger than a critical field strength, H_{crit}. For $H > H_{\text{crit}}$ the magnetization and the applied field are co-linear and the spin wave frequencies increase in a quasi-linear fashion with further increasing external field.

A similar dependence of $\omega(H)$ is obtained if the film exhibits a large uniaxial anisotropy perpendicular to the film which is strong enough to compensate the shape anisotropy and which forces the magnetization direction out of the film plane. If we apply an external field parallel to the film plane the direction of magnetization is tilted into the plane with increasing field until at a critical field strength, H_{crit}, the magnetization is completely forced into the film plane. The azimuthal equilibrium angle θ is obtained by minimizing the free energy, which now contains explicitly the demagnetizing energy:

$$\begin{aligned} E = \; & -M_{\mathrm{s}} H \sin\theta + 2\pi M_{\mathrm{s}}^2 \cos^2\theta \\ & + K_1 \sin^2\theta + K_2 \sin^4\theta - (2k_{\mathrm{s}}/d)\cos^2\theta \,. \end{aligned} \tag{3.71}$$

The first term on the right-hand side is the Zeeman energy, the second term the demagnetizing energy, K_1 and K_2 are the first non-vanishing order anisotropy constants of a perpendicular anisotropy as appropriate for, e.g., hexagonal Co, and k_{s} is the out-of-plane interface anisotropy. The critical field strength is found to be:

$$H_{\mathrm{crit}} = \frac{2}{M_{\mathrm{s}}}(K_1 + 2K_2 + 2k_{\mathrm{s}}/d) - 4\pi M_{\mathrm{s}} \,. \tag{3.72}$$

The calculation of the spin wave frequencies is straightforward albeit an algebraically cumbersome extension of the theory described in Sect. 3.2.2. A full description can be found in [3.52, 3.53, 3.54, 3.82]. Here we give an expression for the spin wave frequency for H not much smaller than H_{crit} with $q_{\parallel} = q_y$ [3.82]:

$$\begin{aligned} \left(\frac{\omega}{\gamma}\right)^2 = & \left(H - H_{\mathrm{crit}} + \frac{4k_{\mathrm{s}}}{M_{\mathrm{s}} d} - 4\pi M_{\mathrm{s}} + \frac{2A}{M_{\mathrm{s}}} q^2\right) \\ & \times \left(H - H_{\mathrm{crit}} + \frac{2A}{M_{\mathrm{s}}} q^2\right) - \frac{16\pi k_{\mathrm{s}} (q_{\parallel}/q)^2}{d} \,. \end{aligned} \tag{3.73}$$

In the case where $q_x \ll q_{\parallel}$, i.e. $\theta \approx \pi/2$, which describes the dipolar-dominated mode, the right-hand side of (3.73) will be positive only if $Aq^2 \approx Aq_y^2 > 2\pi M_{\mathrm{s}}^2$. This states that the dipolar mode will become soft at the critical field unless the exchange energy contained in the mode is greater than the demagnetizing energy. This behavior is illustrated in Fig. 3.2 where the spin wave frequency is plotted as a function of the applied in-plane field for $A = 0$ and for $A = 2.85 \cdot 10^{-6}$ erg/cm. The parameters are appropriate for a 6-Å-thick Co film. Note that the $A = 0$ mode goes soft near $\mathrm{H}_{\mathrm{crit}} = 1.43$ kOe, and there exists a range of fields between about 1.41 and 1.53 kOe, where there is no surface mode. This is due to the influence of the last term in (3.73), which depends both on the perpendicular interface anisotropy and the propagation vector of the mode. When $q_{\parallel} = q_y = 0$ the $A = 0$ mode goes soft exactly at $\mathrm{H}_{\mathrm{crit}}$. The most interesting feature, however, is that the $A \neq 0$ mode has a sharp minimum at $\mathrm{H}_{\mathrm{crit}}$, but does not vanish as the $A = 0$ mode does. Then, according to (3.73), the exchange energy contained in the mode is larger than the demagnetizing energy. Of particular interest is the mode character. It is well known that for a perpendicularly magnetized film no dipolar mode can exist [3.83]. Since near $\mathrm{H}_{\mathrm{crit}}$ the mode energy is strongly affected by exchange interaction the mode character is bulk-mode-like. In the

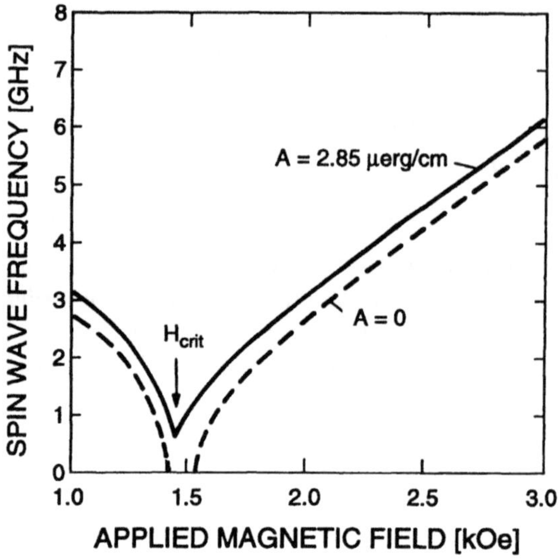

Fig. 3.2. Frequency as a function of the applied field for the surface mode of a 6-Å-thick Co film with (full line) and without (dashed line) exchange. The interface anisotropy constant is $k_s = 0.4\,\mathrm{erg/cm^2}$, the volume anisotropy constant has been set to zero for simplicity. The wavevector is $q_{\parallel} = 1.73 \cdot 10^5\,\mathrm{cm^{-1}}$ (adapted from [3.82])

range of canted magnetization ($H < H_{\mathrm{crit}}$) one finds that a dipolar surface mode exists for some orientation angles $\theta \neq \pi/2$. For fields much larger than $\mathrm{H_{crit}}$ the resulting mode is a surface mode again.

3.2.4 Spin Waves in Magnetic Multilayers

In the case of multilayered structures we have the problem of finding first the static equilibrium orientations of the layer magnetizations before calculating the spin wave frequencies. Due to interface anisotropies and exchange coupling effects (Sect. 3.2.1.2) the static equilibrium direction might differ from the bulk direction, in particular for $A_{12} < 0$, i.e., for antiferromagnetic interlayer coupling. The direction of magnetization can be obtained by solving the equations of motion (3.1) and the magnetostatic Maxwell equations, (3.3) and (3.4), together with the boundary conditions (3.35–39) for time-independent $\boldsymbol{M}$ and $\boldsymbol{H}$. It should be noted that in the general case the direction of the magnetization is a function of the position in each magnetic layer. Once we have solved the static problem, all time-independent terms contained in (3.1–5) cancel. The remaining calculations are straightforward albeit algebraically and/or numerically extensive. A discussion of the problem of the spin wave dispersion in two coupled layers, including the field dependent ground state configuration determined by the competition between anisotropies, interlayer exchange and applied field is reported by *Stamps* [3.84].

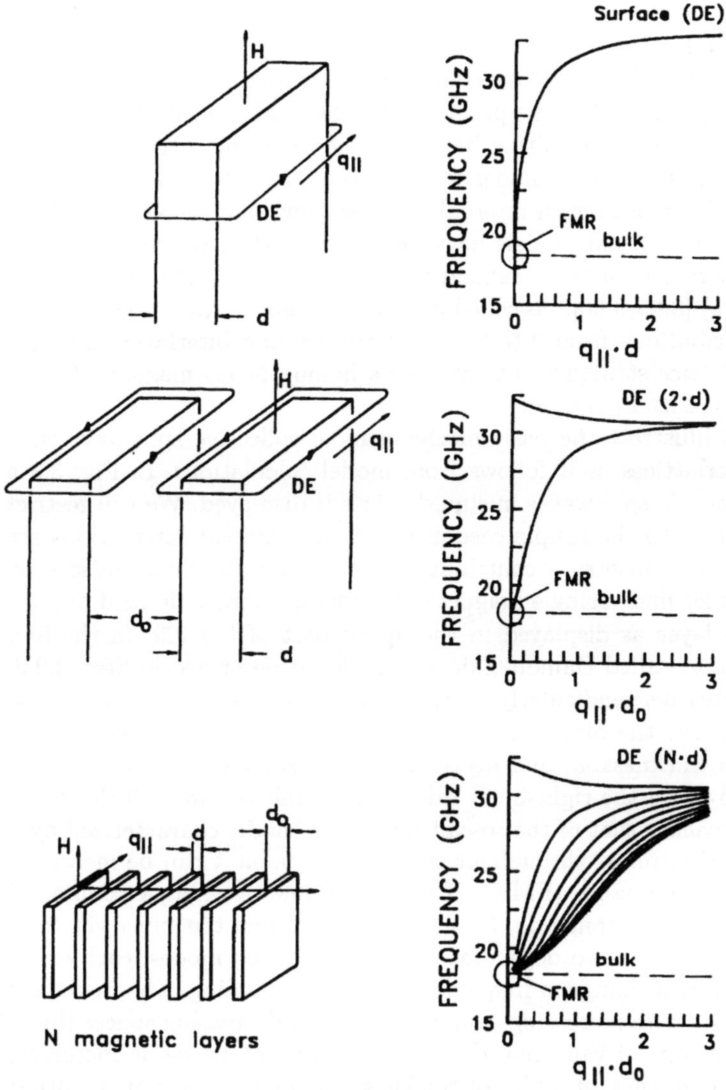

Fig. 3.3. *Top:* The Damon–Eshbach (DE) mode in a single layer. *Middle:* Coupling scheme of dipolar spin waves in a double layer. *Bottom:* Coupling scheme of dipolar spin waves in a multilayer. FMR denotes the $q = 0$ mode frequency, as it can be measured using ferromagnetic resonance. For the multilayer structure $q_{\|} \cdot d = 1$ is assumed

For multilayers the reader is referred to [3.50, 3.51, 3.52, 3.53, 3.54, 3.55] for more details. Since there are six partial solutions for $\boldsymbol{m}(t)$ and $\boldsymbol{h}(t)$ in each magnetic layer and two in each nonmagnetic layer, the dimension of the boundary condition determinant is $((6{+}2)N{+}2)\times((6{+}2)N{+}2)$ for N magnetic layers within the superlattice stack. However, the numerical expense of evaluating these rather large matrices can be reduced in various ways: (i) if only dipolar modes are considered the weak exchange energy contributions can be dropped and the boundary condition matrix reduces in dimension to $((2{+}2)N{+}2)\times((2{+}2)N{+}2)$. The problem can be solved using efficient numerical tools to evaluate the resulting band matrices or using a transfer matrix method; (ii) for systems with modes with not too large exchange energy contributions, i.e. for modes with internal fields varying only slowly from layer to layer, effective-medium models have been proposed in which parameters are averaged across a suitably defined unit cell (e.g., a double layer within the superlattice) [3.85, 3.86]. So-called effective susceptibilities are calculated including contributions from interface anisotropies and interlayer coupling, and the superlattice structure is treated as a homogeneous magnetic film of effective, renormalized parameters.

Let us first illustrate the basic mechanisms of collective spin wave excitations in superlattices as it follows from model calculations. In Fig. 3.3 a sketch of interacting spin waves in superlattices is displayed. We will restrict our considerations to the simple case in which only dipolar interactions are considered and any anisotropy contributions apart from the shape anisotropy are zero. Consider first a single magnetic layer with a magnetic field applied parallel to the layer as displayed in the upper part of Fig. 3.3. In the film there exists the so-called Damon–Eshbach mode, as described in Sect. 3.2.3, which propagates perpendicularly to the applied field with a defined sense of revolution about the film. The frequency ω of this mode decreases with decreasing film thickness, d, and decreasing wavevector parallel to the surface, $q_{\|}$, as shown on the right-hand side. In the limit of $q_{\|}d \to 0$ the mode frequency approaches that of the so-called uniform mode, characterized by a constant precession phase throughout the entire film, as it can be measured by ferromagnetic resonance (FMR). If there are now two magnetic films of same thickness at a distance d_0, thin enough to couple the films via their dipolar stray fields, the two degenerate Damon–Eshbach modes, each existing on each film, will change in frequency due to the coupling. The frequency splitting increases with increasing coupling, i.e., with decreasing spacer thickness, d_0. In the limit of vanishing d_0, the frequency of one mode increases, tending to the frequency of a film of thickness $2d$ whereas that of the other mode tends to the frequency of the uniform mode. Now in the case of a superlattice consisting of N magnetic layers separated by $N-1$ spacer layers (see Fig. 3.3, bottom) the frequency degeneracy of the N Damon–Eshbach modes is lifted, and, in the case of large N a band of so-called collective spin wave modes is formed. Out of the N modes of the band, one mode (highest-frequency mode in Fig. 3.3, bottom) is characterized as a surface mode of the total multilayer stack, with the mode energy (precession amplitude) localized near the surface of the stack, and traveling about the total stack with a well-defined sense of revolution. The remaining modes have, depending on

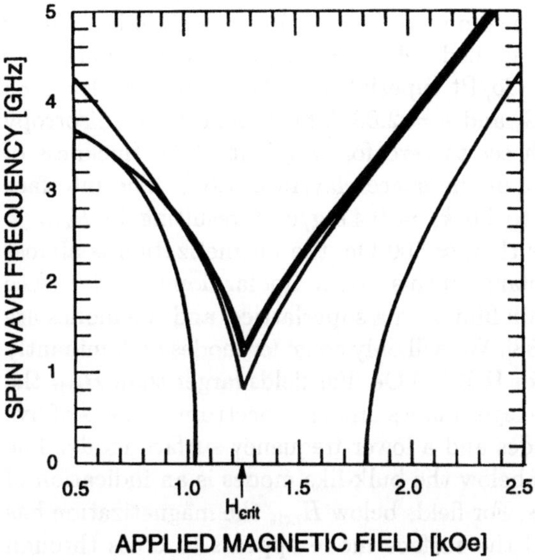

Fig. 3.4. Frequencies of spin wave modes for a six-bilayer stack as a function of the in-plane applied field H. The parameters are appropriate for Co and are given in the text. The Co layers are 8.8 Å thick and the spacer layers are 7.6 Å thick (adapted from [3.53])

the wavevector components perpendicular to the stack, both surface-mode- and bulk-mode-like character to a greater or lesser degree.

3.2.4.1 Large Perpendicular Anisotropies. We will now consider the case that multilayer structures have large perpendicular anisotropies. We assume that they are large enough to compensate for the shape anisotropy and to turn the direction of magnetization out of plane. This is for instance achieved by choosing systems with a large perpendicular interface anisotropy constant, k_s, and with a small magnetic layer thickness, d, such that the effective out-of-plane anisotropy contribution, K_{eff}, is positive [see (3.66–68)].

We will consider the case that the external magnetic field is applied parallel to the layers. With increasing field strength the direction of magnetization, $\boldsymbol{M}_s$, is increasingly tilted into the layer planes until a critical field strength, $H_{\text{crit}} = 2K_{\text{eff}}/M_s$, is reached above which the directions of magnetization and external field are co-linear.

The spin wave frequencies are very dependent on the out-of-plane angle θ between $\boldsymbol{M}_s$ and the $\hat{x}$-axis (stacking axis) [3.53]. With increasing angle θ, caused by an increasing applied field, the spin wave frequencies decrease and some modes may even go soft in the vicinity of the critical field strength, H_{crit}. For $H > H_{\text{crit}}$, i.e. in the regime of $\boldsymbol{M}_s$ parallel to $\boldsymbol{H}$, the spin wave frequencies increase quasi-linearly with further increasing field. This behavior is displayed in Fig. 3.4 where the spin wave frequencies for a six-bilayer stack are shown as a function of the applied field. The magnetic parameters were extracted from experiments on Co/Pt superlattices [3.87], and are $A = 2.85 \cdot 10^{-6}$ erg/cm, $4\pi M_s = 14.5$ kG, and $g = 2.03$. The hcp-Co bulk anisotropy constants K_1 and K_2 are both set to zero for simplicity. The thickness of the Co layers is 8.8 Å and that of the spacer layers is 7.6 Å. The interface anisotropy constant is chosen to be $k_s = 0.4\,\mathrm{erg/cm^2}$ resulting in $H_{crit} = 1.26$ kOe. For applied fields less than ≈ 500 Oe, the magnetization is almost completely normal to the film plane. In this case a dipolar-dominated surface mode cannot exist in a single thin film or in a superlattice, and the modes are mostly of the exchange-type [3.88]. We will only consider modes of dominantly dipolar character, which exist for H $\gtrsim$ 500 Oe. For fields larger than H_{crit} the magnetization is forced into the layer planes and the spectrum consists of five nearly degenerate bulk-like modes and a lower-frequency surface mode. The fact, that the surface mode lies below the bulk-like modes is an indication of large perpendicular anisotropies. For fields below H_{crit} the magnetization has an out-of-plane component and the surface mode appears to cross through and rise above the bulk band as the field is lowered. Near H_{crit} the bulk modes take their minimum values while the surface mode goes completely soft. Such a softening can often be associated with a surface magnetic phase transition in the spin structure [3.89, 3.90]. In the example of Fig. 3.4 this is a strong indication for the direction of magnetization not to be constant but to vary across the stack in order to minimize the net demagnetization energy of the structure.

3.2.4.2 Interlayer Exchange Coupling. So far only dipolar interactions between magnetic layers within the superlattice stack have been considered. We will now discuss the additional contribution of interlayer exchange interactions to the spin wave properties, since they enter the boundary conditions in the spin wave calculations (Sect. 3.2.1.1).

A considerable influence of interlayer exchange interaction of the spin wave frequencies exists only if the spacer layers are thin enough ($\lesssim 10$ Å), since the interaction decays rather fast with increasing spacer thickness. Depending on the spacer material, the interlayer coupling is ferro- or antiferromagnetic or it even oscillates as a function of the spacer thickness as demonstrated further below in this section.

In the presence of interlayer exchange coupling in superlattices, all but the stack surface mode of the dipolar collective modes are converted into exchange modes, which in the full coupling limit become the so-called standing spin waves of the total superlattice stack [3.50, 3.51]. This new type of collective exchange-dominated modes was predicted by the model outlined in Sect. 3.2.1 [3.50, 3.51]. Figure 3.5 shows the calculated frequency dependence of the modes for Co/Pd multilayers of 9 periods as a function of the individual layer thickness, assuming that the thicknesses of all magnetic (d_{Co})

and nonmagnetic ($d_{\rm Pd}$) layers are the same. For this calculation the parameters of Co listed in the figure caption have been used; they were obtained by Brillouin light scattering and SQUID magnetometry measurements on Co/Pd multilayer samples prepared with the same specifications [3.91, 3.92]. For $d_{\rm Co} = d_{\rm Pd} \gtrsim 70\,\text{Å}$ dipolar collective modes are seen to exist in the frequency region between 22 and 28 GHz (see Fig. 3.5). The stack surface mode is well separated from the remaining 8 modes, which form a narrow band of collective excitations. For $d_{\rm Co} = d_{\rm Pd} > 130\,\text{Å}$ the first standing spin wave, which is an exchange mode of each single layer, is obtained (decreasing from 100 GHz to 59 GHz in Fig. 3.5) with its characteristic $1/d_{\rm Co}^2$ dependence. For thinner layers, $d_{\rm Co} = d_{\rm Pd} < 50\,\text{Å}$, all collective dipolar modes, except the highest frequency one (stack surface mode), increase in frequency and cross the stack surface mode due to the onset of interlayer exchange interaction. An analysis of the mode properties shows that these modes are dominated by the exchange energy. In order to model the interlayer exchange coupling strength as a function of the Pd spacer layer thickness, $d_{\rm Pd}$, we have assumed in Fig. 3.5 that A_{12} decreases exponentially with increasing $d_{\rm Pd}$. That is $A_{12}(d) = A_{12}^0 \exp(-d/d_0)$ with $A_{12}^0 = 10\,\text{erg/cm}^2$ and the decay constant $d_0 = 10\,\text{Å}$. In the crossing regime the modes are hybridized, exhibiting a very small mode repulsion which can only be resolved on the scale of Fig. 3.5 for the crossing of the stack surface mode and the lowest-frequency bulk mode.

3.2.5 Nonlinear Excitations.

The equation of motion (3.1) is inherently nonlinear since both the magnetization, $\boldsymbol{M}$, and the effective field, $\boldsymbol{H}_{\rm eff}$, contain time-dependent components $\boldsymbol{m}(t)$ and $\boldsymbol{h}(t)$. Therefore, for high precession amplitudes, terms in the equation of motion, which contain products of components of $\boldsymbol{m}$ and $\boldsymbol{h}$, which cause nonlinear behavior, cannot be neglected. A number of nonlinear phenomena result, which can be tested in BLS experiments. Among them there are parallel pumping [3.85, 3.86, 3.87, 3.93, 3.94, 3.95], subsidiary absorption [3.96, 3.97] spin wave beam shaping [3.98, 3.99] (self-channeling and self-focusing), formation of solitons and two-dimensional spin wave bullets [3.100], and collisions experiments of nonlinear spin waved pulses [3.101]. Most of the nonlinear excitations exist only above a microwave threshold power, $P_{\rm th}$, since external stimulation must overcome spin wave damping.

In the parallel pumping process a microwave field of frequency ω is aligned parallel to the static field. Consider a spin wave of frequency $\omega/2$. The precession is elliptical if the wavevector is not aligned with $\boldsymbol{M}$. This generates a "wobble" in the z-component of the magnetization of frequency ω. The microwave field now can couple to this component. The mode is excited if the microwave input power exceeds the spin wave relaxation rate. The actual threshold power is a function of wavevector and applied field and also a complicated function of anisotropy and damping. Mode coupling, in particular to standing spin waves, further complicates the behavior. In the quantum

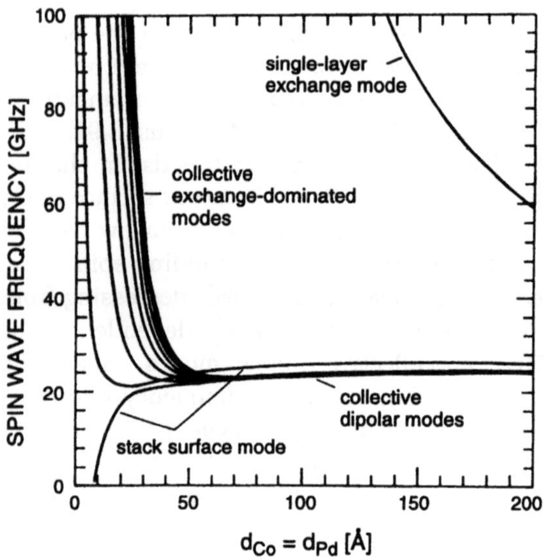

Fig. 3.5. Calculated spin wave excitations in a superlattice of 9 periods. For the magnetic layer the parameters of sputtered Co-films are used (see [3.91]): The saturation magnetization is $4\pi M_s = 14.5\,\mathrm{kG}$, the exchange constant is $A = 2.85 \times 10^{-6}$ erg/cm and the g-factor is $g = 2.03$. For the sum of the two volume anisotropy constants, K_1 and K_2, the value of $K_1 + K_2 = 3.05 \times 10^6$ erg/cm^3 is used and for the interface anisotropy constant $k_s = 0.4\,\mathrm{erg/cm^2}$ is used (from [3.212])

mechanical formulation, a (virtual) magnon of frequency ω is created by a $\boldsymbol{q} = 0$ microwave photon, which decays into magnons with frequency $\omega/2$ and wavevectors $\pm\boldsymbol{q}$. Beam shaping effects appear for propagating plane waves due to the small but nonzero transverse wavevector component modified by nonlinearity. If the input power exceeds a threshold value determined implicitly by a Lighthill criterion [3.102] (compensation of dispersion and/or dissipation by a term proportional to the square modulus of the precession amplitude) self channeling, i.e., the decay of the plane wave front into channels of increased amplitude, wave collapse, i.e., the wave amplitude becomes locally infinite within finite propagation time, or self-focusing, i.e., concave bending of the wave front resulting in one or more focal points may appear.

3.3 The Light Scattering Cross Section

Brillouin light scattering is a spectroscopic method for investigating inelastic excitations with frequencies in the GHz regime. As illustrated in Fig. 3.6 photons of energy $\hbar\omega_L$ and momentum $\boldsymbol{q}_L$ interact with the elementary quanta of spin waves $(\hbar\omega, \boldsymbol{q})$, which are the magnons. The scattered photon gains an increase in energy, $\hbar(\omega_L + \omega)$, and wavevector, $\hbar(q_L + q)$, if a magnon is annihilated. A magnon can also be created by an energy and wavevector transfer

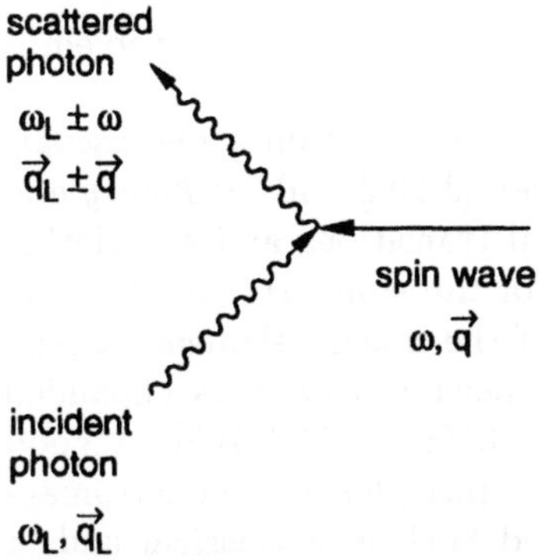

Fig. 3.6. Scattering process of photons from spin wave excitations (magnons)

from the photon, which in the scattered state has the energy $\hbar(\omega_L - \omega)$ and wavevector $\hbar(\boldsymbol{q}_L - \boldsymbol{q})$. For finite temperatures ($T \gg \hbar\omega/k_B \approx 5\,\mathrm{K}$) both processes have about the same probability. In a classic treatment the scattering process can be understood for many materials as follows: Due to the spin-orbit coupling a phase grating is created in the material, which propagates with the velocity of the spin wave. Light is Bragg-reflected from the phase grating with its frequency Doppler-shifted by the spin wave frequency.

In the linear response regime a magnetic material responds to the presence of an external optical electric field by developing an electric polarization, $\boldsymbol{P}$, which is proportional to the electric field, $\boldsymbol{E}$. To terms linear in magnetization, the relation between $\boldsymbol{P}$ and $\boldsymbol{E}$ is

$$4\pi \boldsymbol{P} = (\epsilon_{11} - 1)\boldsymbol{E} + K/M_s(\boldsymbol{E} \times \boldsymbol{M}) \tag{3.74}$$

with the frequency dependent complex coefficients ϵ_{11} (dielectric constant) and K (magneto-optic coefficient), which depend on the actual details of the band structure.

Other mechanisms of inelastic one-magnon light scattering exist. For example, in EuTe a one-magnon Brillouin light scattering process due to s–f exchange interaction has been found [3.103]; it exists in the canted antiferromagnetic state of EuTe for spin wave modes with a longitudinal component of the dynamic magnetization $\boldsymbol{m}(\boldsymbol{r}, t)$.

The calculation of the cross section is now a three-step problem: (i) The distribution of the electromagnetic field inside the interaction volume is calculated; (ii) the intrinsic cross section problem is solved, i.e., the electromagnetic scattered wave is determined, and (iii) the amplitude of the scattered wave outside the sample is calculated. Steps (i) and (iii) pose the problem of solving the Maxwell boundary conditions at the interfaces. For step (ii) the spin wave dispersion and the occupation of the magnon states, which in the classical limit are the precession amplitudes of the spin wave modes, are the input parameters. Since the scattering intensity is proportional to the square of the transverse magnetization components, step (ii) contains most of the salient features of the scattering cross section problem. The main problem is therefore to solve step (ii).

First calculations of the scattering cross section based on a microscopic model were performed by *Shen* and *Bloembergen* [3.104] and by *Fleury* and *Loudon* [3.105] by considering single-ion optical transitions and calculating the corresponding transition matrix elements of the spin–orbit interaction. To overcome the difficulty in the calculation of the matrix elements a phenomenological approach, in which the crystal permittivity, ϵ, is expanded in terms of the magnetization, was proposed [3.106, 3.107, 3.108]. A good review has been given by *Wettling* et al. [3.108]. In it, for a cubic ferromagnet, the light scattering cross section is derived both in a classical and in a quantum-mechanical approach in terms of complex magnetooptic effects including absorption of light within the sample. *Camley* and *Mills* have calculated, for the first time, the light scattering spectrum from surface and bulk spin waves in a semi-infinite ferromagnet including both exchange and dipolar coupling using a Green's function approach [3.109, 3.110, 3.111]. This work was later extended to multilayers [3.112]. The experimentally often observed Stokes/anti-Stokes asymmetry was numerically reproduced by *Camley, Grünberg* and *Mayr* [3.113] by considering the off-diagonal elements of the spin-spin correlation function in the presence of light absorption. *Cottam* calculated the full BLS cross section in thin ferromagnetic films including multiple reflection effects of the incident and the reflected light beams using a Green's function approach [3.114]. A tensorial Green's function theory was developed for perpendicular ferromagnetic films with different boundary conditions at both interfaces by *Cottam* and *Slavin* [3.115, 3.116, 3.117]. A review of their method is given in [3.44]. A different approach for calculating light scattering intensities has been presented by *Cochran* and coworkers. It is based in directly calculating the fluctuating transverse magnetization components from the equipartition law [3.39, 3.118, 3.119, 3.120, 3.121]. BLS results for thin, perpendicularly magnetized films as well as films exchange coupled to a bulk ferromagnet [3.121] were reported.

Spin wave excitations modulate the permittivity, and therefore they create fluctuating terms in the polarization

$$\delta P_i(\boldsymbol{r},t) = 4\pi \sum_j \delta\epsilon_{ij}(\boldsymbol{r},t) E_j^{\mathrm{I}}(\boldsymbol{r},t) \,, \tag{3.75}$$

with $\delta\epsilon_{ij}$ the fluctuating term of the permittivity caused by the spin waves and $\boldsymbol{E}^{\mathrm{I}}$ the incident electric field with frequency ω_{I}. Since $\delta\epsilon_{ij}(\boldsymbol{r},t)$ varies with the spin wave frequency, ω, and $E_j^{\mathrm{I}}(\boldsymbol{r},t)$ varies with the frequency of the incident light, ω_{I}, and since further the intensity of the scattered light is proportional to the square of the second derivative of the polarization with respect to time, terms for the scattered wave with frequency shifts $\omega_{\mathrm{S}} = \omega_{\mathrm{I}} \pm \omega$ are created. The differential scattering cross section $\mathrm{d}^2\sigma/\mathrm{d}\Omega d\omega_{\mathrm{S}}$, i.e., the number of photons scattered into the solid angle $\mathrm{d}\Omega$ in the frequency interval between ω_{S} and $\omega_{\mathrm{S}}+\mathrm{d}\omega$ per unit incident flux density is given by [3.108, 3.122]

$$\frac{\mathrm{d}^2\sigma}{\mathrm{d}\Omega \mathrm{d}\omega_{\mathrm{S}}} = \frac{\omega_{\mathrm{I}}^4}{32\pi^2 c^4} \sum_{ijkl} e_{\mathrm{I},i} e_{\mathrm{S},j} e_{\mathrm{I},k} e_{\mathrm{S},\ell} \left\langle \delta\epsilon_{ij}^*(\boldsymbol{k}) \delta\epsilon_{kl}(\boldsymbol{k}) \right\rangle_\omega , \tag{3.76}$$

with the wavevector $\boldsymbol{k} = \boldsymbol{k}_{\mathrm{I}} - \boldsymbol{k}_{\mathrm{S}}$ and the frequency $\omega = \omega_{\mathrm{I}} - \omega_{\mathrm{S}}$. The correlation function is given by

$$\left\langle \delta\epsilon_{ij}^*(\boldsymbol{k}) \delta\epsilon_{kl}(\boldsymbol{k}) \right\rangle_\omega = \int \mathrm{d}t \mathrm{d}^3(\boldsymbol{r}_2 - \boldsymbol{r}_1)\, \mathrm{e}^{\mathrm{i}(\omega t - \boldsymbol{k}(\boldsymbol{r}_2 - \boldsymbol{r}_1))} \times \left\langle \delta\epsilon_{ij}^*(\boldsymbol{r}_1, t) \delta\epsilon_{kl}(\boldsymbol{r}_2, 0) \right\rangle , \tag{3.77}$$

with $\langle ... \rangle$ the statistical average.

Equations (3.76,3.77) contain all salient features of the scattering process, e.g., the conservation of energy and momentum. The problem is now reduced to the calculation of the appropriate correlation functions. For surface scattering, (3.77) can be used taking into account that the component of $\boldsymbol{k}$ perpendicular to the surface is imaginary and that the volume integration is replaced by an integration across the surface.

Experimentally, an asymmetry in the intensities of the Stokes/anti-Stokes peaks is often observed which can be very large. Possible causes are (i) nonreciprocity in case of Damon–Eshbach type surface waves, (ii) interference between different magnetooptic effects (e.g., those, which are linear and quadratic in magnetization), (iii) contributions of off-diagonal elements of the spin-spin correlation function in the presence of strong optical absorption, and iv) different thermal probability for the Stokes and anti-Stokes process at low ($\lesssim$ 2K) temperatures. Nonreciprocity of the Damon Eshbach mode may cause a Stokes/anti-Stokes asymmetry due to the different precession amplitudes at the film surface of the two modes propagating on the front and counterpropagating on the rear surface, if the light scattering process takes part at the film surface. In particular for light scattering from a semi-infinite medium the Damon–Eshbach mode is only obtained in either the Stokes or anti-Stokes side of the spectrum, depending on the direction of the external field. For ultrathin films and strong light absorption, the asymmetry is usually still obtained due to (iii), although the decay length of the precession amplitude perpendicular to the films is now much larger than the film thickness. The Stokes/anti-Stokes asymmetry, together with the rotation of the polarization plane of the light, is often used to discriminate spin wave excitations from phonon signals in a BLS experiment.

The correlation function can be calculated either by using the fluctuation–dissipation theorem [3.105,3.117] or, alternatively, by using a thermodynamical approach, i.e. the cross section is directly related to the thermal amplitude of the respective normal spin wave mode. Often the problem can be simplified by assuming that the optical fields do not depend on the distance from the film surface, which is valid in particular for scattering from ultrathin films. Using this approach, *Cochran* and coworkers calculate the cross section for a number of geometries [3.39,3.118,3.119,3.120,3.121]. In particular the two limiting cases, that i) the lateral dimensions of the scattering objects in the film are small compared to the wavelengths of the scattering light, λ_S (retardation effects ignored), and ii) the lateral dimensions are large compared to λ_S [3.39]. *Cochran* finds that in both cases approximately the same scattering cross section is found which agrees with experimental results within an order of magnitude. The characteristic difference is that for the first mechanism the mode frequencies are independent of the wavevector.

Although the solution of the scattering cross section problem seems to be straightforward, albeit numerically extensive, a full understanding of this phenomenon has still not been reached. For instance, *Moosmüller* et al. [3.123] report an experiment, in which they find oscillations of the Stokes/anti-Stokes ratio in permalloy films as a function of the film thickness which cannot be reproduced by any of the existing model calculations.

The amount of information which can be extracted from the scattering intensities based on existing knowledge, is often fairly limited. For a quantitative comparison, all the experimental factors which influence the measured intensity need to be determined and the full optical problem must be solved. The Stokes/anti-Stokes ratio is of limited value as well, since a number of different mechanisms, which cannot be easily discriminated against each other, contribute to it.

3.4 Instrumentation

The tandem Fabry–Pérot interferometer developed by J.R. Sandercock can be used as a highly sensitive spectrometer with a frequency resolution in the sub-GHz regime and a contrast of better than 10^{10} [3.32,3.124]. Therefore it is best suited for studying spin wave excitations in layered magnetic structures with monolayer sensitivity.

The frequency selecting element is an etalon consisting of two parallel optical mirror plates (flatness better than $\lambda/200$), of rather high reflectivity (typically 92...96%). The etalon transmits light of wavelength λ if the plate distance is an integer multiple of $\lambda/2$. In conventional interferometry using one etalon the analysis of inelastic excitations is hampered by the ambiguous assignment to the appropriate transmission order, since the transmission is periodic in $\lambda/2$ in the mirror plate spacing. These ambiguities are avoided in a tandem arrangement.

The setup is schematically shown in Fig. 3.7. The light of a frequency stabilized laser ($\Delta\nu$ = 20 MHz), which is typically an Argon$^+$-ion laser (λ = 514.5 nm) is focused onto the sample with an objective lens. The light backscattered from the sample (elastic and inelastic contributions) is collected by the same objective lens and sent through a spatial filter for suppressing background noise before entering the tandem interferometer. The frequency selected light transmitted by the interferometer is detected by a photomultiplier or an avalanche photodiode after passing through a second spatial filter for additional background suppression. A prism or an interference filter

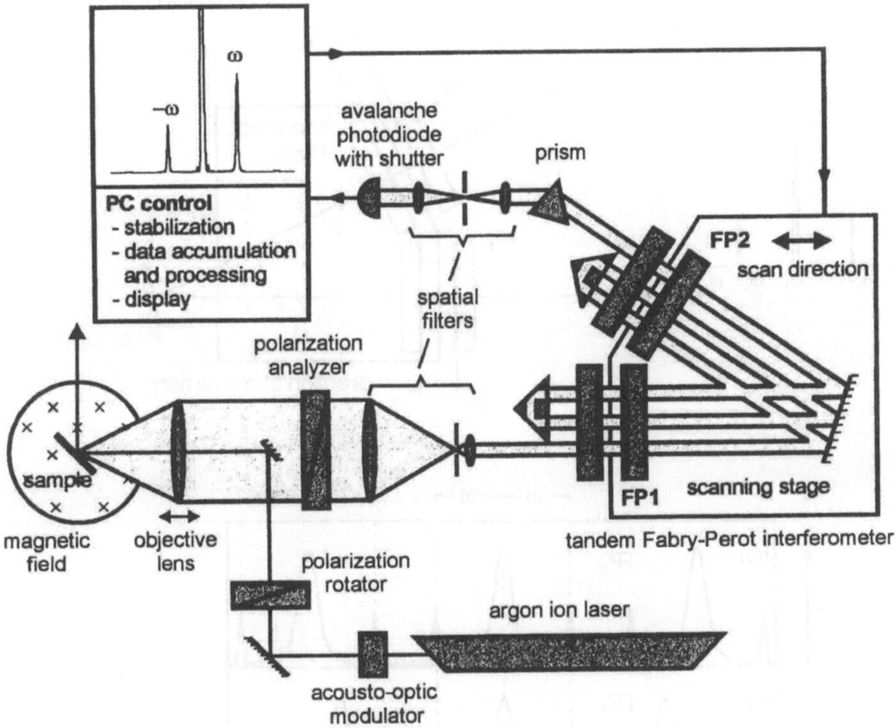

Fig. 3.7. Schematic view of a Brillouin light scattering setup

between the second spatial filter and the detector serve for suppression of inelastic light from common transmisssion orders outside the frequency region of interest. A computer collects the photon counts and displays the data.

The central part of the interferometer is displayed in Fig. 3.8 (top). The light passes in series through two Fabry–Pérot etalons FP1 and FP2. Of both etalons one of the two mirrors is mounted on a common translation stage which is piezo-electrically driven. In order to illustrate the function of the tandem arrangement Fig. 3.8 (bottom) displays schematically the transmission curves of the etalon (a) FP2, (b) FP1 and (c) of both etalons in series (tandem operation) as a function of the mirror separation of the first etalon, L_1. Assuming that for a given value L_1 both etalons transmit, a change of $\lambda/2$ in L_1 puts FP1 into the next transmission order. Due to the common mounting of the movable mirrors, the change in the spacing of the second etalon is smaller by a factor of cos Θ with Θ the angle between the optical axes of the two interferometers as displayed in the figure. Thus FP2 does now not transmit; the transmission maxima of both etalons lie at different values of L_1. The same arguments account for inelastic excitations, which are transmitted only if they belong to the common transmission order. The inelastic signal represents closely the scattering cross section of the sample.

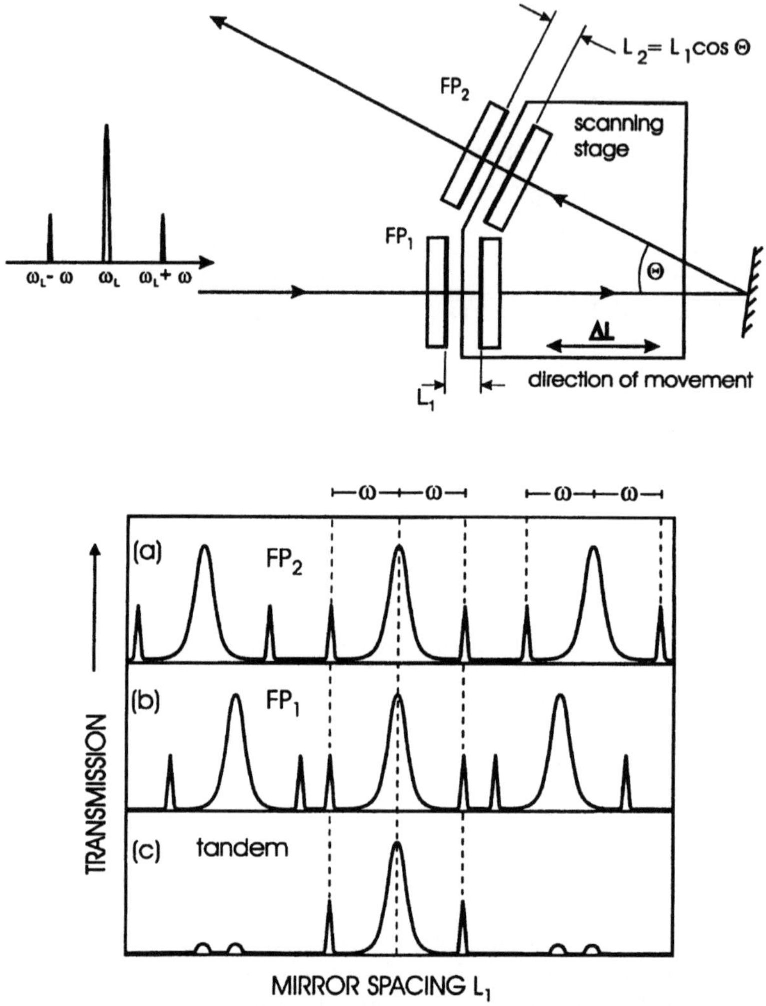

Fig. 3.8. Schematic view of the operation of a tandem Fabry–Pérot interferometer. *Top:* view of the light pass; *bottom:* transmitted intensity of first (FP1) (b), second (FP2) (a) and both etalons in series (c). The inelastic contributions due to an inelastic light scattering process are indicated by the frequency shifts ω

For an experimental realization, a flatness of the mirror surfaces of better than $\lambda/200$ and a parallelism of $\lambda/100$ of the two mirrors of each etalon are necessary. To maintain the latter, a sophisticated active stabilization of the mirror alignment is mandatory: it is performed by analog feedback circuits or by computer control [3.125]. In order to obtain the high contrast necessary to detect the weak inelastic signals, the light is sent through both etalons several times using a system of retroreflectors and mirrors (see Fig. 3.7).

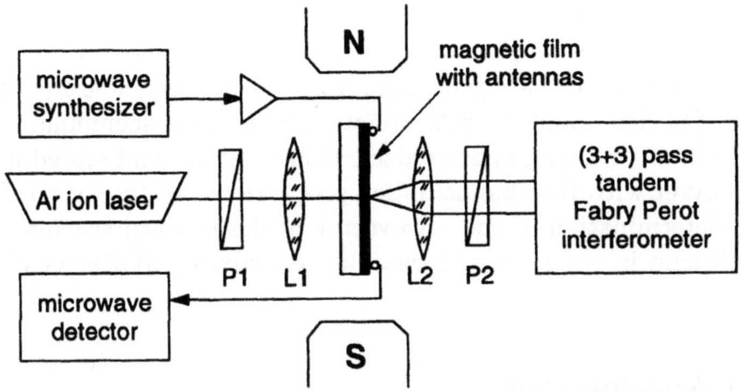

Fig. 3.9. Schematic sketch of the combined BLS-FMR setup. P1, P2: polarizers in crossed orientation, L1, L2: focusing lenses

Modern interferometers [3.32, 3.124] are mostly set up in the (3+3)-pass arrangement. Special measures are taken to protect the detector from overload while scanning through the elastic peak. This is achieved by using an acoustooptic modulator or by a shutter system. Data collection is performed by a personal computer or by a multichannel analyzer.

Apart from studying thermally excited spin wave modes, the interferometer can be used to detect microwave excited modes as well [3.93, 3.94, 3.98, 3.99, 3.100, 3.101, 3.126, 3.127, 3.128, 3.129, 3.130, 3.131, 3.132]. These modes are excited either in a microwave cavity or in a planar structure using microstrip antennas. Figure 3.9 shows a typical setup. The experiments are usually performed in forward scattering geometry, since the wavevector of the excited modes is either zero in ferromagnetic resonance (FMR) geometry (cavity) or very small in planar wave guide geometries ($\approx 10^2\,\mathrm{cm}^{-1}$). The optical detection of spin wave resonances offers advantages over a conventional FMR setup; in particular, a spatial resolution obtainable because of the small size of the laser focus at comparable sensitivity. The dynamic range of such a combined BLS-FMR setup can be very high, 60 dB has been reported [3.99]. The application of a combined BLS-FMR setup for performing the determination of the exchange coupling between two Fe films across a Cr spacer layer at liquid He temperature is reported by *Demokritov* [3.131].

Due to the spatial resolution the mode profile in a cavity [3.94, 3.133, 3.134, 3.135] or in a planar wave guide [3.98, 3.99, 3.126, 3.132] can be determined. Recently *Bauer* et al. reported the construction of a space- and time resolving BLS spectrometer [3.100, 3.101]. Time resolution is achieved by measuring the elapsed time between the launch of a spin wave pulse by applying a microwave pulse to the antenna, and the arrival of an inelastically scattered photon at the detector.

In a number of applications a combined BLS-FMR setup has been used to determine nonlinear spin wave properties (Sect. 3.5.6). Typical phenomena to be studied are parametric excitations, subsidiary absorption and nonlinear anomalies in the spin wave propagation. In order to measure the spin wave intensity as a function of the applied field, in analogy to a field scan in ferromagnetic resonance, a time gate procedure has been implemented in the instrument [3.94,3.126]. The counts from the photomultiplier are accumulated for the inelastic peak, using a gate synchronized with the scan, and stored in a multichannel analyzer with the channel index proportional to the external field. The precise determination of the wavevector of the forward scattered light using a diaphragm in the scattered beam is reported by *Wilber* et al. [3.94].

3.5 Selected Applications

3.5.1 Determination of Magnetic Anisotropies, Reorientation Transitions

One of the fundamental open questions in thin film magnetism is the origin, size and symmetry of magnetic anisotropies, since many of the magnetic properties in ultrathin films are inherently affected by magnetic anisotropies.

As outlined in Sect. 3.2.2 anisotropies reflect, to a high degree, the symmetry as well as symmetry breaking mechanisms present in thin films. However, since anisotropy constants are simply the coefficients of an expansion of the free magnetic energy density with respect to irreducible representations of space symmetries, many mechanisms of sometimes very different origin acting on the direction of magnetization may be subsumed in one common anisotropy constant; an example are the surface and volume anisotropy constants of the same symmetry in a particular film. This makes the analysis and a theoretical modeling of magnetic anisotropies sometimes a difficult task.

3.5.1.1 Fe/W(110): In-Plane Switching of the Easy Axis of Magnetization. We will first discuss a situation often found in surface magnetism, which demonstrates the counteracting influences of interface and volume anisotropies. We consider the case of a (110)-oriented Fe film of thickness d, which has cubic crystallographic symmetry [3.58,3.136,3.137,3.138,3.139]. Therefore, the film exhibits a two-fold symmetry about the film normal. We assume, that the magnetization lies in the film plane. Thus we can neglect the Θ-dependence of the free anisotropy energy.

The free energy of the system is:

$$E = -M_s H \cos(\phi - \phi_H) + \frac{K_1}{4} \sin^2\phi(1 + 3\cos^2\phi) + K_p^{(2)} \cos^2\phi + \frac{2}{d}(k_p^{(2)} \cos^2\phi + k_p^{(4)} \sin^2\phi \cos^2\phi) \,. \quad (3.78)$$

The first term on the right hand side is the Zeeman energy, the second term is the volume anisotropy (3.48), here expressed in a coordinate system

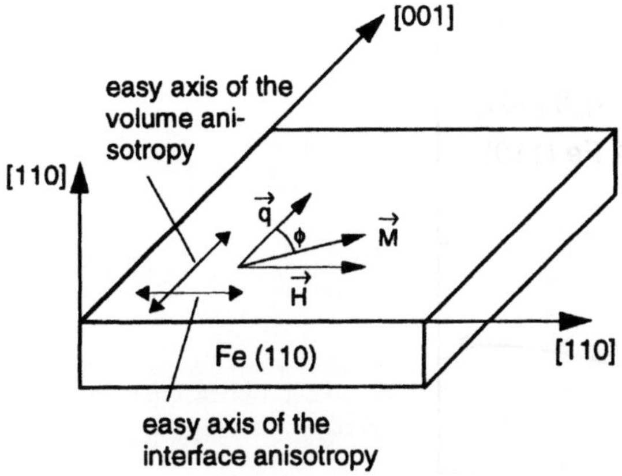

Fig. 3.10. Orientation of the easy axes of the volume and interface anisotropy contributions for Fe(110) films

oriented in the film frame (i.e., $\hat{x}_1 \parallel [001]$, $\hat{x}_2 \parallel [1\bar{1}0]$, $\hat{x}_3 \parallel [110]$), and the fourth term is the in-plane interface anisotropy in the two lowest orders, $k_\mathrm{p}^{(2)}$ and $k_\mathrm{p}^{(4)}$, of two- and fourfold symmetry about the film normal. ϕ_H is the angle between the direction of the external field H and the [001]-direction. For completeness a third term, an in-plane uniaxial volume anisotropy with the constant $K_\mathrm{p}^{(2)}$, has been added. The latter term can be caused by a magneto-elastic interaction.

An interesting case appears if $K_1 - 4K_\mathrm{p}^{(2)} > 0$: Here the in-plane [001]-direction is an easy axis for the volume anisotropy; for the interface anisotropy the perpendicular in-plane $[1\bar{1}0]$ direction is the easy axis. The situation is displayed in Fig. 3.10. Depending on the film thickness the easy axis of the magnetization is either the [001] or the $[1\bar{1}0]$ direction.

Equation (3.78) yields for $H = 0$ a critical thickness,

$$d_\mathrm{c} = \frac{8k_\mathrm{p}^{(2)}}{K_1 - 4K_\mathrm{p}}\,, \tag{3.79}$$

at which the easy axis of magnetization changes from $[1\bar{1}0]$ for $d < d_\mathrm{c}$ to [001] for $d > d_\mathrm{c}$.

Epitaxial Fe(110) films on W(110) substrates were the first epitaxial system investigated in situ in UHV by BLS [3.58]. Preceding investigations using ferromagnetic resonance [3.140], torsion oscillation magnetometry [3.69], and spin and angle resolved photoemission spectroscopy [3.141] indicated the rotation of the easy axis of magnetization in the film plane with increasing film thickness from the $[1\bar{1}0]$ axis into the [001] axis at a critical thickness d_c in the range of 80–100 Å. It was therefore of great importance to determine all relevant anisotropy contributions [3.58, 3.136, 3.139, 3.142].

For Fe(110) on W(110) the following anisotropies were found to contribute: The lowest-order cubic bulk anisotropy constant, K_1, an uniaxial,

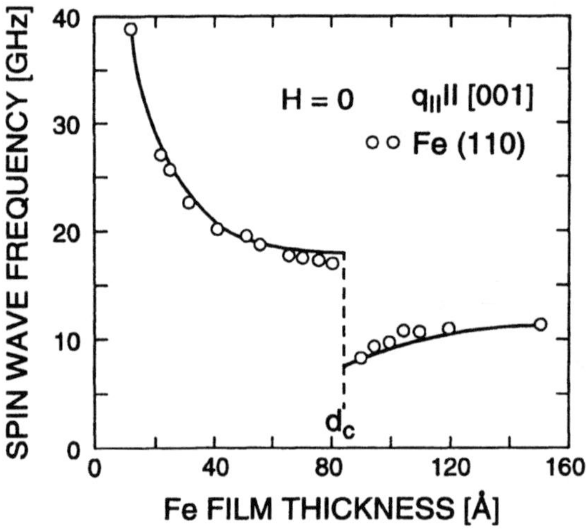

Fig. 3.11. Measured (open circles) and fitted (full line) spin wave frequencies of Fe(110) films on W(110) as a function of the Fe layer thickness (adapted from [3.138])

in-plane anisotropy due to the twofold symmetry axis of the (110) surface, $K_p^{(2)}$, an interface out-of-plane anisotropy constant, k_s, and an in-plane constant of two-fold symmetry, $k_p^{(2)}$. Also, in order to fully describe experiments for fields applied in arbitrary directions within the film plane, the next higher-order in-plane interface anisotropy constant of fourfold symmetry, $k_p^{(4)}$, is considered, see (3.78).

In order to achieve the best possible control over the film quality, all experiments were performed *in situ* in an ultra-high vacuum system. The frequencies of the spin wave modes were measured *in-situ* with BLS and then analyzed in two steps [3.136, 3.137, 3.138]. First, as outlined in Sect. 3.2, the spectra contain information about the direction of magnetization since spin wave frequencies are very sensitive to the orientation of the magnetization relative to the direction of the wavevector and the direction of the applied field. By increasing an external magnetic field applied in a magnetically hard direction, a discontinuous or continuous switching (rotation) of the direction of magnetization from the direction of the easy axis into the direction of the applied field appears and results in discontinuities or minima in the spin wave frequencies. A similar case is displayed in Fig. 3.11, where the spin wave frequencies have been measured for zero applied field with the wavevector parallel to the [001] axis. With increasing film thickness the spin wave frequency first decreases, since the contributions of the interface anisotropies decrease with $1/d$. At a critical thickness, d_c, a discontinuity is observed. Here the easy axis of magnetization switches from $[1\bar{1}0]$ to [001]. From 3.78 the ratio of the in-plane interface anisotropy constant to the volume constants is obtained.

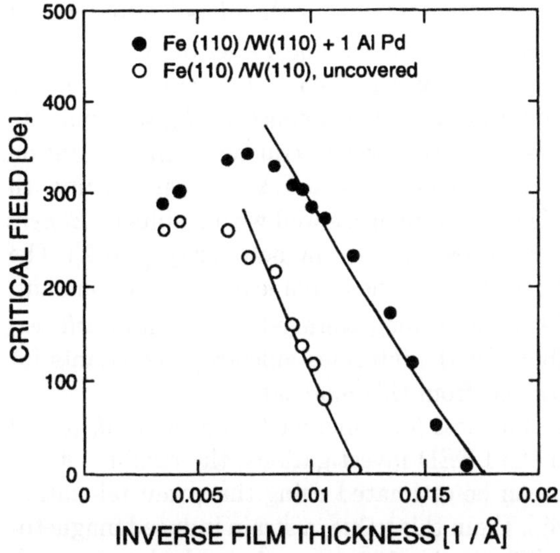

Fig. 3.12. Critical magnetic field as a function of the inverse film thickness for Fe(110) films on W(110) substrates uncovered (∘) and covered (•) by 1 ML of Pd(110). The solid lines are a model fit to the experimental data in the nearly linear regime of data points; see text (adapted from [3.136])

In a series of measurements of spin wave modes propagating in $[1\bar{1}0]$ and [001] directions the characteristic critical magnetic field at which the direction of magnetization switches, H_{crit}, was determined as a function of the film thickness, d. In Fig. 3.12, H_{crit} is plotted as a function of the inverse film thickness for Fe(110) films on W(110) both uncovered and covered with 1 monolayer (ML) of Pd. From the plots in Fig. 3.12 estimates for $K_{\text{p}}^{(2)}$, $k_{\text{p}}^{(2)}$ and $k_{\text{p}}^{(4)}$ are obtained as follows:

The intersection of the curves with the $1/d$ axis, i.e. at $H_{\text{crit}} = 0$, determines $d_{\text{c}} = 84 \pm 1$ Å for the uncovered film and $d_{\text{c}} = 58 \pm 2$ Å for the film covered with 1 ML Pd. Fitting the linear part of the data to the model, as indicated by the full lines in Fig. 3.12, yields $K_{\text{p}}^{(2)} = -8.0 \times 10^5\,\text{erg/cm}^3$ from the slope of the curve. The large slope indicates that a large amount of uniaxial bulk anisotropy is involved. From the maximum in the critical field strength near $d = 250$ Å for the uncovered film, and $d = 140$ Å for the covered film, it is concluded that only up to these thicknesses any substrate-induced elastic strains might contribute to the observed uniaxial anisotropy.

From a fit to the experimental data the results are as follows for Fe(110) films on W(110) [3.58,3.136,3.137,3.138]: The volume anisotropy constant, $K_1 = (4.1 \pm 0.3) \times 10^5\,\text{erg/cm}^3$ is close to the literature value of $4.5 \cdot 10^5\,\text{erg/cm}^3$. For the uniaxial anisotropy, a value of $K_{\text{p}}^{(2)} = -8.0 \times 10^5\,\text{erg/cm}^3$ is obtained, favoring the in-plane [001] direction as the easy axis for this anisotropy. The in-plane interface anisotropy constant $k_{\text{p}}^{(2)} = (0.34 \pm 0.04)\,\text{erg/cm}^2$ for the uncovered films and $k_p^{(2)} = (0.26 \pm 0.03)\,\mathrm{erg/cm^2}$ for the films covered with 1 ML Pd with the $[1\bar{1}0]$ direction as the easy axis. The values of $k_p^{(4)}$ are $(0.04 \pm 0.01)\,\mathrm{erg/cm^2}$ for the uncovered films and $(-0.03 \pm 0.01)\,\mathrm{erg/cm^2}$ for the covered films. The out-of-plane interface anisotropy constant, k_s, was found to be $(0\pm0.3)\,\mathrm{erg/cm^2}$. From the fits to the spin wave frequencies, the saturation magnetization, $4\pi M_s$, was determined to be (21.0 ± 0.5) kG, in agreement with the literature value of bulk Fe. These results agree well with results of *Elmers* and *Gradmann* using *in situ* torsion oscillation magnetometry [3.142]. The 1-ML thick Pd overlayer has a large effect on the in-plane anisotropy constant, $k_p^{(2)}$, of Fe(110)/W(110) by reducing its value (averaged over both interfaces) by about 24%. Individual numbers for the interface anisotropy constants for each Fe interface cannot be obtained from this analysis.

From the critical maximum thickness for coherent film growth, $d_{crit} \approx 2$ ML = 4.04 Å, as obtained from the LEED investigations, the residual strain for an, e.g., 100 Å thick Fe film can be estimated using the strain relaxation model (Sect. 3.2.2.3) to be $\approx 0.4\%$. From this value, a strain induced magneto-elastic anisotropy constant of $K_p^{me} = -4 \times 10^5\,\mathrm{erg/cm^3}$ is calculated, which is in agreement within a factor of two with the above listed value for $K_p^{(2)}$. For thicknesses larger than 250 Å (140 Å) the critical field starts to decrease, indicating a relaxation of this uniaxial anisotropy contribution. In the limit of bulk Fe, i.e. $1/d \to 0$, a much smaller value of $H_{crit} = 80$ Oe caused by the volume anisotropy would be expected. The rather simple estimate of the elastic strain fields, neglecting any details of the misfit dislocations (e.g. anisotropic relaxation) and the influence of an overlayer, cannot serve for a more quantitative investigation, in particular the thickness dependence of H_{crit}. It would be of interest to follow in detail the strain relaxation as a function of film thickness, as well as its depth dependence, in order to better characterize the implied magneto-elastic contributions in this system.

3.5.1.2 Co/Cu(001) and (1 1 13): Stabilization of Ferromagnetic Order by In-Plane Anisotropies, Induced Uniaxial Anisotropy and Step Anisotropy . We address next a problem of great interest: Will magnetic anisotropies stabilize ferromagnetic order in ultrathin films? According to the theorem of *Mermin* and *Wagner* [3.143] a ferromagnetic ground state cannot exist at finite temperatures in an isotropic two-dimensional non-Ising system with short-range interactions. Several mechanisms have been proposed to account for the experimentally observed existence of ferromagnetic order in two-dimensional systems [3.144], among them the stabilization of ferromagnetic order due to dipolar interactions or to magnetic anisotropies. BLS is best suited to address this problem. In-situ experiments permit the determination of all contributing anisotropies. Measurements in remanence, i.e., at $H = 0$, allow one to determine the presence of a ferromagnetic ground state as a function of film thickness by testing the existence of long-wavelength spin wave excitations.

In the following we discuss the determination of all relevant magnetic volume and interface anisotropy contributions at room temperature [3.145, 3.146]. We show evidence of the fact that in this system magnetic in-plane anisotropy contributions stabilize ferromagnetic order.

Ferromagnetic order is observed for a film thickness, d, larger than $d_c = 1.6 \pm 0.3$ ML for uncovered Co films and $d_c = 1.9 \pm 0.3$ ML for Cu covered Co films [3.145, 3.146]. The onset of ferromagnetic order is identified by the existence of a remanent magnetization resulting in a coercive field, which is tested by the magneto-optic Kerr effect (MOKE), as well as by the existence of spin waves at zero applied fields. The obtained coercive field is (81 ± 14) Oe for the uncovered film and (66 ± 11) Oe for the covered film.

For ultrathin Co films of two- or fourfold symmetry about the surface normal, F_{ani} is expressed in lowest order as

$$\begin{aligned} F_{\text{ani}} = & \left(K_{\text{p}}^{(2)} + \frac{2}{d}k_{\text{p}}^{(2)}\right)\cos^2(\phi - \phi_0)\sin^2\theta \\ & + \frac{1}{4}\left(K_{\text{p}}^{(4)} + \frac{2}{d}k_{\text{p}}^{(4)}\right)\sin^2(2\phi)\sin^4\theta \\ & - \left(K_{\text{s}}^{(2)} + \frac{2}{d}k_{\text{s}}^{(2)}\right)\cos^2\theta\,, \end{aligned} \tag{3.80}$$

where θ and ϕ are the polar and azimuthal angle of the direction of magnetization, with ϕ measured against the in-plane [100]-axis, and ϕ_0 is the angle of the symmetry axis of the uniaxial in-plane anisotropy with respect to [3.100]. The volume and interface terms are often combined and plotted multiplied by the film thickness, d, vs. d, resulting in a straight line (cf. (3.52)) with the slope equal to the volume anisotropy constant and the intercept equal to twice the interface anisotropy constant:

$$K_{\text{in-plane}}^{(n)} d = K_{\text{p}}^{(n)} d + 2k_{\text{p}}^{(n)} \tag{3.81}$$

with $n = 2{,}4$ and

$$K_{\text{out-of-plane}} d = K_{\text{s}}^{(2)} d + 2k_{\text{s}}^{(2)}\,. \tag{3.82}$$

In situ MOKE measurements of the remanent field as a function of the in-plane direction of the external field show that the acting in-plane anisotropies are of fourfold symmetry about the film normal in lowest order, i.e., the first r.h.s. term in (3.80), and thus $K_{\text{p}}^{(2)}$ and $k_{\text{p}}^{(2)}$ are zero.

By applying the external magnetic field along the magnetic hard [100]-axis, one can probe the magnetic anisotropies by studying the rotation of the magnetization with increasing field into the direction of the applied field via the corresponding change in the spin wave frequency: Upon increasing the applied field the spin wave frequency first decreases, due to the change in ϕ, until a critical field strength H_{crit} is reached. For $H > H_{\text{crit}}$ the magnetization and the applied field are collinear and the spin wave frequency increases nearly linearly with further increasing field. $H_{\rm crit}$ is a measure of the in-plane anisotropy.

For the film considered here the product of the film thickness d and the wavevector $q_{\|}$ is small compared to unity. Therefore we may use (3.70) for calculating the spin wave frequency of the Damon–Eshbach mode. From a least-squares fit of (3.70) to the measured spin wave frequencies as a function of the applied field and the Co film thickness the saturation magnetization and the anisotropy constants $K_{\rm p}^{(4)}$, $k_{\rm p}^{(4)}$ and $k_{\rm s}$ are determined. For all studied films with $d > d_{\rm c}$ the saturation magnetization does not deviate from the bulk value of 17.9 kG within the limit of accuracy of 5%. The result is corroborated by finding a linear increase of the Kerr rotation angle with increasing film thickness measured in situ on the same films as used for the light scattering studies.

Figure 3.13 shows the obtained anisotropy values as a function of the film thickness for uncovered Co films (open symbols) and Co films covered with 2 ML Cu (closed symbols). In the upper part the total in-plane anisotropy, $K_{\text{in-plane}}^{(4)}$, multiplied with the film thickness, d, and in the lower part the corresponding out-of-plane anisotropy contribution, $K_{\text{out-of-plane}}d$, as defined in (3.81,3.82) are shown. From a fit to the experimental data a value of $K_{\rm s}^{(2)} = 0$ is found both for the uncovered and covered films. The average value of $k_{\rm s} = (-0.46 \pm 0.09)\,\text{erg/cm}^2$ for uncovered Co films changes to $k_{\rm s} = (0.15 \pm 0.04)\,\text{erg/cm}^2$ upon covering the Co films by 2 ML Cu. The negative sign indicates that the surface normal is a magnetic hard axis for this anisotropy contribution.

Of particular interest are the properties of $K_{\text{in-plane}}$. From the slope and the intercept of the straight lines in Fig. 3.13 with the ordinate the volume ($K_{\rm p}^{(4)}$) and the interface ($k_{\rm p}^{(4)}$) contributions are obtained. They are: $K_{\rm p}^{(4)} = (-2.3 \pm 0.15) \cdot 10^6$ erg/cm^3 and $k_{\rm p}^{(4)} = (0.034 \pm 0.004)\,\text{erg/cm}^2$ for the uncovered films, and $K_{\rm p}^{(4)} = (-2.2 \pm 0.15) \cdot 10^6\,\text{erg/cm}^3$ and $k_{\rm p}^{(4)} = (0.031 \pm 0.003)\,\text{erg/cm}^2$ for the Co layers covered with 2 ML Cu. Due to their opposite signs, the contributions of $K_{\rm p}^{(4)}$ and $k_{\rm p}^{(4)}$ to $K_{\text{in-plane}}$ cancel each other at $d_{\rm c}^{+} = (1.7 \pm 0.3)$ ML for the uncovered films and at $d_{\rm c}^{+} = (1.6 \pm 0.3)$ ML for the Cu covered films. It should be pointed out that $d_{\rm c}^{+} = d_{\rm c}$ within the experimental uncertainty for both covered and uncovered Co films, and both quantities are determined by independent experiments.

Assuming that the Co films covered by 2 ML Cu have symmetric interfaces with the same anisotropy constants, the Co interface anisotropy constants are independently accessible for the vacuum and the Cu side by comparing the data of the uncovered to the covered Co films. Since $K_{\text{in-plane}}$ does not significantly change upon covering the Co layer by Cu, both the Co/Cu and Co/vacuum surface have the same value of $k_{\rm p}^{(4)} = (0.032 \pm 0.003)\,\text{erg/cm}^2$. The out-of-plane anisotropy constant, $k_{\rm s}$, was found to be $k_{\rm s} = (-1.06 \pm 0.17)\,\text{erg/cm}^2$ for the Co/vacuum interface and $k_{\rm s} = (0.15 \pm 0.04)\,\text{erg/cm}^2$

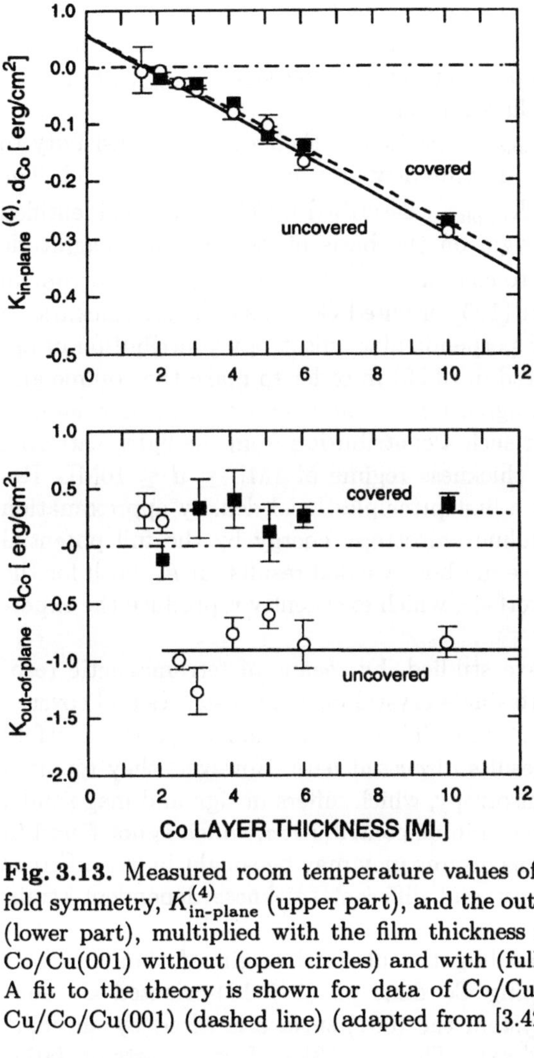

Fig. 3.13. Measured room temperature values of the in-plane anisotropy of four-fold symmetry, $K^{(4)}_{\text{in-plane}}$ (upper part), and the out-of-plane anisotropy, $K_{\text{out-of-plane}}$ (lower part), multiplied with the film thickness d, as a function of d in ML for Co/Cu(001) without (open circles) and with (full circles) a 2 ML Cu cover layer. A fit to the theory is shown for data of Co/Cu(001) (full line) and for data of Cu/Co/Cu(001) (dashed line) (adapted from [3.42])

for the Co/Cu interface, i.e., the two sides of the Co film have opposite signs in k_s.

From the observed agreement between the critical thickness for ferromagnetic order, d_c, with the thickness d_c^+, at which the contributions to the in-plane anisotropy cancel, it is concluded that the symmetry breaking interaction for stabilizing ferromagnetic order in Co(001) films at room temperature is indeed given by the magnetic in-plane anisotropy contribution [3.145, 3.146]. It should be pointed out that this argument is backed by the fact that d_c^+ may also be obtained from an extrapolation of $K_{\text{in-plane}}(d)$ from data with film thicknesses significantly larger than d_c, thus ruling out structural and/or magnetic percolation effects near d_c^+. The out-of-plane anisotropy, described by the thickness independent parameter k_s, does not support the stabilization: The data clearly indicate that, in contrast to the critical thickness, d_c, and to the in-plane anisotropy constant $K_{\text{in-plane}}$, k_s is very sensitive to the presence of a Cu overlayer.

The origin of k_s and $K_{\text{in-plane}}$, in particular their different sensitivity to the presence of a Cu overlayer, is presently only partially understood. The thickness-independent part of $K_{\text{in-plane}}$, described by $K_p^{(4)}$, can be identified to be of magnetocrystalline origin on the basis of its sign and magnitude (see below). However, in order to compare $K_p^{(4)}$ to a crystalline anisotropy of cubic symmetry appropriate for (100)-oriented Co films of larger thickness, a thickness-independent uniaxial perpendicular anisotropy contribution of appropriate size must be considered in (3.80) in order to make the volume anisotropy contribution invariant against all cubic symmetry transformations. No evidence has been found for such a contribution from the light scattering data for the investigated film thickness regime of $1\text{ML} \leq d \leq 10\text{ML}$. Recent calculations [3.147], using a first principles local density approximation of the interface magnetocrystalline anisotropy energy by the full potential linearized augmented plane wave method, yielded results for k_s, both for the Co/Cu and the Co/vacuum interface, which excellently reproduce the experimental values.

Heinrich and coworkers have studied, by means of ferromagnetic resonance, Co(001) films grown onto single-crystalline Cu substrates and covered by 6–11.5 ML Cu [3.38, 3.148, 3.149]. Their results agree for the fourfold in-plane anisotropy with the results discussed here. However, they report a large uniaxial perpendicular anisotropy, which differs in sign and magnitude. In particular, they report a large volume contribution, which is not found in the BLS study. The cause for this discrepancy may be sought in the different thicknesses of the Cu cover layer and different thickness dependent strain contributions.

Additional information regarding the physical origin of thin film anisotropies is obtained by studying Co films on (1 1 13)-oriented Cu substrates [3.146, 3.150]. In this system the (001)-surface is tilted by an angle of 6.2° about the in-plane $[1\bar{1}0]$ axis. The (1 1 13)-surface consists of (001)-oriented terraces with an average width of 6.5 atomic distances, separated by monoatomic steps aligned with the $[1\bar{1}0]$-axis. Here, due to the induced twofold in-plane symmetry, an additional in-plane uniaxial anisotropy contribution is found. The easy axis was found to lie along the steps ($\phi_0 = 45°$ in (3.80)) [3.151].

By use of (3.80) the measured spin wave frequencies were fitted using (3.70) with the anisotropy constants as fit parameters. Performing the same type of analysis as in the case of Co(100) films, critical thicknesses are deduced from the data, at which the corresponding volume and interface anisotropies cancel. They are $d_c^{(2)} = (2.9 \pm 0.5)$ ML for the uniaxial anisotropy and $d_c^{(4)} = (2.2 \pm 0.4)$ ML for the fourfold anisotropy. It is found that, within the experimental error, $d_c^{(2)} = d_c^{(4)} = d_c$ with d_c the independently determined minimum thickness of 2 ML for the onset of ferromagnetic order [3.152]. It should be pointed out that, although the film growth mode is very different for the (1 1 13) surface orientation compared to the (001)-orientation [3.145, 3.150, 3.153], the critical thickness for the onset of ferromagnetic order is the same within the error margins for both orientations.

From the data analysis the following anisotropy constants were obtained: the twofold in-plane anisotropy constants are $K_p^{(2)} = (6.0 \pm 0.7) \times 10^5$ erg/cm^3 and $k_p^{(2)} = (-0.009 \pm 0.002)$ erg/cm^2, and the fourfold contributions are $K_p^{(4)} = (-6.5 \pm 0.2) \times 10^5$ erg/cm^3 and $k_p^{(4)} = (0.012 \pm 0.002)$ erg/cm^2.

For $d > d_c^{(2)}$ the $[1\bar{1}0]$ direction (parallel to the steps) is the easy axis for $K_{\text{in-plane}}^{(2)}$, and for $d > d_c^{(4)}$ the in-plane $\langle 13\ 13\ \bar{2} \rangle$ directions (approximate $\langle 110 \rangle$-directions) are the easy axes for $K_{\text{in-plane}}^{(4)}$. For $K_{\text{out-of-plane}}$ no interface dependent contribution, k_s, was found, whereas the volume part was found to be $K_s = (-5.0 \pm 0.6) \times 10^6$ erg/cm^3. The negative sign indicates that the surface normal is a magnetic hard axis.

We now show that both $K_p^{(2)}$ and $K_s^{(2)}$ are caused by magnetoelastic interaction due to the elastic strain field originating from the lattice mismatch at the Co/Cu interface. In a continuum approach we assume a smooth film, i.e., we neglect the stepped surface structure. This approach is valid since the step distance of 6.5 atomic distances is much smaller than the static coherence length over which the magnetic moments might vary in direction. In the (1 1 13)-orientation the magnetoelastic tensor, b_{ijkl}, must be rotated from the crystallographic reference frame into the film coordinate system with the cartesian axes aligned along the $[1\bar{1}0]$-, $[13\ 13\ \bar{2}]$- and [1 1 13]-axes which are the approximate $[1\bar{1}0]$-, [110]- and [001]-axes (i.e. here: tilted by 6.2° about the $[1\bar{1}0]$-axis). In the film coordinate system only the diagonal strain components ϵ_{ii} are nonzero. The tensor components of the magnetoelastic tensor are obtained [3.70] from the magnetostriction constants λ_{100} and λ_{111} of fcc-Co extrapolated from Co-rich CoPd alloys [3.154], as well as from the elastic constants of bulk fcc Co [3.155].

Figure 3.14 shows the obtained results. The values of the magnetoelastic anisotropy constant, $K_p^{(2)}$, are plotted as a function of the tilt angle, α, by which the surface is rotated about the $[1\bar{1}0]$-axis. For $\alpha = 0$, i.e. for the (001)-orientation, the in-plane constant, $K_p^{(2)}$, is zero for symmetry reasons. With increasing tilt $K_p^{(2)}$ increases. The experimental value of $K_p^{(2)}$ for the (1 1 13)-orientation ($\alpha = 6.2°$) is shown as well. It agrees with the calculated value within a factor of two, a rather good agreement in view of the uncertainty in the estimates of the magnetostriction constants as well as the use of bulk elastic constants for the film. We conclude that this anisotropy contribution is indeed caused by magnetoelastic interaction.

A careful investigation shows that the value and sign of $K_p^{(2)}$ depend very sensitively on a possible strain relaxation: A relaxation of the lattice

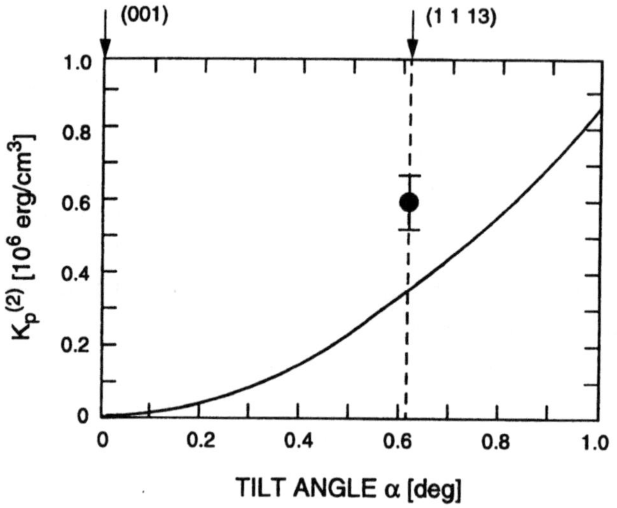

Fig. 3.14. Calculated in-plane ($K_p^{(2)}$) magnetoelastic anisotropy contribution as a function of the tilt angle α, by which the surface is rotated about the in-plane $[1\bar{1}0]$-axis out of the (001)-orientation. The (1 1 13)-orientation is marked by a dashed line. The experimental value for the (1 1 13)-orientation is shown (adapted from [3.42])

parameter in the in-plane direction perpendicular to the steps by 3% with no relaxation parallel to the steps would cancel $K_p^{(2)}$ [3.152], and a further relaxation would reverse its sign. From the LEED data we estimate that this type of relaxation is not present in the films although a relaxation of 3% is at the limit of resolution. It should be emphasized that in addition to the magnetoelastic contribution the break in symmetry by the atomic steps is evidenced by the in-plane uniaxial surface anisotropy contribution (step anisotropy), $k_p^{(2)}$.

3.5.1.3 Co/Cu(110): Strain-Induced Suppression of the Magnetocrystalline Anisotropy. In thick films of cubic symmetry only the fourth-order magnetocrystalline anisotropy is expected to be the leading contribution beside a possible growth induced contribution. On the other hand, in the thin film regime surface and strain-induced anisotropies will dominate the magnetic properties. The transition between these two cases as a function of film thickness, d, is of considerable interest.

In order to clarify this point (110)-oriented Co films of fcc structure prepared on Cu(110) were studied by BLS [3.156,3.157,3.158]. The (110)-surface contains the [001]-, $[1\bar{1}0]$- and $[1\bar{1}1]$-axes. Any in-plane surface anisotropy contribution would have either the [001]- or the $[1\bar{1}0]$-axis as the symmetry axis for symmetry reasons. On the other hand, in fcc-Co the magnetocrystalline bulk anisotropy favors the $\langle 111\rangle$-axes as the easy axes of magnetization. A

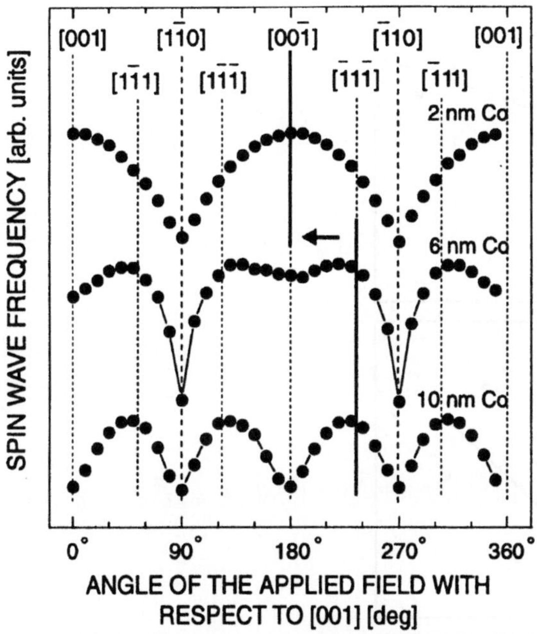

Fig. 3.15. Measured room temperature spin wave frequencies for epitaxial Co(110) layers on Cu(110) for various film thicknesses as a function of the in-plane angle ϕ_H between the external applied field and the [001]-axis. The external field strength is 3 kOe. The in-plane crystallographic directions and the change of easy axis (arrow) are indicated (adapted from [3.45])

study of the spin wave frequencies as a measure of the free energy of the system vs. the in-plane direction of the applied field ϕ_H and the film thickness, should provide information about the contributing anisotropies.

Figure 3.15 shows the measured spin wave frequencies for various film thicknesses as a function of the in-plane angle, ϕ_H, of the external applied field with the [001]-axis. Maxima indicate easy directions of the magnetization. From Fig. 3.15 it is evident that the easy axes of magnetization switch from $\langle 001 \rangle$ for $d < 50\,\text{Å}$ to $\langle 111 \rangle$ for $d > 50\,\text{Å}$, indicating the transition from the dominance of surface and strain anisotropies to magnetocrystalline anisotropy. The shape anisotropy causes the magnetization to lie in the film plane for the investigated Co thickness range of 8–110Å.

A detailed analysis is performed, using the anisotropy expression

$$\begin{aligned} E_{\text{ani}} &= K_1(\alpha_{x'}^2\alpha_{y'}^2 + \alpha_{y'}^2\alpha_{z'}^2 + \alpha_{z'}^2\alpha_{x'}^2) \\ &\quad + K_{\text{in-plane}}\alpha_x^2 - K_{\text{out-of-plane}}\alpha_z^2 \,, \end{aligned} \tag{3.83}$$

with x', y' and z' in the crystallographic reference frame. A simultaneous fit to the data of $K^{(2)}_{\text{in-plane}}$, $K^{(2)}_{\text{out-of-plane}}$ and the cubic magnetocrystalline bulk anisotropy constant, K_1, yields the anisotropy constants.

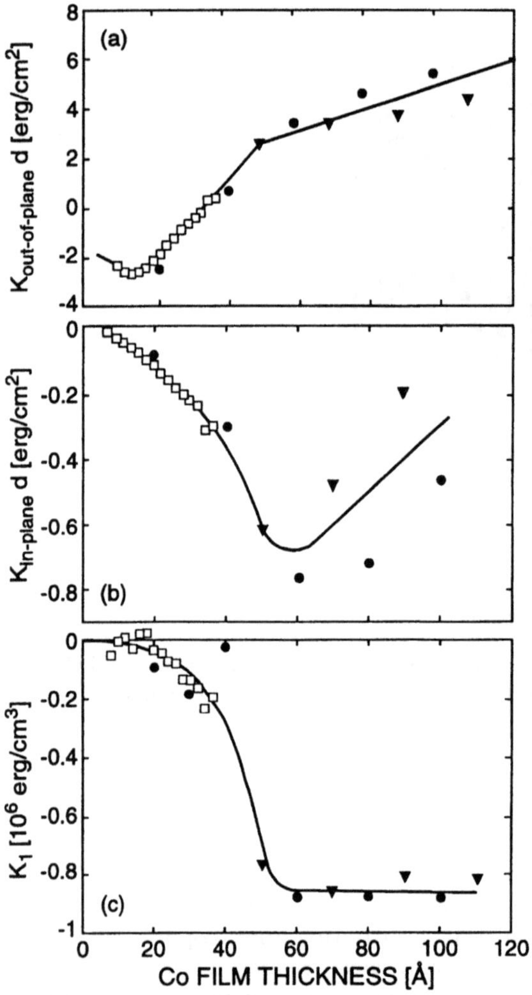

Fig. 3.16. (**a**): Effective out-of-plane anisotropy constant, $K_{\text{out-of-plane}}$, and (**b**): effective in-plane anisotropy constant, $K_{\text{in-plane}}$, multiplied by the Co film thickness, d. (**c**): magnetocrystalline anisotropy constant, K_1, as a function of d, measured at room temperature. The solid lines are guides to the eye (adapted from [3.45])

Figure 3.16a shows the obtained effective out-of-plane anisotropy constant, $K_{\text{out-of-plane}}$, multiplied by the Co film thickness, d, as a function of d. Such a plot yields the bulk anisotropy contributions as slopes and the interface anisotropy contributions as the (extrapolated) intercepts with the y-axis. Three different thickness regimes can be identified. First there is a thickness region up to 13Å indicated by a negative slope. This region is attributed to a coherently grown Co film with a large bulk magnetoelastic anisotropy contribution. In the region of 13Å to near 50Å a positive slope is formed, indicating a large thickness dependent magnetoelastic anisotropy contribution due to progressive, anisotropic strain relaxation. These anisotropy mechanisms are in agreement with the observed LEED patterns and scanning tunneling microscopy investigations [3.159]. Finally, for $d > 50$ Å we find a reduction in slope which is interpreted as the onset of complete elastic relaxation, as expected for larger film thicknesses. In this regime we find that the anisotropy remains constant and non-zero; it must be due to either a morphology induced anisotropy contribution from residual strains or a three-dimensional dislocation formation.

Figure 3.16b shows the effective in-plane anisotropy constant, $K_{\text{in-plane}}$, multiplied by d, vs. d. For d near about 50–70Å we find a break in slope for the effective in-plane anisotropy. As observed by LEED patterns and by STM it coincides with a continuous transition from anisotropic to isotropic in-plane strain relaxation.

New insight is gained into the magnetocrystalline anisotropy constant, K_1, as displayed in Fig. 3.16c as a function of the film thickness, d. For d larger than 50Å, a thickness-independent value of K_1 is found which is comparable with the value of the high-temperature bulk fcc-Co phase [3.160]. For Co film thicknesses smaller than 50Å a sudden breakdown in magnetocrystalline anisotropy to almost zero is observed.

Here we have the surprising result that the presence of a uniaxial strain strongly suppresses the cubic anisotropy. We now outline a phenomenological approach which provides some insight into the relationship between these second and fourth order anisotropies [3.158, 3.161]. The model gives a natural dependence of the cubic anisotropy on symmetry breaking effects which produce second-order anisotropies.

The existence of a uniaxial surface anisotropy was predicted by *Néel* as a consequence of the electronic symmetry breaking which occurs at the surface due to lower atomic coordination [3.65]. Recently, band theoretical methods have made considerable progress in the calculation of anisotropies which result from Néel effects [3.147, 3.162, 3.163]. Notably *ab initio* calculations for a Co(001) film were reported by *Wang* et al. [3.147], and a tight-binding calculation, in which only d-states are considered for a Co(110) film by *Cinal* et al. [3.162]. However, the calculation of forth-order anisotropies requires an energy resolution beyond the scope of current computational methods and, consequently, tends to be limited to second-order phenomena.

We start from a simple crystal field Hamiltonian to discuss changes in the symmetry of the system. The wave functions relevant to our analysis are $x'y'$, $y'z'$, $x'z'$, $x'^2 - y'^2$ and $3z'^2 - r^2$ with x', y' and z' the Cartesian coordinates of the electrons in the crystallographic reference frame. The surface normal is along $z = (x' + y')/\sqrt{2}$. We consider a Hamiltonian in terms of Stevens's operators [3.164] in the form

$$H = A(l_{x'}^4 + l_{y'}^4 + l_{z'}^4) + X(l_{x'} + l_{y'})^2 \,, \tag{3.84}$$

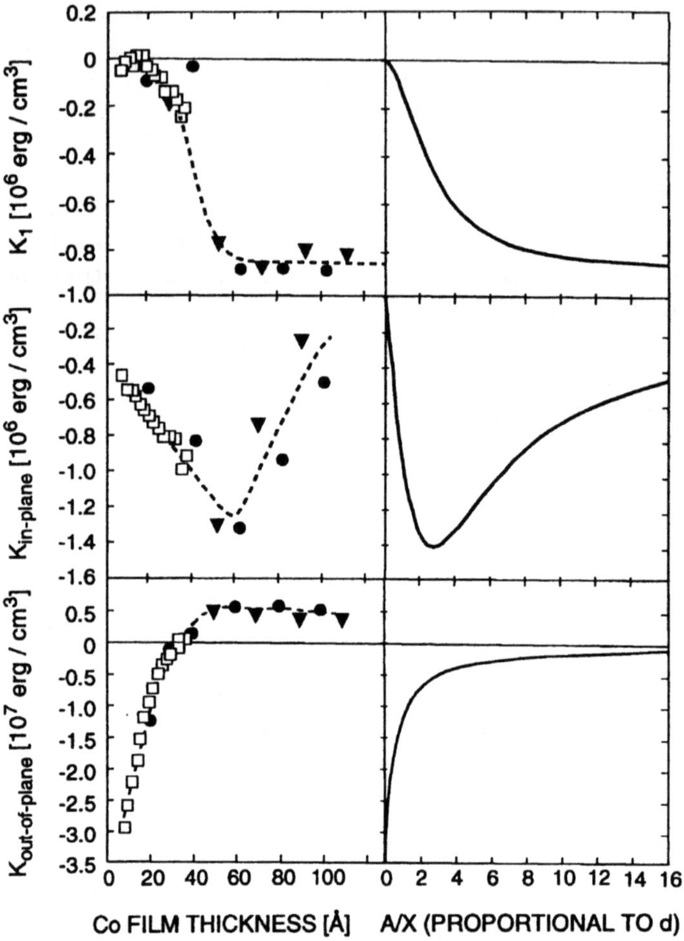

Fig. 3.17. *Left:* Experimental anisotropy data as in Fig. 3.16. *Right:* Calculated cubic anisotropy constant, K_1, (*top*), in-plane anisotropy constant, $K_{\text{in-plane}}$, (*middle*), and out-of-plane anisotropy constant, $K_{\text{out-of-plane}}$, (*bottom*) as a function of the ratio of the cubic and uniaxial energy parameter, A/X (adapted from [3.158])

where A and X are the cubic and uniaxial energy parameters. From (3.84) it is apparent that we have added a strain beyond cubic symmetry along the growth direction. Anisotropies are calculated in the usual way by including the spin–orbit coupling as a perturbation [3.165]. Details of the calculation are reported in [3.158, 3.161]. The results are shown in Fig. 3.17. The uniaxial energy parameter, X, is by definition proportional to the misfit strain, ϵ. *Chappert* and *Bruno* [3.77] and *den Broeder* et al. [3.166] argue that ϵ is inversely proportional to the film thickness and so we plot the anisotropies as a function of $A/X \propto d$. The results are shown on the r.h.s. in Fig 3.17. An overall good agreement between experiment (Fig. 3.16 and l.h.s. of Fig. 3.17)

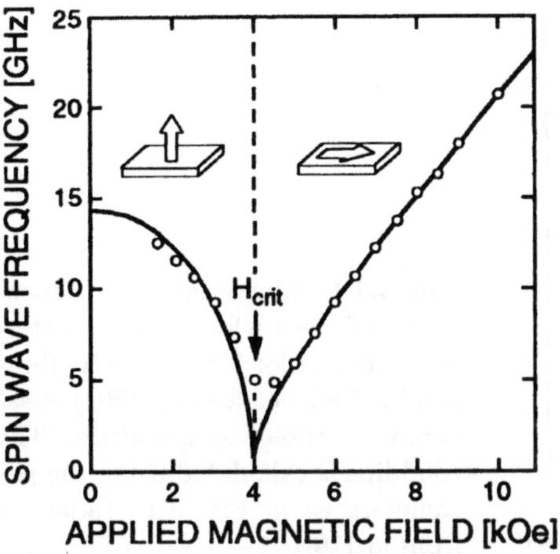

Fig. 3.18. Room temperature spin wave frequencies as a function of the magnetic field applied parallel to the film plane for 3 ML thick Fe(001) films sandwiched between Cu(001). The solid line has been calculated using the following parameters: effective magnetization $4\pi M_s - 2K_1/M_s = -4.0\,\text{kG}$, g-factor $g = 1.95$, second order uniaxial anisotropy field $4K_2/M_s = 1.17\,\text{kOe}$, Gilbert damping parameter $G = 7.0 \times 10^7$ Hz and resistivity $\rho = 1.0 \times 10^{-5}\,\Omega\,\text{cm}$ (adapted from [3.119])

and the model calculations is obtained. In particular, we see that K_1 is suppressed with increasing strain.

The data in Fig. 3.17 have been obtained by adjusting the cubic and tetragonal energy parameters, A and X, and the spin–orbit coupling constant, ξ, such that K_1 approaches its experimental value for large thicknesses and that we obtain the correct values for $K_{\text{out-of-plane}}$ in the limit $X \gg A$. The agreement of the magnitude of the measured value of $K_{\text{in-plane}}$ is not fully satisfactory. However, the fact that such a simplistic approach yields qualitative agreement is in itself important.

3.5.2 Perpendicularly Magnetized Films: Fe/Cu(001)

Of great current interest are ultrathin magnetic films with large perpendicular anisotropies which tend to pull the magnetization out of the film plane. *Dutcher* et al. have investigated the Brillouin light scattering spectra of 3-ML thick Fe(001) films epitaxially grown on Cu(001) substrates and covered with a Cu(001) overlayer [3.119,3.167]. This system is perpendicularly magnetized for zero applied field. The sample preparation is described in [3.168,3.169].

Figure 3.18 shows the measured spin wave frequencies as a function of a magnetic field, applied parallel to the films. With increasing field H, the spin wave frequencies first decrease while the direction of magnetization is increasingly tilted into the layer plane. At $H_{\text{crit}} = 4\,\text{kOe}$ the magnetization lies in-plane. For further increasing fields, the spin wave frequencies increase in a nearly linear fashion, as characteristic for in-plane magnetized samples.

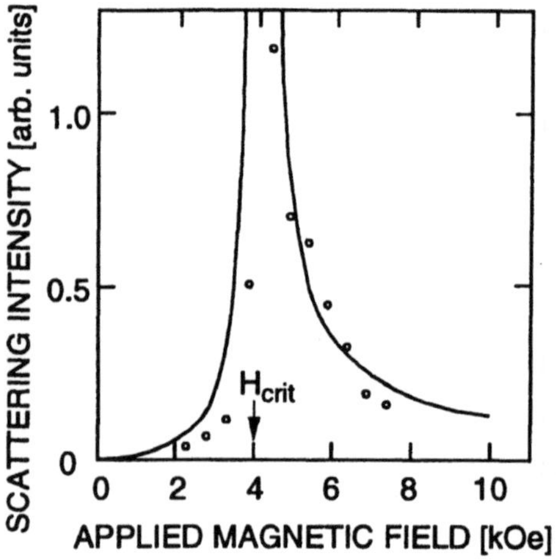

Fig. 3.19. Magnetic field dependence of the surface mode scattering intensity for 3 ML Fe(001) films sandwiched between Cu(001) measured at room temperature. The solid line is calculated using the parameters as in Fig. 3.18 (adapted from [3.119])

The data were fitted by a model which neglects volume exchange contributions but is otherwise equivalent to the theory outlined in Sect. 3.2, yielding a good agreement with the experimental data [3.119, 3.167, 3.170]. The fit is shown in Fig. 3.18 as a solid line. In order to achieve a better agreement with experiment, the authors introduced a higher order out-of-plane anisotropy contribution described by the anisotropy constant K_2 in (3.71) [3.171].

Of particular interest are the spin wave properties near H_{crit}. Here the torques from the shape anisotropy and the out-of-plane anisotropy contributions (K_1, K_2 and k_s) acting on the magnetization cancel each other. As shown in Sect. 3.2, the minimum frequency should then be determined by the exchange interaction. The volume exchange constant can be derived in principle from the frequency minimum at H_{crit} in Fig. 3.18. In practice, however, it can be shown, that already a local variation of 1% in k_s in (3.71) over the laser spot region is sufficiently large to "wash out" the minimum [3.171].

Dutcher et al. showed [3.119] that due to the torque cancellation effect near H_{crit} the spin wave frequencies become very small and the thermal fluctuations in M and thus the BLS cross section increase by several orders of magnitude for H approaching H_{crit}. This is demonstrated in Fig. 3.19 where the light scattering cross section of the Damon–Eshbach mode is plotted as a function of applied field. The divergent behavior near H_{crit} is very pronounced, both in the calculation (solid line in Fig. 3.19) and in the experimental data (circles in Fig. 3.19).

3.5.3 Multilayered Structures With Dipolar Coupling

Many of the phenomena discussed for single magnetic films apply as well to multilayered and superlattice structures. For instance, BLS has successfully been used for determining magnetic anisotropies of multilayered structures. Due to the periodicity of the stacking sequence, however, superlattice struc-

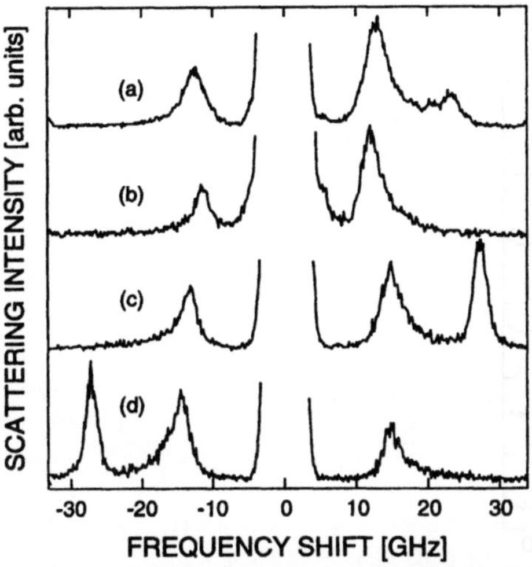

Fig. 3.20. Room-temperature Brillouin light scattering spectra of Fe/Pd superlattices in an applied magnetic field of 1 kOe: **(a)** $d_{\mathrm{Fe}} = 21.9\,\text{Å}$ and $d_{\mathrm{Pd}} = 24.3\,\text{Å}$, **(b)** $d_{\mathrm{Fe}} = 41.7\,\text{Å}$ and $d_{\mathrm{Pd}} = 138.7\,\text{Å}$, **(c)**, **(d)** $d_{\mathrm{Fe}} = 41.0\,\text{Å}$ and $d_{\mathrm{Pd}} = 9.1\,\text{Å}$. In **(d)** the direction of the applied field has been reversed compared to **(c)**. The number of repeated bilayers is 90 (adapted from [3.29])

tures exhibit novel collective spin wave phenomena. They are the subject of this section.

3.5.3.1 Fe/Pd: Demonstration of Dipolar Coupling Effects. We first consider collective spin wave excitations formed by dipolar coupling from the Damon–Eshbach modes of the individual magnetic layers. We demonstrate the properties of the collective spin wave band with some sample spectra of Fe/Pd superlattices [3.29]. The samples were prepared on single-crystal sapphire substrates using an rf sputtering technique [3.139, 3.172, 3.173]. As shown by Bragg and wide-film Debye–Scherrer x-ray diffraction the layers grew with a preferred orientation of bcc Fe(110) planes and fcc Pd(111) planes, with no preferred in-plane orientation. The samples exhibited long-range structural coherence of at least 300 Å perpendicular to the layers.

In Figs. 3.20 and 3.21, typical Brillouin spectra of spin wave excitations in Fe/Pd superlattices are displayed. The measured scattering intensities are plotted as a function of frequency shift, $\nu = \omega/2\pi$, with respect to the laser frequency. The magnetic field applied parallel to the layers is 1 kOe in Fig. 3.20. In Fig. 3.20a the thickness $d_{\mathrm{Fe}} = 21.9\,\text{Å}$ of the magnetic material is close to that of the spacer material $d_{\mathrm{Pd}} = 24.3\,\text{Å}$. The band of collective spin wave excitations can clearly be identified in the right-hand part of the spectrum by its specific asymmetric shape: The density of states is largest at small frequency shifts and decreases in an asymmetric fashion towards the upper band edge. At the latter, a few discrete spin wave modes can still be

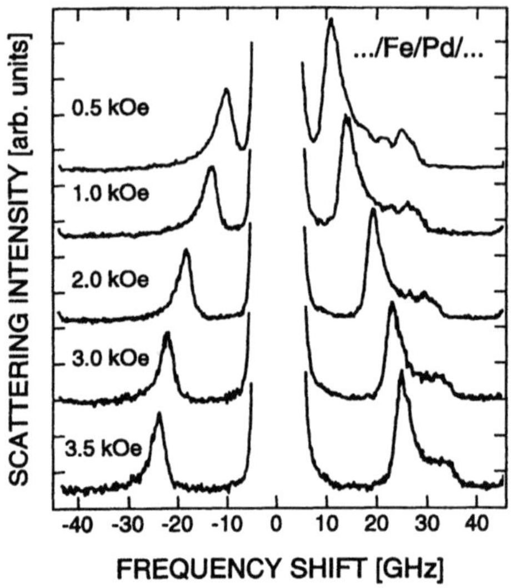

Fig. 3.21. Room-temperature Brillouin light scattering spectra of a Fe/Pd superlattice consisting of 49 bilayers with $d_{\mathrm{Fe}} = 89.4\,\text{Å}$ and $d_{\mathrm{Pd}} = 99.0\,\text{Å}$ for different applied magnetic fields, as indicated in the figure (adapted from [3.29])

resolved due to the small layer thicknesses and the still finite number of bilayers (ninety) [3.139, 3.173]. The large Stoke/anti-Stokes asymmetry identifies them as surface-mode-like spin waves. On the other hand, the modes near the lower edge of the spin wave band are found to be bulk-mode-like from the much smaller Stokes/anti-Stokes asymmetry. If we neglect the discrete modes near the upper band edge the shape of the spin wave excitation band is qualitatively very similar to the calculated Brillouin scattering cross section for the semi-infinite superlattice system Mo/Ni [3.112].

In Fig. 3.20b we show the Brillouin spectrum of an Fe/Pd superlattice with $d_{\mathrm{Fe}} = 41.7\,\text{Å}$ and a much larger spacer thickness $d_{\mathrm{Pd}} = 138.7\,\text{Å}$. In this case the spin wave band becomes narrower due to the reduced coupling across the spacer layers. A very different spectrum is found for the case of the spacer thickness, d_{Pd}, (9.1 Å) much smaller than d_{Fe} (41.1 Å), as shown in Fig. 3.20c. Here a very intense discrete mode is found near 27.7 GHz in the anti-Stokes spectrum apart from the band of collective modes near $\pm$ 15 GHz. This superlattice surface spin wave mode, which travels about the total superlattice stack, is allowed to exist beside the collective spin wave band. It would merge with the latter for $d_{\mathrm{Fe}} = d_{\mathrm{Pd}}$ [3.174].

The effect of reversing the direction of the applied magnetic field is demonstrated in Fig. 3.20d. Since the direction of the applied field defines the sense of revolution of each surface spin wave mode about each magnetic layer, a reversed field causes the Stokes and anti-Stokes parts of the spectrum to be exchanged.

Figure 3.21 shows Brillouin spectra of an Fe/Pd superlattice with $d_{Fe} = 89.4$ Å and $d_{Pd} = 99.0$ Å for different applied magnetic fields. With increasing field the spin wave frequencies increase in a quasi-linear fashion, accompanied by a slight narrowing. Fitting the measured peak positions of Fe/Pd superlattice samples like in Fig. 3.21 with the model outlined in Sect. 3.2 assuming for the bulk anisotropy constant the bulk value of $K_1 = 4.5 \times 10^5$ erg/cm^3, the following parameters are obtained [3.29]: $4\pi M_s = 17 \pm 2$ kG, $k_s = 0.15 \pm 0.03$ erg/cm^2.

We would like to point out a specific difference between data obtained from Brillouin light scattering and those obtained with standard magnetometry. Measurements of the magnetization of the same Fe/Pd superlattice samples using a SQUID magnetometer yielded enhanced values which exceeded for some samples the bulk value of 21 kG contrary to the BLS analysis [3.29, 3.175]. These magnetization values were obtained by dividing the measured total magnetic moment of the sample by the Fe volume. A comparison of this result to the Brillouin light scattering data provides clear evidence that the additional magnetic moment must be attributed to the Pd spacer layers, which are therefore magnetically polarized. On the contrary, an enhanced moment of the Fe atoms would result in increased spin wave frequencies, which is not experimentally observed. Simulations showed that an additional moment of the Pd layers would not significantly change the spin wave properties [3.29]. A combination of BLS and magnetometry might therefore serve for characterizing spacer layer polarization effects. The opposite effect, namely formation of magnetically dead Fe layers at the interfaces, has been found by Brillouin light scattering in sputtered Fe/Ti superlattices [3.176].

3.5.3.2 Co/Pd, Co/Au: Large Perpendicular Anisotropies. Multilayered systems with large perpendicular magnetic anisotropies are of high interest due to their potential applicability for magneto-optic recording. If the effective out-of-plane anisotropy, K_{eff}, which is the sum of all anisotropy contributions including the shape anisotropy with the film normal as their symmetry axis (see (3.66–68), is larger than zero the system is perpendicularly magnetized. The critical field strength of an in-plane applied field, which is needed to force the direction of magnetization into the layer planes, $H_{\text{crit}} = 2K_{\text{eff}}/M_s$, can be as large as 50 kOe.

Brillouin light scattering is a useful tool for investigating these large anisotropy contributions. However, we would like to point out a few limitations of the technique:

(i) In order to observe dipolar modes the magnetization must have a large component parallel to the film planes. It is therefore a prerequisite that H_{crit} be smaller than the experimentally available magnetic field strength, since otherwise a well-defined in-plane magnetization state cannot be achieved. Thus, systems with extremely large out-of-plane anisotropies, such as state-of-the-art Co/Pt (111)-oriented superlattices with optimized out-of-plane anisotropies, are beyond reach for Brillouin light scattering experiments.

(ii) The line width of the observed spin wave excitations is very sensitive to the distribution of the magnetization directions. If the magnetization has an out-of-plane component, the multilayers often are in a multidomain state which results in broad spin wave excitations with reduced scattering cross sections. A quantitative analysis of the frequencies of these modes is not possible.

(iii) Large positive perpendicular anisotropies reduce the BLS cross section, since they cause the magnetization precession to be largely elliptical with the larger amplitudes perpendicular to the layers. The light couples mostly to the parallel precession components, which are weak, and therefore large accumulation times in a BLS experiment have to be planned.

In order to determine anisotropy constants BLS may be used. Often, however, standard magnetometry methods are more appropriate, since some of the specific characteristics of the Brillouin light scattering method, in particular its monolayer sensitivity and easy implementation into UHV, are not as advantageous as in the case of ultrathin single films. Nevertheless, there are a number of applications which demand for the specific advantages of the BLS method, for instance in the case of interlayer exchange coupling for the characterization of spatially varying anisotropies. In the following subsection we demonstrate the applicability of BLS to the determination of anisotropies and show the specific advantages of the method.

3.5.3.3 Co/Pd Superlattices. Co/Pd superlattices can be prepared on GaAs substrates by MBE using appropriate buffer layers with high crystallographic perfection for layer orientations of (100), (110) and (111) [3.177,3.178, 3.179,3.180]. The sample growth and quality can be monitored by RHEED and LEED, as well as by ex-situ studies using a scanning tunneling microscope and Rutherford backscattering. The Co and Pd layers are of fcc structure for the (100) and (110) oriented layers. For the (111) oriented layers the stacking sequence (fcc or hcp) could not be uniquely determined. Of particular interest are (111)-oriented superlattices, due to their large perpendicular anisotropy, as well as (110)-oriented superlattices, since here the twofold symmetry in the layer planes introduces an additional uniaxial in-plane anisotropy [3.70].

Figure 3.22 shows spin wave spectra of a series of (111)-oriented Co/Pd superlattices with a Pd layer thickness of $d_{\mathrm{Pd}} = 12\,\text{Å}$, and a Co bilayer thickness, d_{Co} varying between 8 Å and 20 Å. The strength of the in-plane applied external field is 10 kOe. With decreasing Co layer thickness (top to bottom in Fig. 3.22) the spin wave frequencies decrease due to the increasing contribution from interface anisotropies to the effective fields. For $d_{\mathrm{Co}} = 8\,\text{Å}$ a very broad spin wave spectrum is obtained. Here the applied field is not strong enough to force the magnetization into the layer planes. The two peaks

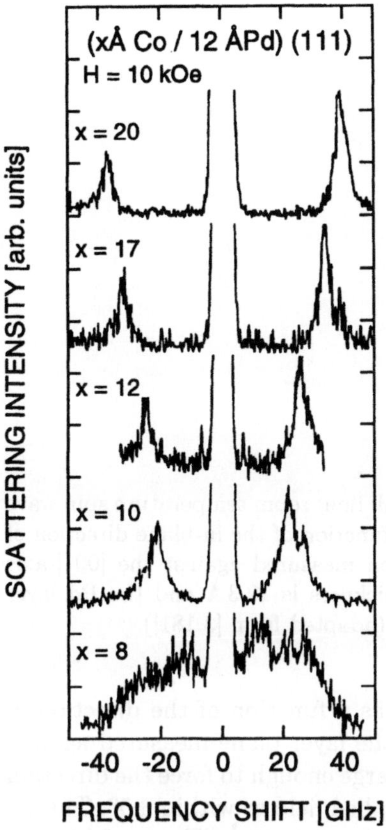

Fig. 3.22. Room temperature Brillouin light scattering spectra of MBE-prepared (111)-oriented Co/Pd superlattices with varying Co layer thickness as indicated (adapted from [3.181])

at 12 GHz and 26 GHz might be attributed to different well-defined directions of domain magnetizations, as implied by closure domains. This assignment however, is somewhat speculative.

The frequency positions of the spin wave bands displayed in Fig. 3.22 were fitted by adjusting the saturation magnetization, the hcp-Co bulk anisotropy constant, $K_1 + K_2$, as defined in (3.49), and the interface anisotropy constant, k_s. The results are: $4\pi M_s = 17.6\,\mathrm{kG}$, which is close to the Co bulk value, $K_1 + K_2 = 7.4 \times 10^6\,\mathrm{erg/cm^3}$. These results are in excellent agreement with measurements performed with a vibrating-sample magnetometer [3.179, 3.180] which yielded $K_1 + K_2 = 7.0 \times 10^6\,\mathrm{erg/cm^3}$ and $k_s = 0.6\,\mathrm{erg/cm^2}$. The fits reproduce the peak positions, but not the experimentally observed rather broad line widths (Sect. 3.5.5).

In the Co/Pd(111) system no clear separation between magnetocrystalline and magneto-elastic anisotropies, and perhaps other additional anisotropy contributions, can be made. The situation is clearer for (110)-oriented superlattices [3.181]. Here a large in-plane anisotropy contribution is found. Figure 3.23 shows the measured (open circles) and fitted (full lines) spin

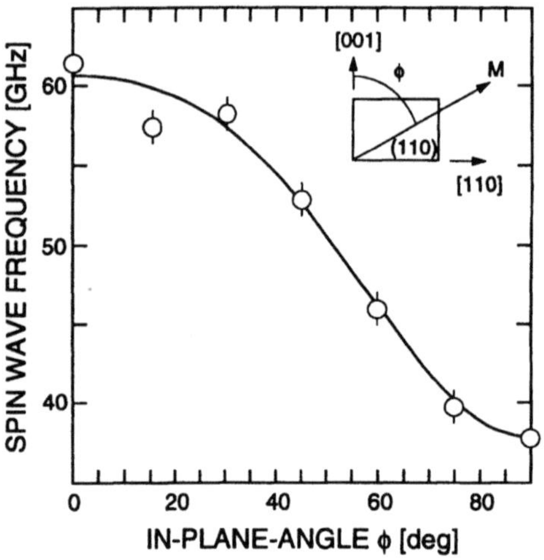

Fig. 3.23. Measured (circles) and calculated (full line) room temperature spin wave frequencies for a Co/Pd(110) superlattice as a function of the in-plane direction of the magnetic field as well as the magnetization measured against the [001]-axis in an applied field of 12 kOe. The Co layer thickness is 12.3 Å and the Pd layer thickness is 10 Å. The number of bilayers is 10 (adapted from [3.181])

wave frequencies for a Co/Pd(110) sample as a function of the direction of the applied field and the magnetization in the layer plane measured against the [001]-axis. The applied field of 12 kOe is large enough to force the direction of magnetization into the direction of the applied field to within ±5°. The Co layer thickness is 12.3 Å and the Pd layer thickness is 10 Å. The twofold symmetry of the in-plane anisotropy is well observed. The spin wave frequencies are largest for $\phi = 0$, i.e., for M_s parallel to the [001]-axis. Thus the in-plane [001]-axis is an easy axis for the magnetization in the multilayer stack in contrast to bulk Co. The obtained anisotropy constants are for the perpendicular anisotropy contribution (3.67) $K_{\text{eff}} = -9.5 \times 10^6$ erg/cm^3 and for the uniaxial in-plane volume anisotropy of second order $K_{\text{p}} = 3.5 \times 10^6$ erg/cm^3. The negative sign of K_{eff} indicates that the sample is magnetized in the film plane. The in-plane anisotropy is found to be magnetoelastic in origin: Using the results of Sect. 3.2.2, first the strain within the Co layers is estimated to be $\epsilon_1 = 3.5\%$, $\epsilon_2 = 3.2\%$ and $\epsilon_3 = -2.4\%$. Please note that $\epsilon_1 \neq \epsilon_2$ since the [001]- and the [1$\bar{1}$0]-directions are elastically not equivalent. The in-plane second order magnetoelastic anisotropy constant $K_{\text{p}} = 18.9 \times 10^6$ erg/cm^3 is calculated using the fcc-Co magnetostriction constants $\lambda_{100} = 130 \times 10^{-6}$ and $\lambda_{111} = -65 \times 10^{-6}$ extrapolated for Co rich fcc Pd-Co alloys [3.154]. Its value agrees with the experimental value within an order of magnitude. In particular, the sign is the same as that which determines the easy direction for this anisotropy contribution. Taking into account that for both the strains in the Co layers and the magnetostriction constants of fcc Co only estimates can be made, this agreement is rather good. Using the results of recent x-ray measurements of the in-plane and out-of-plane strains in Co/Pt multilayers yielded an even better agreement of the calculated magnetoelastic anisotropy and the experimental data [3.182].

3.5.3.4 Co/Au Superlattices. Co/Au multilayered structures have spurred strong interest since they exhibit unusual structural and magnetic properties. Magnetic properties, such as the strength of perpendicular anisotropy, are known to depend largely on the chemical and topographic quality of the interfaces involved. The interfaces are very sensitive to the preparation conditions and, because of the immiscibility of Co and Au in the bulk, can be modified in this system by postannealing the samples [3.183]. For samples prepared by ion beam sputtering techniques, as reported by *den Broeder* et al. [3.183], the easy axis of magnetization can be turned out of the film plane in a postannealing process for $d_{Co} < 14$ Å. This is interpreted by assuming "interface sharpening", i.e., the thickness of the intermixing zones at the interfaces in the as-prepared samples is reduced due to backdiffusion of Co and Au at elevated temperatures because both constituents are mutually insoluble below 420° C [3.184]. This assumption has been tested by x-ray diffraction studies, in which the expected increase in peak intensities of Bragg diffraction from the multilayer periodicity has been clearly observed [3.183, 3.185]. In the following we show how the spin wave spectra and the anisotropies are modified by the backdiffusion process [3.186]. The samples were prepared on epipolished sapphire substrates by ion-beam sputtering in an ultra-high vacuum system. The preparation and characterization is described elsewhere [3.185, 3.186]. The samples consist of 70 bilayers with a Co layer thickness of 8.8 Å and a Au layer thickness of 7.5 Å.

Figure 3.24 shows the measured dependence of the spin wave frequencies on the external field applied parallel to the layers [3.186]. In the top part of Fig. 3.24 the as-prepared sample is measured, and in the bottom part the sample has been annealed for 1 hour at 200°C prior to the measurement. The full lines represent a data fit assuming that the magnetization lies parallel to the layers for all applied field strengths. The good agreement between the data of the as-prepared sample with the fit proves that the sample is in-plane magnetized for all field strengths. The values obtained for the saturation magnetization and the out-of-plane uniaxial anisotropy, K_{eff}, are $4\pi M_s = 15.7 \pm 0.1$ kG and $K_{\mathrm{eff}} = (-2.58 \pm 0.2) \times 10^6$ erg/cm^3. For the annealed sample a good agreement between the experimental data and the fit is achieved for H $\gtrsim$ 4 kOe. The obtained values are $4\pi M_s = 14.5 \pm 0.2$ kG and $K_{\mathrm{eff}} = (-0.85 \pm 0.2) \times 10^6$ erg/cm^3. The value of K_{eff} is very close to zero albeit still negative, indicating that although the sample is supposed to be in-plane magnetized for zero field an additional small amount of anisotropy would turn the direction of magnetization out of plane. At $H = 3.8$ kOe the lowest frequency mode goes soft indicating that for $H < 3.8$ kOe the assumed magnetization ground state,

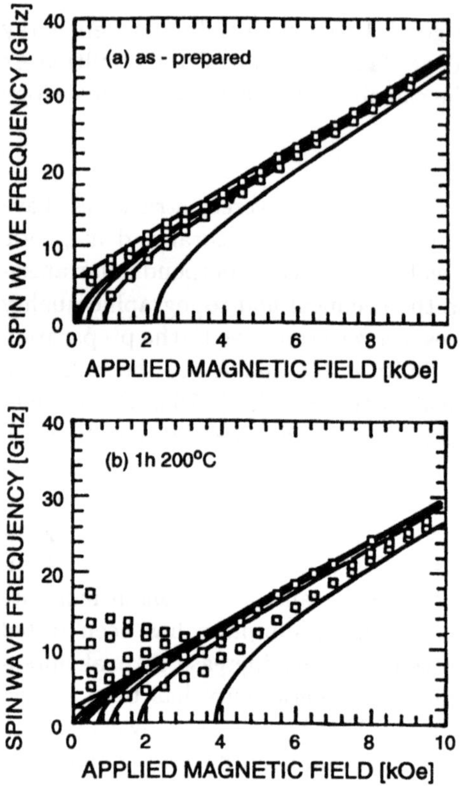

Fig. 3.24. Measured (squares) and fitted (full lines) spin wave frequencies as a function of the applied field for the as-prepared Co/Au multilayer sample (*top*) and the same sample annealed for 1 hour at 200° C (*bottom*). The Co (Au) layer thickness is 8.8 Å (7.5 Å), the number of bilayers is 70 (adapted from [3.186])

which is a homogeneous, in-plane aligned magnetization state, is unstable. Experimentally, for $H < 4\,\text{kOe}$, the frequency increase observed in Fig. 3.24 (bottom) of some spin wave modes with decreasing field strength indicates that the magnetization turns out of plane, at least for some domains. This phenomenon, which points to a distribution of magnetization directions in this field range, is discussed in Sect. 3.5.5.1.

3.5.4 Interlayer Exchange Coupling

Antiferromagnetic interlayer exchange coupling between ferromagnetic layers in sandwich and superlattice structures has become one of the most discussed phenomena in magnetism of layered structures. Although the existence of antiferromagnetic coupling is already very surprising in itself, the observation of exchange coupling oscillating in strength and sign as a function of spacer layer thickness is even more exciting. *Grünberg* et al. [3.26] first discussed experimental evidence for antiferromagnetic interlayer exchange coupling in Fe/Cr/Fe sandwich structures, followed by the discovery of oscillatory interlayer exchange as a function of the spacer material thickness in different multilayered structures by *Parkin* et al. [3.187]. Since then, many other systems with antiferromagnetic or oscillatory interayer coupling have been found [3.188, 3.189, 3.190, 3.191, 3.192, 3.193, 3.194, 3.195], even with two oscillation periods [3.196, 3.197, 3.198, 3.199, 3.200]. The interest in these phenomena was further boosted by the discovery of the so-called "giant magnetoresistance" effect in the antiferromagnetically coupled regimes [3.188, 3.201], making the effect a promising subject for, e.g., designing magnetoresistive reading heads for magnetic storage devices.

The most pronounced interlayer exchange coupling effect is due to indirect exchange through the conduction electrons of the nonmagnetic spacer layers (Ruderman–Kittel–Kasuya–Yoshida (RKKY-) interaction). It is of long range order. The oscillatory coupling arises from the oscillating spin polarization of the conduction electrons. The periodicity (periodicities) is (are) connected with the caliper distance(s) of the Fermi surface along the film normal. Apart from this dominantly bilinear interlayer exchange coupling mechanism other coupling mechanisms, such as biquadratic coupling caused by interface roughness and frustrated bilinear coupling, dipolar antiferromagnetic coupling due to interface roughness, and nontrigonometric coupling across antiferromagnetically ordered spacer layers are reported. For reviews see [3.202, 3.203].

The bilinear interlayer exchange coupling strength, A_{12}, can be obtained in the antiferromagnetic regimes from magnetometry measurements of the saturation field [3.187]. For the ferromagnetic regimes, so-called exchange-biased [3.204] and spin-engineered [3.205] layered structures were investigated. As has been pioneered by *Grünberg* for magnetic sandwich structures, an easier access to A_{12}, both in the ferro- and the antiferromagnetic regime, is provided by Brillouin light scattering [3.20, 3.51, 3.80, 3.118, 3.200, 3.206, 3.207].

In a phenomenological classification the interlayer exchange coupling is of bilinear, biquadratic or of non-trigonometric type [3.203]. With 2α the angle between the magnetizations of the two coupled, in-plane magnetized layers the corresponding free coupling energies are:

$$E_{\text{bilinear}} = -A_{12} \cos 2\alpha \; , \tag{3.85}$$

$$E_{\text{biquadratic}} = B_{12} \cos^2 2\alpha \; , \tag{3.86}$$

$$E_{\text{non-trig.}} = C_+ (2\alpha)^2 + C_- (2\alpha - \pi)^2 \; , \tag{3.87}$$

with A_{12} (B_{12}) the bilinear (biquadratic) coupling constant and C_+ and C_- the ferromagnetic and antiferromagnetic coupling constants. Minimizing the free energy expressions (3.85–87) together with Zeeman and anisotropy energy terms yields the canting angle 2α as a function of the applied field. Bilinear coupling is considered in the Hoffman boundary conditions (3.39). In the case of biquadratic or non-trigonometric coupling, (3.86) and (3.87) need to be modified by including out-of-plane contributions of the layer magnetizations due to the moment precessions. The resulting boundary conditions are rather complex, they can only be reduced to simple expressions in the case of parallel or antiparallel alignment of the layer magnetizations. Results are reported for the biquadratic coupling case by *Macciò* et al. [3.208] and for the non-trigonometric case by *Tschopp* et al. [3.209]. Although the type of coupling determines the remagnetization curves to a large extent, the dynamic properties depend to a much higher degree on the underlying coupling mechanism. In the case of nontrigonometric coupling the layer magnetizations show only an asymptotic approach to saturation with increasing field. Therefore the optic spin wave mode does not soften, contrary to the case of bilinear and biquadratic coupling [3.209]. With a fully developed theory BLS can serve as a very sensitive tool for characterizing the interlayer exchange coupling mechanism.

3.5.4.1 Fe/Au/Fe and Fe/Cr/Fe Trilayers. The Fe/Cr/Fe trilayer system is the best studied system for oscillatory interlayer exchange coupling. The reader is referred to [3.202, 3.203] for a discussion of the coupling mechanisms. Here we only discuss the determination of the interlayer exchange coupling strength by Brillouin light scattering.

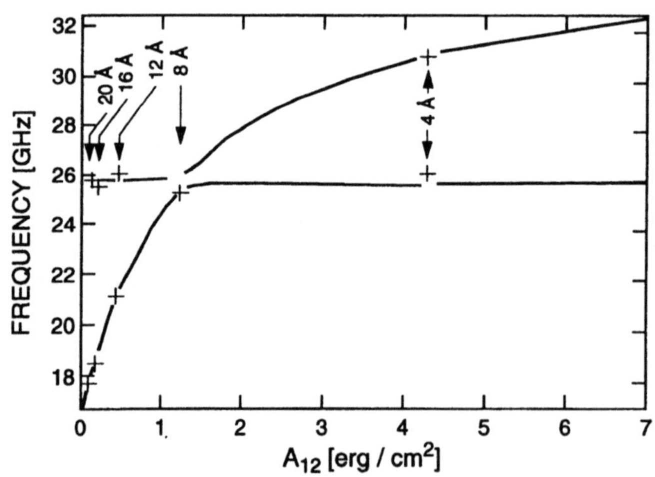

Fig. 3.25. Theoretical (solid lines) and experimental data (crosses) for spin wave frequencies in the Fe/Cr/Fe system at room temperature. The applied field is 3 kOe applied along the easy in-plane axis. For the calculation the following parameters are used: $q_{\|} = 1.73 \times 10^5\,\mathrm{cm}^{-1}$, $4\pi M_s = 20.7\,\mathrm{kG}$, $A = 1.6 \times 10^{-6}\,\mathrm{erg/cm}$, $K_1 = 4.0 \times 10^5\,\mathrm{erg/cm^3}$, $g = 2.1$. The Fe layers are both 106 Å thick. The Cr layer thickness is indicated by arrows (adapted from [3.206])

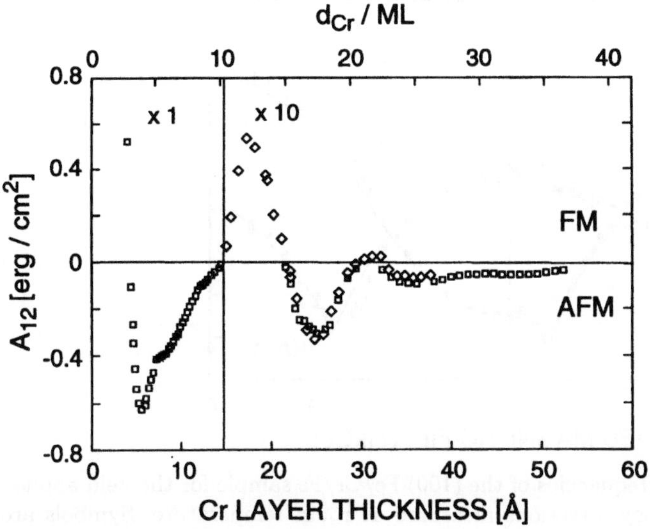

Fig. 3.26. Interlayer constant A_{12} as a function of d_{Cr} at room temperature, obtained from BLS (◇) and $M(H)$ (□) measurements (adapted from [3.43])

The calculation of the spin wave frequencies in the antiferromagnetic coupling case is governed by the equations of motion (3.1) and the Hoffmann boundary conditions (3.39). The calculations are, however, complicated by the presence of the rotation of the two magnetization vectors in the two magnetic layers as a function of the external field and/or in-plane anisotropies, which involves algebraically intensive matrix rotations. Details of the calculation are given in [3.206, 3.207, 3.208, 3.209, 3.210, 3.211].

In the Fe/Cr/Fe trilayer systems two low frequency modes exist (see Fig. 3.3). One is the Damon–Eshbach mode of the combined system, which is independent of interlayer exchange coupling, since the precessions of the spins in both layers are in phase. The other mode, the so-called optic mode, is sensitive to interlayer exchange coupling. It shifts to larger (smaller) frequency values with increasing (decreasing) interlayer exchange coupling strength both in the positive and negative regimes of the coupling constant. Figure 3.25 shows one of the first measurements of the two modes by *Barnaś* and *Grünberg* [3.206]. By comparing the measured spin wave frequencies with those calculated as a function of the interlayer exchange coupling constant, A_{12}, the value of A_{12} is determined from the experiment. Figure 3.25 demonstrates the independence of the Damon–Eshbach mode on A_{12} and the large dependence of the optic mode, in particular for small A_{12}. A detailed study of the interlayer coupling was performed by *Demokritov* et al. [3.196]. Figure 3.26 demonstrates the oscillatory behavior of A_{12} as a function of the Cr-spacer thickness, measured by means of BLS. The values of A_{12} are determined both in the antiferro- and in the ferro-magnetic coupling regimes. Current state-

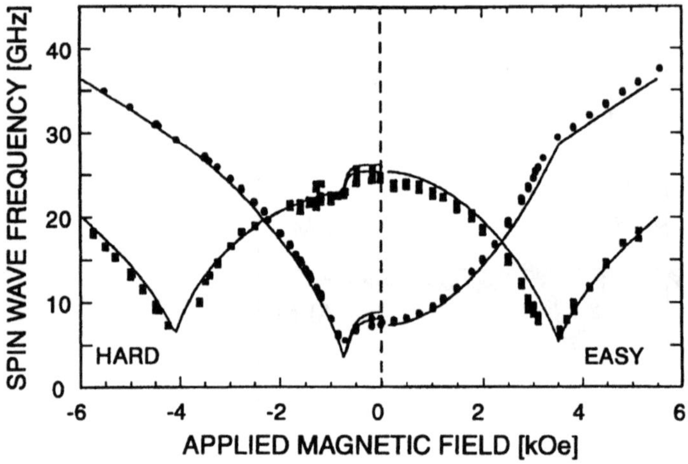

Fig. 3.27. Spin wave frequencies of the (100) Fe/Cr/Fe sample for the field applied along the hard and easy directions measured at room temperature. Symbols are experimental points, the lines represent a fit. For clarity the hard axis results have been plotted along the negative field axis (adapted from [3.211])

of-the-art results [3.211] on Fe/Cr/Fe(100) trilayers are shown in Fig. 3.27. Here a fourfold in-plane anisotropy acts on the direction of magnetization in addition to the interlayer exchange coupling. Measurements are shown for the external field applied along the in-plane easy direction (positive field values) and along the in-plane hard direction (plotted as negative field values). The full lines are a fit to the experimental data (dots and squares) with the bilinear and biquadratic exchange coupling constant as fit parameters. The results are $A_{12} = 0.57 \pm 0.02$ erg/cm^2 and $B_{12} = 0.003 \pm 0.003$ erg/cm^2. The saturation magnetization of $4\pi M_s = 19.8$ kG and the in-plane anisotropy of 2.2×10^5 erg/cm^3 were determined independently. The data shows very clearly the properties of the acoustic and the optic spin wave modes in the aligned state ($| H | > 3.5$ (4.0) kOe in the easy (hard) direction and in the canted state. At a field of 2.4 kOe (easy direction) both modes cross. For measurements along the hard direction (negative field values in Fig. 3.27) a characteristic change in frequency at -0.7 kOe is observed, which is due to the compensation of the in-plane anisotropy by the applied field. The excellent agreement between the fitted curves and the experimental data demonstrates the power of the BLS technique for determining exchange coupling strengths and anisotropies.

3.5.4.2 Exchange-Dominated Collective Spin Waves. For magnetic multilayers only dipolar interactions between magnetic layers within the superlattice stack have been discussed so far. We will now discuss experimental results for the additional influence of interlayer exchange interaction on the spin wave properties.

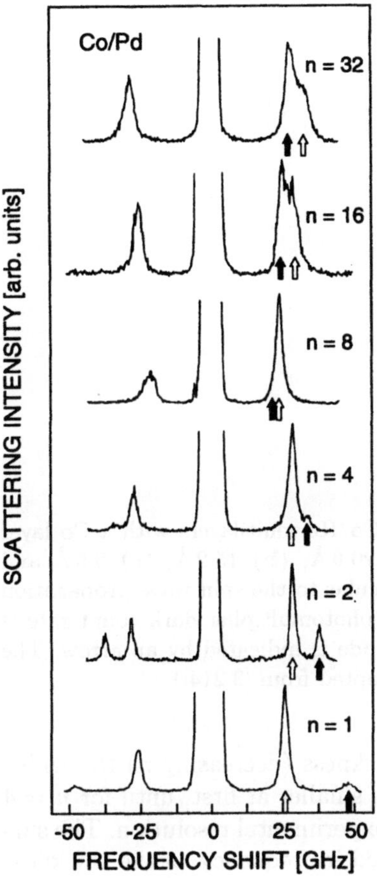

Fig. 3.28. Room temperature Brillouin light scattering spectra of Co/Pd multilayers at $H = 5\,\mathrm{kOe}$ applied parallel to the layer planes and perpendicular to the scattering plane. The number of atomic layers per magnetic and nonmagnetic layer is indicated by n. The number of bilayers is $N = 30$. The open arrows denote modes which are predominantly surface-mode-like in character, the full arrows denote those which are mainly bulk-mode-like in character (adapted from [3.212])

We first turn to the experimental demonstration of the numerical findings of the collective exchange modes [3.212,3.213]. The samples used are magnetically enhanced triode sputtered (111) c-axis textured Co/Pd superlattices of 30 bilayers. The samples were grown with the number of atomic layers, n, in each Co and Pd layer the same, varying for different samples between 1 and 32. The sample preparation and characterization is reported in [3.91].

Figure 3.28 shows the Brillouin light scattering spectra of a series of Co/Pd multilayer samples in an external magnetic field of $H = 5\,\mathrm{kOe}$, applied parallel to the layer planes and perpendicular to the scattering plane. The applied field is large enough to generate a single magnetic domain with the direction of magnetization lying in-plane. The number of bilayers is $N = 30$. For the sample with thickest layers, $n = 32$, we obtain a spectrum typical for collective dipolar spin wave excitations. The mode character changes from surface-mode-like at the upper band edge (indicated by the open arrows in Fig. 3.28) to bulk-mode-like at the lower band edge (full arrows in Fig. 3.28) as determined by the characteristic Stokes/anti-Stokes

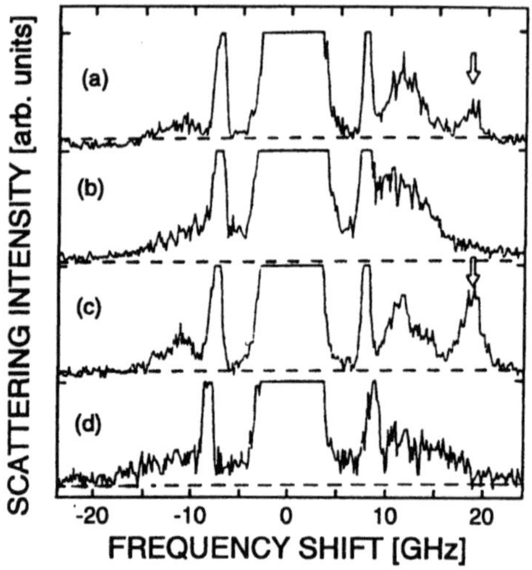

Fig. 3.29. Room temperature BLS spectra of Co/Ru multilayers with a Co layer thickness of 20 Å and a Ru thickness of **(a)** 20.9 Å, **(b)** 15.2 Å, **(c)** 9.5 Å and **(d)** 3.8 Å. The magnetic field applied perpendicular to the spin wave propagation direction is 1 kOe. The background due to the photomultiplier dark count rate is indicated by dashed lines. The stack surface mode is indicated by an arrow. The maxima near ± 8 GHz are surface phonons (adapted from [3.214])

asymmetry [3.29]. With decreasing layer thickness (decreasing n) the width of the band of collective excitations becomes smaller at first, until for $n = 4$ the observed band width comes close to the experimental resolution. The surface mode then remains essentially unchanged in frequency as the thickness is decreased further. The frequencies of bulk modes, however, move above the stack surface mode, and show a large increase in frequency with decreasing n. Here the bulk modes contain, to a large degree, interlayer exchange energy, and thus they are identified as collective exchange-dominated spin wave modes.

3.5.4.3 Spin Waves in Antiferromagnetically Coupled Multilayers. We now demonstrate the influence of antiferromagnetic as well as oscillating interlayer exchange coupling on the spin wave frequencies in magnetic superlattice structures for the case of sputtered Co/Ru superlattices [3.214,3.215]. For this system A_{12} oscillates as a function of the ruthenium thickness with a periodicity of 11.5 Å [3.187,3.214,3.215]. Figure 3.29 shows four spectra of Co/Ru multilayers with a Co layer thickness of 20 Å and a Ru layer thickness of (a) 20.9 Å, (b) 15.2 Å, (c) 9.5 Å and (d) 3.8 Å measured with an applied magnetic field of 1 kOe parallel to the layers. The peaks at ± 8.5 GHz correspond to the surface phonon (Rayleigh mode) of the system and are not further considered here. In all spectra we observe a band of collective spin wave excitations in the frequency range between 10 GHz and 20 GHz. Near

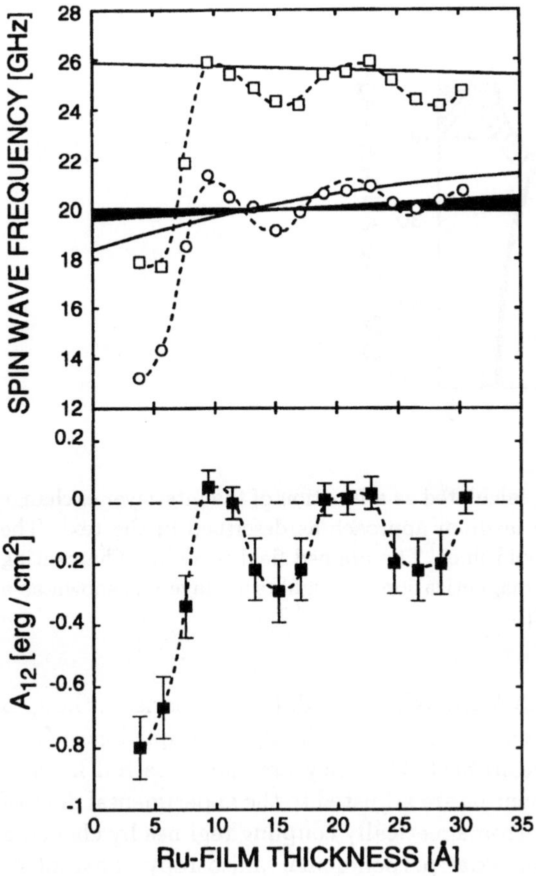

Fig. 3.30. *Upper part:* Room temperature spin wave frequencies of the stack surface mode (squares) and the bulk modes (circles) as a function of the Ru layer thickness of Co/Ru multilayers measured at an applied field of 3 kOe. The Co layer thickness is 20 Å and the number of bilayers is $N = 20$. For comparison the spin wave frequencies calculated for zero interlayer exchange coupling are shown as full lines. *Lower part:* Experimentally determined values of the interlayer exchange constant, A_{12}, as a function of Ru layer thickness (adapted from [3.214])

19 GHz the stack surface mode (marked in Fig. 3.29 with an open arrow) is identified in Fig. 3.29a,c by its characteristic Stokes/anti-Stokes intensity asymmetry. This pronounced mode is only observable in the regimes of $d_{Co} =$ 10..14 Å and 20..24 Å, which are identified as the ferromagnetically coupled regimes. Otherwise (cf. Fig. 3.29b,d) the mode is shifted to lower frequencies and merges with the other band modes.

In the upper part of Fig. 3.30 the frequency positions of the stack surface mode (squares) and the center of the bulk modes (circles), measured in an applied field of 3 kOe, are plotted as a function of the Ru layer thickness.

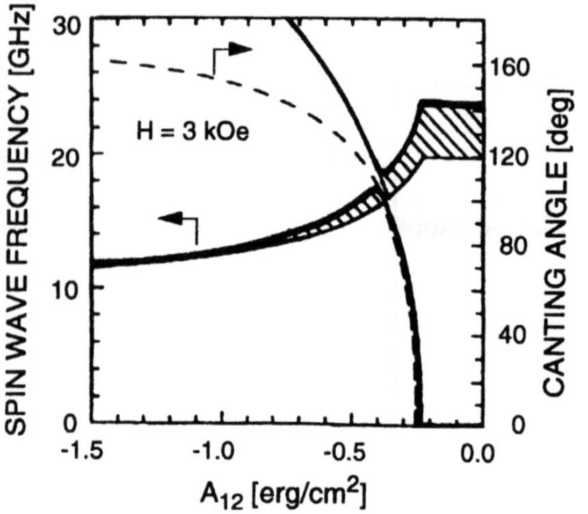

Fig. 3.31. Spin wave frequencies calculated as a function of the interlayer exchange constant, A_{12}, using an effective medium approach as described in the text. The surface modes are indicated by bold lines. The applied field is 3 kOe. The canting angle 2α between the saturation magnetization of neighboring layers is shown as a dashed line (adapted from [3.214])

Oscillations with a period of 11.5 Å are well resolved. For comparison, the spin wave frequencies have been calculated for the exchange-uncoupled case ($A_{12} = 0$) using the model described in Sect. 3.2. They are shown as full lines in Fig. 3.30, upper part. The frequencies are adjusted to the experimental data of the stack surface mode in the ferromagnetically coupling regimes by choosing an appropriate value for the uniaxial perpendicular anisotropy constant of hexagonal symmetry $K_1 = 4.7 \times 10^6\,\mathrm{erg/cm^3}$, which is typical for Co layers. The regimes of Ru thicknesses exhibiting reduced spin wave frequencies are identified as the antiferromagnetically coupling regimes as described further below. In particular for $d_{\mathrm{Ru}} < 6$ Å a very large antiferromagnetic coupling is revealed by the large frequency decrease.

We will now discuss the spin wave properties in the antiferromagnetic coupling regimes. For not too large external fields the magnetizations of neighbouring layers are canted with respect to each other. The canting angle depends on the (negative) value of A_{12} and the strength of the applied field. Calculations of the spin wave frequencies in this regime have been performed based on an effective medium model [3.85, 3.86]. The total multilayer stack is treated as a ferromagnetic film with effective susceptibilities that include exchange coupling. The susceptibilities are calculated assuming that the electromagnetic fields vary only slightly across each bilayer period. Therefore this model allows us only to calculate the frequencies of the stack surface mode and the first few bulk modes, but these are the modes that contribute mostly to the light scattering cross section.

In Fig. 3.31 the calculated spin wave frequencies are plotted as a function of the interlayer exchange constant, A_{12}. For comparison, the calculated canting angle 2α of the saturation magnetization between neighboring magnetic layers is shown as a dashed line. For $A_{12} > -0.25\,\mathrm{erg/cm^2}$, i.e. for zero canting angle, the spin wave frequency of the stack surface mode is independent of A_{12}, since here the net magnetization is constant.

Canting of the magnetization occurs in Fig. 3.31 for $A_{12} < -0.25\,\mathrm{erg/cm^2}$. Here the spin wave frequencies display a more complicated behavior. There are now two surface modes, indicated in the figure by bold lines, and we can observe that one of these surface modes goes soft when the magnetizations lie parallel to one another at $A_{12} = -0.25\,\mathrm{erg/cm^2}$. The bulk spin wave bands are shown as hatched areas. For $d_{\mathrm{Co}} = d_{\mathrm{Ru}}$, as assumed in Fig. 3.31, the surface modes are not well defined and exist at the top of the bulk bands, thus forming the upper frequency limit of the dipolar bulk modes.

There are now two bands. One band is the continuation of the collective spin wave band of ferromagnetically coupled layers. Its frequencies decrease with increasing negative value of A_{12}, i.e. with increasing canting angle. This band is crossed by a new collective band of bulk modes, which is reminiscent of the "optic" high-frequency spin wave mode of antiferromagnetic bulk material and goes soft for $A_{12} \geq -0.25\,\mathrm{erg/cm^2}$ [3.216]. The behavior of the former band, apart from the crossing regime, can be easily understood.

The modes respond to the net magnetization in the direction of the field. As the canting angle increases, the net magnetization in the direction of the field decreases approximately according to

$$M_{\mathrm{s}}(\alpha) = M_{\mathrm{s}} \cos\alpha = -\frac{HM_{\mathrm{s}}^2 d}{4A_{12}}\,, \tag{3.88}$$

where M_{s} is the net magnetization of each film and α is the canting angle [3.86]. In the simple case of a semi-infinite multilayer structure without anisotropies, the frequency of the surface mode is well described by

$$\frac{\omega}{\gamma} = H + 2\pi M_{\mathrm{s}} \cos\alpha\,. \tag{3.89}$$

Similarly, the bottom of the associated bulk band is given by

$$\frac{\omega}{\gamma} = [H(H + 4\pi M_{\mathrm{s}} \cos\alpha)]^{1/2}\,. \tag{3.90}$$

Thus, measurements of the frequencies are measures of the interlayer exchange constant A_{12} and of the canting angle α, which are interrelated by (3.88). We emphasize that modes at these frequencies make the largest contribution to the light scattering cross section. For multilayer structures of finite thickness with anisotropies, however, the frequencies can be determined only numerically.

Fitting this model to the experimental data, values for the interlayer exchange coupling constant, A_{12}, are obtained. They are displayed in the lower part of Fig. 3.30. Although the error bars are rather large due to the experimentally observed large linewidth of the modes, the oscillations between ferro- and antiferromagnetic coupling are clearly observed. However, the error bars are too large to allow the determination of the decay in oscillation amplitude with increasing Ru spacer thickness d_{Ru}.

3.5.5 Systems with Spatial Inhomogeneities

In the preceding sections we discussed systems characterized by homogeneous material properties. We will now focus on systems where the internal field shows spatial variations. The equation of motion (3.1) is then only locally defined and new models are required to discuss the spin wave propagation across the regions of inhomogeneity.

A full development of the theory is still lacking, in particular for the Damon–Eshbach mode. *Stamps* et al. [3.217] discuss the propagation of the Damon–Eshbach mode across an inhomogeneous model film consisting of alternating areas with two values of the internal field. They find, that (i) lifetime shortening due to scattering from imperfections contributes only very weakly to a mode broadening, and (ii) the dominant effect is the creation of new spin wave states resulting in a large mode broadening effect. For this the wavelength of the spin wave may be much larger than the spatial variation length. In addition, an inhomogeneous line width broadening is obtained if the spatial variations of the internal field are large with respect to the spin wave wavelength, but still small compared to the laser focus diameter in a BLS experiment.

Spatial variations of the internal field are caused by variations in material properties like the composition, variations of the direction of magnetization if the film consists of domains, changes in anisotropy energy caused by surface anisotropy and thickness variations, spatially varying interface anisotropy caused by, e.g., the exchange bias coupling mechanism, and more. The measurement of the resulting line width broadening provides information about the degree and to some extent about the origin of the inhomogeneity.

3.5.5.1 Co/Pt, Co/Au Multilayers: Spatially Varying Anisotropies. We will first discuss the case of non-homogeneous superlattices. This case will be illustrated for Co/Pt superlattices as displayed in Fig. 3.32 [3.218]. For a superlattice with $d_{\mathrm{Co}} = d_{\mathrm{Pt}} = 5\,\text{Å}$, the collective exchange modes are well separated in frequency from the dipolar stack surface mode due to large ferromagnetic interlayer exchange coupling [3.212, 3.213, 3.218]. The observed width of the stack surface mode of 14 GHz is still much larger than the experimental resolution. This broadening is attributed to spatially varying anisotropies caused by thickness variations of the Co layers as follows: In Fig. 3.32 the frequency of this mode is calculated as a function of the Co layer thickness, d_{Co}, for an applied magnetic field of 8 kOe using for $4\pi M_s$

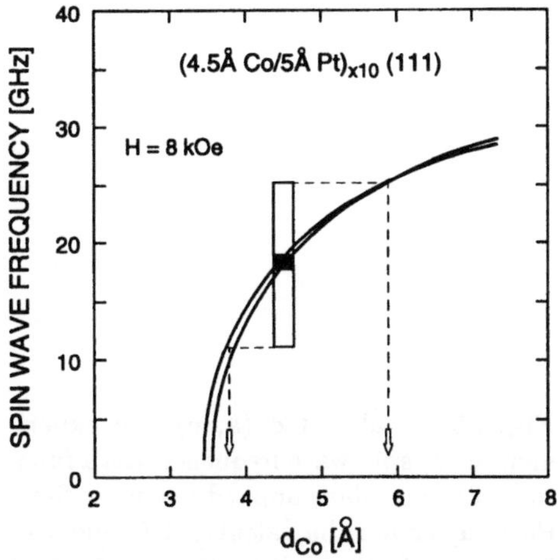

Fig. 3.32. Calculated spin wave frequencies of a Co/Pt multilayer structure with 10 bilayers of varying Co thickness and of 5 Å Pt thickness in an external field of 8 kOe as a function of the Co thickness. The experimental, broad mode is shown as a bar with the intensity maximum marked with a black square. From the frequency spread of the mode (range of the bar) the corresponding change in Co layer thickness is estimated as indicated by the dashed lines (from [3.218])

and the uniaxial anisotropy constant, K_1, the results obtained from a fit to the experimental data, as reported elsewhere [3.219]. By varying d_{Co}, the contribution of the interface anisotropy field to the internal field varies with $1/d_{Co}$. The frequencies go to zero at $d_{Co} \cong 3.4$ Å, indicating a perpendicular magnetized state for smaller values of d_{Co}. The experimentally observed line width of the peak is indicated as a bar in Fig. 3.32, with the center of the peak as a black square. From the length of the bar the range of spatial variations of d_{Co} of 3.8...5.8 Å is deduced, as illustrated in the figure. This would translate into variations in the interface anisotropy constant of $k_s = (0.21..0.32)\mathrm{erg/cm^2}$ on assuming flat interfaces.

We will now discuss the case of multilayer structures, in which the individual layer thicknesses vary from layer to layer. This might happen, e.g., due to changing deposition rates in the sample fabrication process. But even for samples with nominal identical thicknesses of all magnetic layers, the "local" thickness, i.e. the thickness on a length scale of the wavelength of the spin waves, may vary from layer to layer in the same manner like the thicknesses vary laterally as discussed above.

We will assume a system with large interface anisotropy values, which therefore exhibits a large dependence of spin wave frequencies on the layer thickness [3.181, 3.186]. Without an external field the uniaxial anisotropy of each layer is assumed to be large enough to force the direction of magnetization perpendicular to the layer planes. Figure 3.33a shows the calculated spin wave frequencies of a "perfect" multilayer structure of 8 bilayers with the

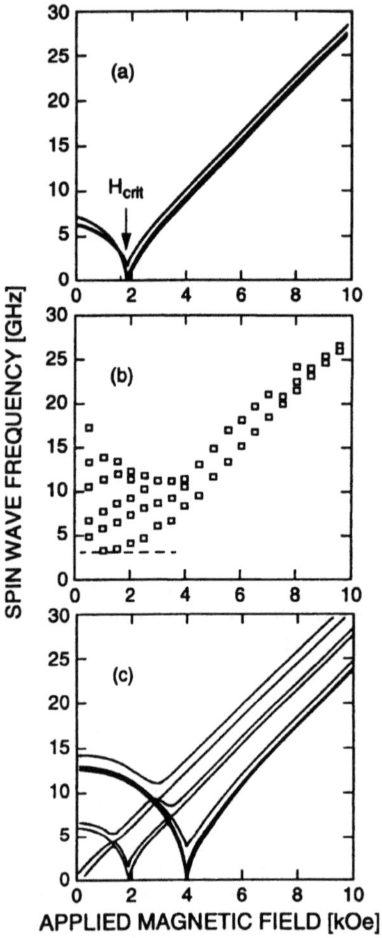

Fig. 3.33. Calculated (**a**, **c**) and experimental (**b**) spin wave frequencies as a function of the in-plane applied magnetic field. Part (**a**) shows the calculated frequencies for a "perfect" multilayer consisting of 8 bilayers of same parameters. The thickness of magnetic and nonmagnetic layers is 10 Å. In (**b**) experimental room temperature data of Co/Au multilayers with 70 periods and with nominal thicknesses of $d_{Co} = 8.8$ Å and $d_{Au} = 7.5$ Å are shown. The dashed line marks the threshold, below which spin wave observation is inhibited in the Brillouin light scattering experiment due to elastically scattered light. In (**c**) the layer thicknesses of the 8 magnetic layers of the multilayer are assumed to be 10, 9, 9, 11, 10, 9, 11 and 9 Å. The other parameters are as in (**a**) (adapted from [3.181])

same thickness of magnetic and nonmagnetic layers at 10 Å [3.181,3.186]. For the magnetic layers the bulk parameters of Co are assumed. With increasing in-plane applied magnetic field the spin wave frequencies first decrease while the direction of magnetization is increasingly tilted towards the layer planes, lying in the layer planes at and above a critical field strength, $H_{\rm crit} = 2$ kOe (Sect. 3.2). For $H > H_{\rm crit}$ the spin wave frequencies increase approximately linearly with further increasing external field. Near $H_{\rm crit}$ the calculated spin wave frequencies show a sharp minimum with some modes going soft.

Figure 3.33b shows experimental data for a Co/Au superlattice sample consisting of 70 bilayers of 8.8 Å Co and 7.5 Å Au, as discussed in Sect. 3.5.3.4. Above 4 kOe the spin wave frequencies increase approximately linearly with increasing applied field, a fact which is indicative of the saturation magnetization lying in-plane. Near $H_{\rm crit} \approx 3.5$ kOe the spin wave frequencies show a broad minimum, and they are rather widely spread below $H_{\rm crit}$. The observed behavior is in only rough qualitative agreement with the calculated field dependence of the spin wave modes shown in Fig. 3.33a. We will now show that by allowing the individual layer thicknesses to have a distribution about the mean thickness value the calculated spin wave properties resemble much better the experimental data.

Figure 3.33 shows calculated spin wave frequencies for the case, where the nonmagnetic layer thickness is fixed at 10 Å, but the magnetic layer thicknesses are 10, 9, 9, 11, 10, 9, 11 and 9 Å, respectively [3.40]. For this calculation A_{12} has been set to zero. For each layer, first the critical field, H_{crit}, is calculated as well as the direction of the magnetization as a function of the applied in-plane field. H_{crit} varies from layer to layer due to different layer thicknesses. Then the spin wave frequencies of the multilayer stack are calculated using an effective medium approach, using the static orientation of the layer magnetizations as input data [3.40]. The spin wave modes show zero frequencies at $H_{\mathrm{crit}} = 0$, 1.9 and 4 kOe. The values of H_{crit} correspond to the chosen thickness values of the magnetic layers of 11, 10, and 9 Å, respectively. The obtained spin wave mode distribution resembles the experimentally observed mode spectrum (Fig. 3.33b) in a much better way then does the calculation assuming the same parameters for each layer, as shown in Fig. 3.33a. Please note that spin wave modes with frequencies smaller than about 3 GHz (dashed line in Fig. 3.33b) are not accessible in the Brillouin light scattering experiment due to the overlap with elastically scattered laser light.

For a "real" multilayer structure, both thickness variations from layer to layer as well as thicknesses varying laterally due to, e.g., a mosaic spread, contribute to the effect. The Co/Au sample, of which the spin wave data are shown in Fig. 3.33b, was prepared by postannealing the sample in order to gain atomically sharp interfaces for maximizing interface anisotropies [3.183]. On the other hand, evidence has been found that the postannealing process introduces interface corrugations, which might be responsible for local, layer-to-layer thickness variations [3.220]. The pronounced difference in the spin wave properties between a "perfect" structure (Fig. 3.33a) and a "realistic" structure as described above (Fig. 3.33c) is already obtained for a corrugation of ± 1 Å of each layer.

3.5.5.2 Fe/Cu(001): Spin Waves in Films with Growth Domains. An excellent case for the study of spin wave propagation in a system with a finite number of internal fields is the system of 12–16 ML Fe on Cu (001) single crystal substrates grown at room temperature, as shown by *Scheurer* et al. [3.221]. In this thickness range Fe undergoes a phase transition from fcc to bcc with increasing thickness. As revealed by LEED studies the Fe films consist of four types of coexisting equivalent elongated bcc domains, as shown in Fig. 3.34. Measured in-plane angular dependencies of the coercive field by means of the magnetooptic Kerr-effect show a fourfold in-plane symmetry.

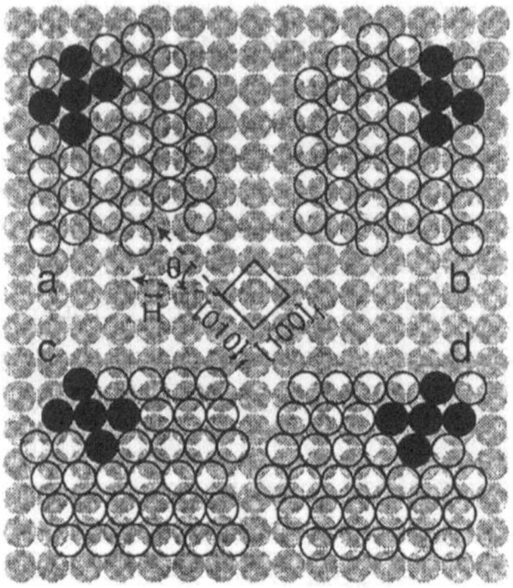

Fig. 3.34. The four growth domains of bcc Fe (open circles) on fcc Cu (shaded circles) drawn to scale, each with its bulk lattice constant. The solid circles illustrate the single rectangular unit cells of each domain (from [3.221]

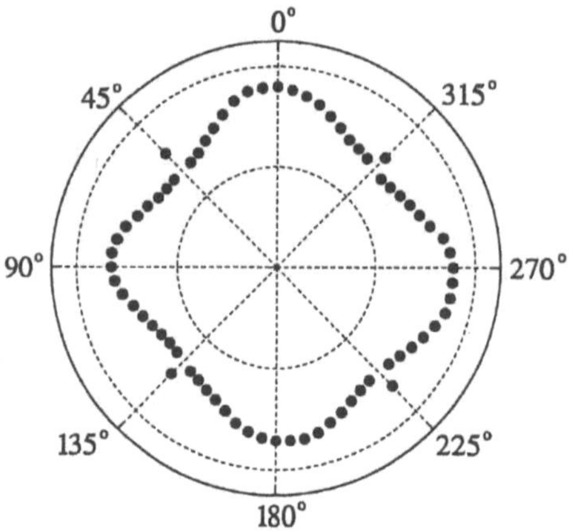

Fig. 3.35. The coercivity field, H_c, as a function of the in-plane direction of the applied field with respect to the $[3.100]_f$ direction, as measured by MOKE. The equifield line distance is 50 Oe. The Fe film is 25 ML thick and coated with 20-ML Cu (from [3.221])

Surprisingly very narrow maxima along the four hard directions are superimposed on broad minima along the in-plane $\langle 110 \rangle_f$ axes (the index f denotes those directions lying in the film plane). This is shown in Fig. 3.35. From Fig. 3.34 it is inferred that, without coupling, the easy and hard axes of the four growth domains are canted away from the $\langle 100 \rangle_f$-easy and $\langle 100 \rangle_f$-hard axes of the resulting film by $\pm$ 9.7°.

The unusual narrow maxima along the $\langle 110 \rangle_f$ axes can be understood by magnetically coupled twin domains, which are the domains (a) and (b), or the domains (c) and (d) in Fig. 3.34. The coupling generates a small, narrow local minimum in the combined anisotropy energy along the $\langle 110 \rangle_f$ axes caused by a small angle between the magnetizations $\boldsymbol{M}_a$ and $\boldsymbol{M}_b$ in the domains (a) and (b). The local minimum in the free energy gives rise to the observed intermediate easy axes.

Brillouin light scattering can be used to detect the different values of internal fields caused by the growth domains, since each value (modulus and

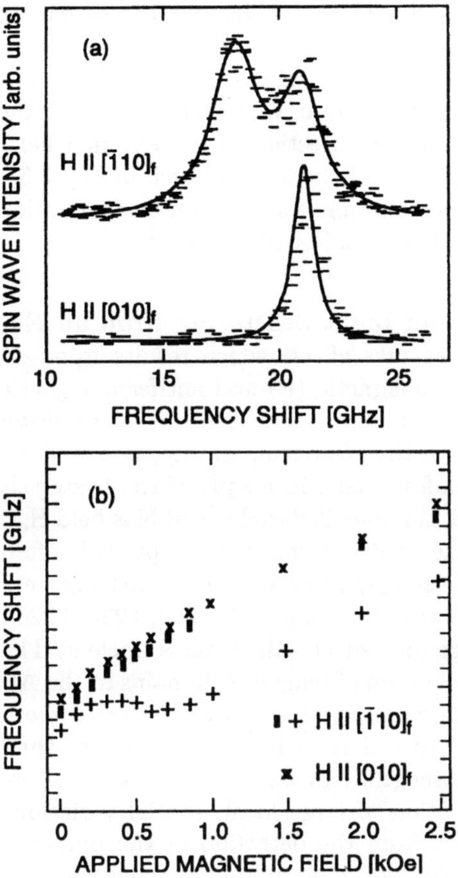

Fig. 3.36. (**a**) Spin wave spectra obtained for a 25-ML Fe film for H parallel to $[010]_f$ and H parallel to $[\bar{1}10]_f$ with $H = 2.5\,\mathrm{kOe}$. (**b**) Spin wave frequencies vs. applied magnetic field H for a 18-ML Fe film (adapted from [3.221])

direction) of internal field will result in a separate mode, apart from cases of mode degeneracy for symmetry reasons. As outlined by *Stamps* et al. [3.217] the size of the regions with constant internal fields, i.e., the size of the growth domains, may be smaller than the spin wave wavelengths, and still separate modes will be observed. This is due to the fact that spin waves corresponding to one value of the internal field can propagate through small regions of different internal fields without too much scattering [3.217]. In the presence of coupled twin growth domains, only one spin wave mode may exist for propagation along the easy $\langle 100 \rangle_f$ axes, since the internal fields for the (a, b)- and (c, d)-twin domains are symmetric with respect to the propagation direction. This is what is observed, as displayed in Fig. 3.36. Note that without domain coupling, two modes should be observable, since only the modes from (a) and (d), or those from (b) and (c) would be degenerate but not all four. In case of propagation along the $[3.110]_f$-direction, the degeneracy of the (a, b) and (c, d) twin domain fields no longer holds, and two modes are observed (see Fig. 3.36). This is very clearly seen in the field dependence of the spin wave frequencies of the two modes: the magnetization of the domain couple (a, b) lies in the frustrated state between the two hard axes of the two domains, which leads to the high frequency branch. For the domain couple (c, d) the field points along a hard direction. Here, with increasing field, the direction of magnetization is tilted into the direction of the external field, giving rise to the observed field dependence of the lower branch in Fig. 3.36. The data in Fig. 3.36 can be reproduced assuming a saturation magnetization of 13 kG and an anisotropy constant of $K_1 = 2.5 \times 10^5\,\mathrm{erg/cm^3}$.

3.5.5.3 FeNi/FeMn(110): Brillouin Light Scattering from an Exchange Bias System. The exchange bias effect, which results from exchange coupling between adjacent ferromagnetic (F) and antiferromagnetic (AF) layers which are either deposited in a magnetic field or cooled down in a magnetic field after heating above the Néel temperature, is a well established phenomenon [3.222]. It manifests itself in a shift of the hysteresis loop along the axis of the applied field, the so-called exchanged bias field H_{eb}, and it is often described as an in-plane unidirectional anisotropy. It is often accompanied by large changes in the magnetic anisotropies. Although exchange bias systems have been extensively studied [3.223, 3.224, 3.225, 3.226], the microscopic origin of the exchange bias effect still remains unclear. The most advanced models propose the formation of magnetic domains in the AF layer causing a macroscopic exchange coupling strength of the experimentally observed size [3.227], which is two orders of magnitude smaller than the atomic F-AF exchange coupling strength. Key issues to be tested experimentally are (i) to prove the assumption of large local, spatial variations of the F-AF coupling, and, (ii), apart from the detection of the unidirectional anisotropy, to search for possible additional anisotropy contributions of higher order induced by the exchange coupling mechanism. For both tasks, Brillouin light scattering is well suited. The first observation by BLS of the change in magnetic anisotropies caused by the exchange-bias effect was reported by *Ercole* et al. [3.228]. *Mathieu* et al. performed a detailed study of the exchange-bias effect by BLS, which is summarized in the following [3.229]. Here (110)-oriented FeNi/FeMn-bilayers were chosen. The (110)-oriented interface, which has two-fold symmetry, allows for an easy decomposition of the relevant anisotropy contributions.

The samples were grown by MBE onto Cu (110) single crystal substrates. The NiFe film thickness is 18 Å, 24 Å, 37 Å and 90 Å. Half of the sample was covered by an 80 Å thick FeMn layer. At this thickness the exchange bias effect is saturated. The sample was covered with a 30 Å thick protective Au layer. The preparation is described elsewhere [3.229]. During growth a magnetic field of $\approx$ 250 Oe was applied in the filme plane along the $[1\bar{1}0]$ direction to induce exchange bias. The (110)-oriented FeMn surface has an uncompensated spin structure with a resultant in-plane magnetization along the $\pm$ [001]-directions.

The Brillouin light scattering experiments were performed with the external field applied in the film plane and in magnetic saturation to ensure a one-domain state of the F-film. From the spin wave dispersion curves measured as a function of the in-plane angle of the external field with respect to the [001]-direction, ϕ_H, the free energy density, F_{ani}, was obtained. It is expressed as

$$\begin{aligned} F_{\rm ani} = & -K_{\rm s}^{(2)} \cos^2\theta + K_{\rm p}^{(1)} \cos(\phi - \phi_{\rm uni}) \sin\theta \\ & + K_{\rm p}^{(2)} \cos^2\phi \sin^2\theta \\ & + K_{\rm p}^{(4)} \cos^2\phi \sin^2\phi \sin^4\theta \, , \end{aligned} \tag{3.91}$$

with $K_{\rm s}^{(2)}$ the perpendicular anisotropy constant. The in-plane anisotropy constants $K_{\rm p}^{(1)}$, $K_{\rm p}^{(2)}$ and $K_{\rm p}^{(4)}$ are of unidirectional, uniaxial and fourfold symmetry, respectively, and ϕ is the in-plane angle of the direction of magnetization. $\phi_{\rm uni}$ describes the reference direction of the unidirectional anisotropy. All in-plane angles are measured relative to the [001] direction. θ is the out-of-plane polar angle.

First we discuss the uncovered $Ni_{80}Fe_{20}$ staircase-shaped samples for reference. The measured spin wave frequencies as a function of ϕ_H are displayed in Fig. 37a for $d_{\rm NiFe} = 18$ Å. For all F-layer thicknesses a nearly identical spin wave dependence on ϕ_H is obtained with a large uniaxial anisotropy contribution. The spin wave maxima which are present for the $\pm$[001] directions indicate that the easy axis of magnetization is along [001]. The corresponding anisotropy field of about 600 Oe is much larger than the value of 5 Oe usually found in polycrystalline $Ni_{80}Fe_{20}$ films grown in an external field. Its origin is still unclear though it may well be of magneto-elastic origin and caused by the $Ni_{80}Fe_{20}$ growth mode of long (500 Å), narrow islands with a length-to-width ratio of approximately 10 lying along the magnetically hard $[1\bar{1}0]$ direction, as has been shown by scanning tunneling microscopy.

Entirely different results are obtained upon covering the $Ni_{80}Fe_{20}$ layers with thicknesses from 18 Å to 90 Å with an 80 Å thick $Fe_{50}Mn_{50}$ layer as displayed in Fig. 37b–e. For the $Ni_{80}Fe_{20}$ layer thicknesses of 18 Å to 37 Å, the spin wave maxima and thus the easy axes of magnetization are shifted by 90°. Moreover, for the 18 Å thick $Ni_{80}Fe_{20}$ layer, and to a lesser degree for the 24 Å and the 37 Å thick films, the two maxima at ϕ_H equal to 90° and 270° do not agree in their spin wave frequencies as indicated in Fig. 37b by the dashed horizontal lines. The difference in frequencies provides a measure of the unidirectional anisotropy.

From a model fit using (3.91) all anisotropy constants were determined. The results are displayed in Fig. 3.38 for the uncovered and FeMn-covered $Ni_{80}Fe_{20}$ films. Again, first we discuss the uncovered samples (Fig.3.38a,b,c). The in-plane fourfold anisotropy contribution is more or less zero. A large uniaxial in-plane anisotropy of $K_{\rm p}^{(2)} = (-3 \pm 1) \times 10^5\ {\rm erg/cm^3}$ is obtained nearly independent of the $Ni_{80}Fe_{20}$ thickness within the error margins. As ex-

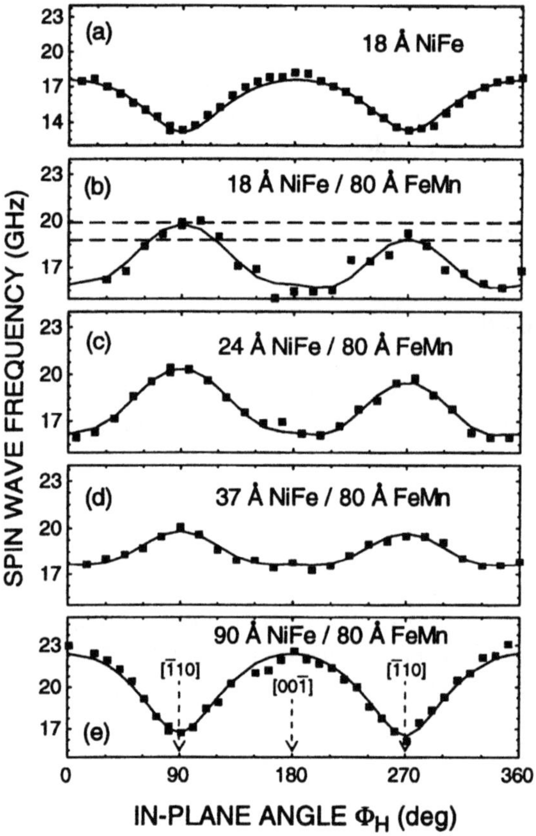

Fig. 37a–e. Spin wave frequencies as a function of the angle of the in-plane applied field, ϕ_H, with the in-plane [001] direction for the Cu(110)/$Ni_{80}Fe_{20}$/(80 Å $Fe_{50}Mn_{50}$)/Au staircase-shaped sample with $Ni_{80}Fe_{20}$ layer thicknesses of 18 Å, 24 Å, 37 Å and 90 Å. The full lines are least squares fits. The difference in frequency of the spin wave maxima for the covered $Ni_{80}Fe_{20}$ layer of 18 Å thickness, representing the unidirectional anisotropy contribution, is indicated by the dashed horizontal lines. The applied field was 3 kOe (from [3.229])

pected the unidirectional anisotropy parameter, $K_p^{(1)}$, was found to be nearly zero within the error margins.

The anisotropy behavior of the covered $Ni_{80}Fe_{20}$ films is displayed in Fig. 3.38d–f. All three in-plane anisotropy contributions decrease with increasing Ni_0Fe_{20} thickness. For large $Ni_{80}Fe_{20}$ thicknesses the respective anisotropy values of the uncovered and the covered $Ni_{80}Fe_{20}$ films converge, indicating an interface effect. Within the given limited accuracy the differences in the respective anisotropy values are consistent with a $1/d_{NiFe}$ scaling law for all three in-plane contributions. For the fourfold in-plane anisotropy, $K_p^{(4)}$, the data of the second sample agree only within a factor of two with

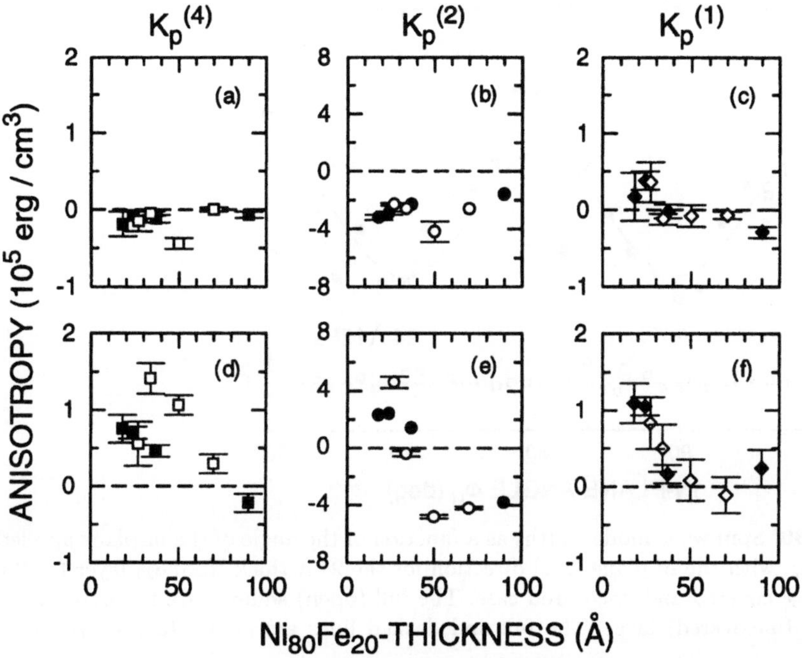

Fig. 3.38. Anisotropy constants obtained for the two staircase-shaped samples (open and closed symbols for the first and second sample, respectively) as a function of the $Ni_{80}Fe_{20}$ layer thickness for the uncovered layers (**a,b,c**) and the layers covered by 80 Å $Fe_{50}Mn_{50}$ (**d,e,f**) (from [3.229])

those of the first sample. This is probably due to a lesser quality in the film-substrate interface due to problems in the sputter cleaning process [3.229]. However, both samples show the same systematic decrease of this anisotropy contribution with increasing $Ni_{80}Fe_{20}$ thickness. For the uniaxial in-plane anisotropy, $K_p^{(2)}$, a change in sign is obtained for the first sample near (35 ± 5) Å and for the second sample near (50 ± 10) Å.

For the unidirectional anisotropy constant, $K_p^{(1)}$, we obtain for NiFe film thicknesses smaller than about 40 Å (i.e. the range were reliable conclusions can be made) an angle for the reference direction of $\phi_{\text{uni}} = 90°$, i.e. the easy direction of the unidirectional and the easy axis of the uniaxial anisotropy contributions are colinear.

While the origin of the unidirectional anisotropy contribution ($K_p^{(1)}$) is likely found in the exchange bias machanism, the cause of the large modifications of the two other in-plane anisotropies ($K_p^{(2)}$, $K_p^{(4)}$), compared to the uncovered $Ni_{80}Fe_{20}$ films, needs to be discussed. Since these modifications are identified as interface contributions it is very likely that they are also induced by the exchange coupling interaction and compete with the respective intrinsic contributions of the uncovered layer. Indeed, *Jungblut* et al. have recently

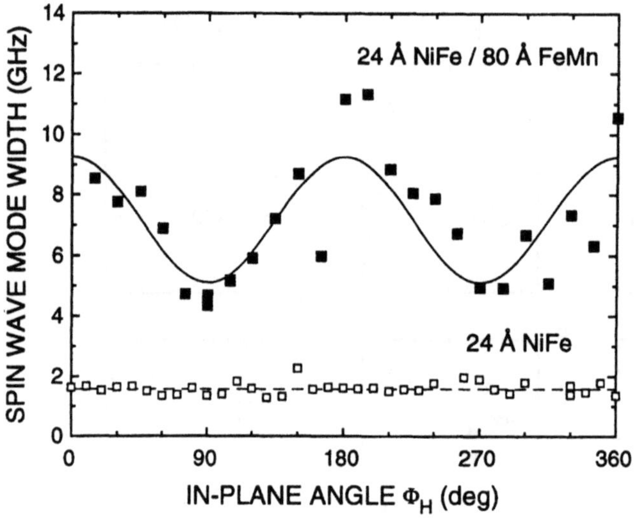

Fig. 3.39. Spin wave mode widths as a function of the angle of the in-plane applied field, ϕ_H, with the in-plane [001] direction for the 24 Å thick $Ni_{80}Fe_{20}$ layer for the $Fe_{50}Mn_{50}$ covered and uncovered case. The full (open) squares are the data of the covered (uncovered) layer. The full and dashed lines are guides to the eye (from [3.229])

shown that the exchange bias effect is causing the additional contribution in $K_p^{(2)}$ [3.230].

Of particular interest are the measurements of the spin wave line widths. Mode broadening for propagating spin waves is obtained if the internal fields vary locally on a length scale which is of the order of the spin wave wavelength (3000 Å). In the experiments the spin wave modes show a large mode broadening of more than a factor of six upon covering the NiFe layers by FeMn for the lower NiFe layer thicknesses. This is displayed in Fig. 3.39 for the 24 Å thick NiFe layer for the covered and uncovered case. For the covered sample the mode width varies within a factor of two as a function of the azimuthal angle, ϕ_H. As can be seen from Figs. 37 and 3.39 the maximum and minimum values correspond to the hard [001] and easy [1$\bar{1}$ 0] direction of magnetization, respectively. The line width is strongly decreasing with increasing NiFe layer thickness converging to the width of the uncovered NiFe films. This is also characteristic for an interface effect.

The large spin wave mode broadening and its dependence on the in-plane angle of the external field can be understood as follows: We assume variations in the local F–AF exchange field on a scale of the atomic terrace width, which is of the order of 50 Å [3.230]. From the local exchange field the macroscopic, averaged exchange bias field, or equivalently, the unidirectional in-plane anisotropy, are generated. The variations in the F–AF exchange field will cause a broadening of the spin wave line width, which therefore is a characteristic fingerprint of the variations [3.55, 3.217, 3.181]. The broadening is largest, if not only internal field contributions (here the exchange coupling field) vary, but also the direction of magnetization. The latter occurs if the easy axis of the spatially varying internal field is not collinear with the magnetization [3.231]. By changing the direction of the external field minima and maxima in the line width appear at the easy and hard axes of the dominating, exchange coupling induced uniaxial anisotropy.

3.5.6 Light Scattering from Microwave Excited Spin Waves

Since the beginning of the development of high performance interferometers for optical spectroscopy in the GHz regime there has been great interest in detecting microwave induced spin waves optically. It has been shown that Brillouin light scattering is an adequate method for the detection of the uniform mode under ferromagnetic resonance (FMR) conditions [3.232, 3.233, 3.234]. However, although the BLS technique is comparable in sensitivity to a standard FMR setup, this combination has not found widespread use due to the comparatively large experimental effort. The method found its place however for spatially resolved FMR measurements and for investigations of nonlinear phenomena.

Nonlinear, parametrically excited spin waves have first been observed by *Khotikov* and *Kreines* [3.95]. *Wettling* et al. observed nonlinear spin wave phenomena, in particular they determined the critical threshold power of the microwave field for generation of nonlinear excitations as a function of the wavevector, q, of the spin waves [3.93, 3.94]. *Srinivasan* et al. first observed $q \neq 0$ propagating magnetostatic spin waves in microwave device structures [3.128]. They have observed the parametric decay of the magnetostatic surface mode ($q \approx 10^2 \mathrm{cm}^{-1}$) into two backward volume spin waves ($q \approx 10^4 - 10^5\,\mathrm{cm}^{-1}$). Backward volume spin waves are excitations with negative dispersion such that the wavevector and the Poynting vector have opposite directions. In these experiments the spin wave modes could only be observed by BLS. *Kabos* et al. have investigated the case of perpendicular pumping, i.e. the microwave field is perpendicular to the direction of magnetization [3.97].

Brillouin light scattering as a detection method of spin waves has, compared to microwave techniques, the great advantage that spatially resolved measurements (laser spot diameter $\approx 30\mu\mathrm{m}$) can be performed at comparable sensitivity. This method is potentially suited for measurements of mode profiles of spin wave excitations. This has been performed first by *Azevedo* and *Rezende*, who have determined the mode profile of a ferromagnetic resonance mode in a microwave cavity [3.135]. This is shown in Fig. 3.40. The successful use of a combined Brillouin light scattering and ferromagnetic resonance technique at low temperatures for the determination of the magnetization and the coupling between two metallic magnetic layers has been shown by *Demokritov* on samples with a spatially varying spacer layer [3.131]. Recently, nonlinear phenomena in the propagation of spin waves in two-

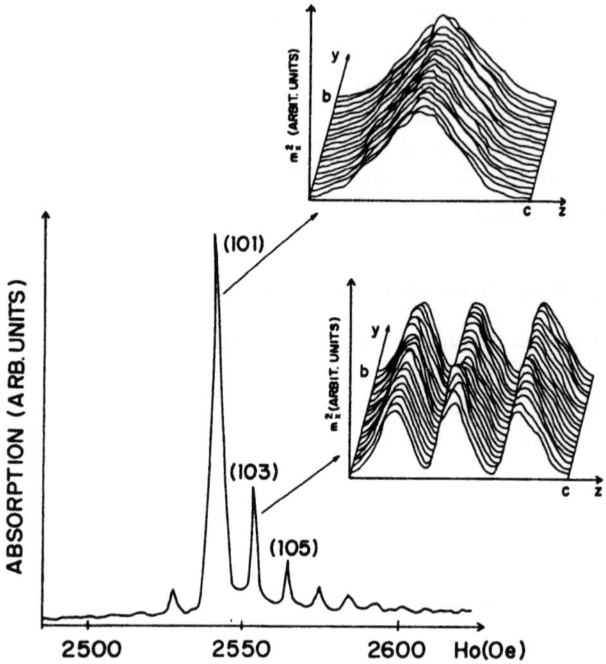

Fig. 3.40. Microwave absorption spectrum in a thin YIG slab at 9.4 GHz. Insets show the spatial variations of the rf magnetization squared for the (101) and (103) cavity modes measured with BLS (from [3.135])

dimensional films have been investigated by BLS. *Boyle* et al. showed first that the two-dimensional profile of nonlinear wave propagation in ferromagnetic films can be investigated by Brillouin light scattering [3.98]. *Bauer* et al. report about the first identification of self-focusing and initial stages of wave collapse for dipolar backward volume spin waves in garnet film media [3.99]. They show that for film wave guides of finite width the evolution of the wave beam, that is modulationally instable in both in-plane directions, does not lead to a collapse as predicted for an infinite film. Using a time resolving technique, the propagation of nonlinear spin wave pulses (solitons in one dimension and so-called spin wave bullets in an infinite film) [3.100] and collisions thereof [3.101] were observed.

3.5.7 Spin Waves in Corrugated and Patterned Films

Not much Brillouin light scattering work has been carried out so far on spin wave propagation in periodically corrugated or in patterned films. A number of calculations exist about this topic [3.235, 3.236, 3.237, 3.238, 3.239, 3.240, 3.241, 3.242, 3.243]. *Elachi* calculated the propagation of a magnetic wave in a periodic medium, in which the dielectric constant shows a periodic variation along the direction of the magnetization. The essential re-

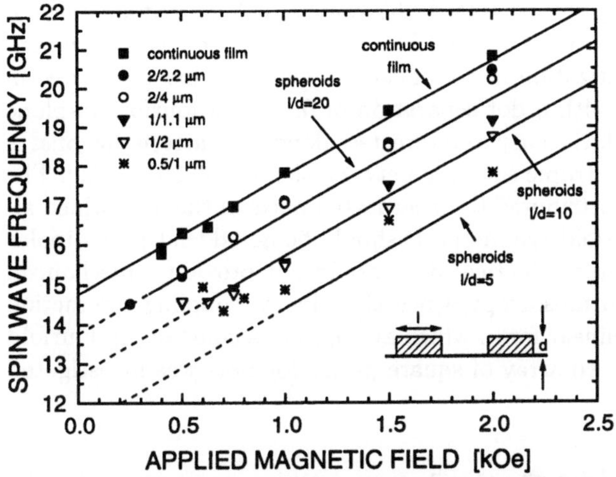

Fig. 3.41. Measured spin wave frequencies as a function of the strength of the applied external field, H. The symbols denote the measured data with the first number the dot diameter and the second number the dot periodicity. The dot thickness is 1000 Å. The full lines are fits to the data using aspect ratios of the spherical dot shapes as indicated in the figure (adapted from [3.250])

sult is that, due to the back-folding of the dispersion curve at the wavevector $k = 2\pi/\Lambda$, with Λ the period, hybridization of modes in the crossing regimes appears. The properties are calculated in this and in most following papers by use of Floquet's theorem, expressing the solutions of the magnetic potentials in an infinite number of space harmonics. The dispersion properties in periodically corrugated YIG films has been extensively studied [3.236, 3.237, 3.238, 3.239, 3.240, 3.241, 3.242], recently with the inclusion of propagation losses [3.243]. BLS experiments from these structures have not been performed so far. *Gurney* et al. report on Brillouin light scattering from ferromagnetic patterned submicron structures [3.244]. The structures consist of rf-diode sputtered, 300 Å thick $Ni_{80}Fe_{20}$ layers patterned into 1 µm wide lines with 1 µm spacing. They observe the Damon–Eshbach mode, with its frequency shifted to lower values due to the demagnetizing field, as well as the first standing spin wave. For the wavevector $\boldsymbol{q}$ perpendicular to the stripes a splitting of the Damon–Eshbach mode into several additional modes separated by about 1 GHz is observed. This is attributed to finite size effects in the spin wave band structure due to the loss of translational symmetry and confinement within the stripe. A full study of spin waves in these wire structures including a discussion of the light scattering cross section is reported by *Mathieu* et al. [3.245] and *Hillebrands* et al. [3.246]. *Ercole* et al. investigated FeNi wires, however, without finding quantization effects [3.247, 3.248]. *Chérif* et al. performed similar studies in Co wires [3.249]. *Hillebrands* et al. have studied the propagation of the Damon–Eshbach mode across a square lattice of magnetic $Ni_{80}Fe_{20}$ dots of 1000 Å thickness of 0.5–2 μm diameter and 1–4 μm periodicity [3.250, 3.251, 3.252]. The spin wave frequencies are very sensitive to the demagnetization factor of each single dot. This is depicted in Fig. 3.41. For a sample with a dot separation of 0.1 μm, a fourfold in-plane magnetic anisotropy with the easy axis directed along the pattern diagonal is observed, indicating anisotropic coupling between the dots [3.251, 3.252]. The origin of the coupling is attributed to unsaturated parts of the dots which allow for a coupling of fourfold symmetry. It should be noted that pure dipolar coupling between saturated, identical dots would not provide a mechanism for generating a fourfold anisotropy, since the dipolar coupling interaction can be expressed by a bilinear form, which averages to a constant on fourfold symmetry [3.251, 3.252]. An array of square permalloy dots was investigated by *Cherif* et al. [3.253].

3.6 Conclusions and Outlook

We want to conclude by adding some general remarks about the phenomenon of Brillouin light scattering from spin waves in films, multilayered structures and superlattices. Although the penetration depth of light for typical metallic materials may be as small as 100 Å, the information depth is given by the perpendicular coherence length of spin waves, which is in most cases the total thickness of the film or the multilayer stack, and which can be up to at least a few thousand Å. For multilayer structures this is because the collective spin wave excitations are coherent throughout all magnetic layers. Therefore by probing them in the first few layers of the multilayer stack, the complete spin wave information of the total stack can be obtained although the first few layers may even have modified parameters due to, e.g., corrosion.

The light scattering cross section is proportional to the net fluctuating part of the dipolar moment of the precessing spins within the light scattering interaction volume. Thus pure exchange-type spin wave modes contribute to the cross section only very weakly. In order to study exchange interaction, in particular interlayer exchange interaction, a fair amount of dipolar coupling is necessary. This is the case if modes are studied which in frequency are not too much separated from the dipolar surface mode or if the net fluctuating part of the dipolar moment averaged over the light penetrated region is sufficiently large.

Brillouin light scattering is a local probe. Its sensitivity is comparable to a high-sensitivity superconducting quantum interference device (SQUID) instrument (2 ML of Fe or Co and a sampling area of 30 μm diameter provide at room temperature for a spectrum with good signal-to-noise ratio).

The local character can be utilized in various ways. The BLS technique can be used to scan across a sample to measure its homogeneity, or the dependence of thickness dependent magnetic parameters in the case of wedge type films. Localized size-effect modes in μm-wide bars have been observed [3.244], as well as line width broadening effects.

The broad line widths observed in many experiments seem not to be correlated with intrinsic damping mechanisms of spin waves [3.217]. The line broadening is caused by spatial inhomogeneities on a length scale comparable to and larger than the spin-wave wavelength ($\approx$ 3000 Å) and by sampling over many areas with different local properties within the laser spot, which is typically 30–50 μm in diameter. Loss mechanisms due to direct scattering of spin waves at, e.g., inhomogeneities, are weak, since there are usually no scattering channels available.

Not all areas of Brillouin light scattering in artificially layered structures could be covered in this review. Contrary to magnetometry, spin wave frequencies in layered systems composed of different magnetic materials are mostly sensitive to the magnetically stiffest material. Thus, by comparing Brillouin light scattering results with magnetometric investigations, access is gained to the characterization of atomic interface layers with reduced or increased magnetic moments, like magnetically dead layers, or, on the contrary, on magnetically polarized spacer layers [3.29, 3.176]. The same applies to superlattice structures composed of two magnetic materials, which then are strongly exchange coupled [3.50, 3.51, 3.254]. Here a new type of collective exchange modes exists. The collective modes are composed of exchange modes of each magnetic layer of one kind of material and they are exchange coupled through the intervening magnetic layers of the other kind.

The field of magnetic layers and superlattices is advancing very fast. Due to its potential, the Brillouin light scattering technique certainly will be of central importance in understanding some of the scientific surprises which forthcoming studies of ultrathin layered magnetic structures are bound to reveal.

Acknowledgements. I would like to thank all those who have contributed to this work through numerous collaborations, discussions, and advice. I cannot name them all here since they were too many. In particular I would like to thank G. Güntherodt for his major continuing support, R.L. Stamps for many discussions and advice about spin-wave theory, J.R. Cochran, S.O. Demokritov and G. Gubbiotti for a careful reading of the manuscript, and M. Bauer, S. Müller and I. Wollscheid for technical help, typing the manuscript and drawing the figures.

3.A Appendix: Summary of Experimental Brillouin Light Scattering Work

In the preceding section, a number of selected applications were presented. We will conclude this chapter with a survey made in the fall of 1996 of reported Brillouin light scattering work and updated in winter 1998/99. Tables 3.A.1–3.A.7 list the systems studied. The preparation method, the year of publication and short comments on the work are listed for each system.

Comments: The substrate is always on the right hand side. If the film orientation is the same as the orientation of the underlying layer or substrate its orientation is not listed. Papers of similar content and by the same group of authors are listed on one line. Preparation methods are denoted as: MBE: molecular beam epitaxy, sp: sputtering (without further specification), rfs: rf sputtering, rfms: rf magnetron sputtering, dcs: dc sputtering, dcms: dc magnetron sputtering, ibs: ion beam sputtering, ev: evaporation including e^--beam evaporation, LPE: liquid phase epitaxy.

Table 3.A.1. Fe, Ni and Fe, Ni alloy films

System	Prep.	Reference	Year	Comment
Fe/Cu(001)	MBE	[3.255]	1989	3 ML thick fcc Fe films. Study of perpendicular anisotropy
		[3.119, 3.167]	1989	Perpendicular magnetized films, cross section study
		[3.256, 3.257]	1991	ditto
		[3.221]	1993	Spin waves in coupled growth domains
Fe/Ag(001)	MBE	[3.169, 3.255, 3.258]	1988-89	Fe/Ag, and Fe/vacuum interface anisotropies
		[3.259]	1993	Wave vector dependence, spin wave correlation length
		[3.260]	1996	Fe/Ag, Fe/Cr and Fe/vacuum interface anisotropies
Fe/Au/Ag/MgO(001)	MBE	[3.261]	1996	In-plane – out-of-plane reorientation transition
Pd/Fe/Ag(001)	MBE	[3.262]	1996	Pd-induced interface anisotropies
Fe/Ni/Ag(001)	MBE	[3.80, 3.255, 3.263]	1988-89	Large fourfold in-plane anisotropy caused by reconstructed Ni layer
Fe/W(110)	MBE	[3.58, 3.264, 3.137, 3.139]	1987–91	In situ experiment, in-plane reorientation transition, Fe/W and Fe/Pd interface anisotropy
Fe/Pd/MgO(001)	MBE	[3.265]	1994	
Fe/Al(001)	MBE	[3.266]	1994	
Fe/GaAs(001)	MBE	[3.267, 3.268, 3.269]	1994–95	Reorientation transition
Fe/GaAs(110)	MBE	[3.61, 3.270, 3.271]	1983-84	First determination of magnetic anisotropies by BLS
		[3.268, 3.269]	1995	Reorientation transition

Table 3.A.1. (Continued)

System	Prep.	Reference	Year	Comment
Fe/sapphire	ev	[3.113]	1980	First experiments on Fe films
	ev	[3.24, 3.272]	1982	First determination of magnetic parameters in Fe
	MBE	[3.133]	1984	Magnon branch crossover
	MBE	[3.273]	1984	Anomalous q-dependence of spin wave frequency
	MBE	[3.274]	1986	
FeB and related metglasses		[3.275, 3.276, 3.277]	1978-79	First observation of standing spin waves
FeNiB		[3.278]	1992	Comparison BLS-FMR
FeCo alloy	MBE	[3.279]	1994	
Fe-polyethylene	ev	[3.280]	1992	
FeSi	sp	[3.281]	1989	Determination of stiffness constant and magnetization
Ni/sapphire	ev	[3.272]	1982	First determination of magnetic parameters by BLS
Ni/Cu/Si(111)	MBE	[3.282]	1997	Determination of (111)-in-plane anisotropy
Ni/Si(001)	ev	[3.283, 3.284, 3.285]	1995	
Ni/GaAs(111)	ev	[3.283]	1995	
Ni/Al_2O_3(0001)	ev	[3.283, 3.284, 3.285]	1995	
NiFe/Cu(110)	MBE	[3.229]	1997	
FeMn/NiFe/Cu(110)	MBE	[3.229]	1997	Exchange bias effect and modified anisotropies
NiFe/a-Al_2O_3/Si	rfs	[3.286]	1997	
NiFe/Si	sp	[3.287]	1987	Study of critical angle behavior
		[3.123]	1991	Oscillations in the Stokes/anti-Stokes ratio as function of wavevector
NiFe/sapphire	sp	[3.272]	1982	

Table 3.A.2. Co and Co alloy films

System	Prep.	Reference	Year	Comment
Co/Cu(001)	MBE	[3.288]	1989	
		[3.289]	1991	Microscopic spin wave theory for this system
	MBE	[3.145, 3.146], [3.42]	1992–94	In situ experiment, stabilization of ferromagnetic order due to anisotropies
Co/Cu(1 1 13)	MBE	[3.146, 3.150], [3.42]	1993–94	Magnetoelastic in-plane anisotropy, step anisotropy
Co/Cu(110)	MBE	[3.157, 3.158], [3.42]	1994–96	Suppression of magnetocrystalline bulk anisotropy in thin, stressed films
Co/Cu(111)	MBE	[3.42]	1994	
Au/Co/Au	MBE	[3.290, 3.291], [3.292, 3.293]	1999	Perp. anisotropy, effect of strain
Co/GaAs(001)	MBE	[3.294]	1994	hcp Co film with fourfold anisotropy
	MBE	[3.295]	1995	ditto
	MBE	[3.296, 3.297]	1995–96	bcc, fcc and hcp Co films
Co/GaAs(110)	MBE	[3.296, 3.297], [3.298, 3.299]	1991–96	bcc, fcc and hcp Co films
Co/Cr(211)/MgO(110)	MBE	[3.300]	1997	B-axis oriented hcp films
Co/CoO		[3.228, 3.301], [3.302]	1996–97	
Co/sapphire	ev	[3.303]	1984	
CoZr	rfms	[3.304]	1989	
CoNbZr	rfms	[3.305]	1994	
CoCr alloy	ibs	[3.306]	1986	
	sp	[3.307]	1991	Perp. anisotropy
CoPt alloy	rfs	[3.308, 3.309]	1991–94	
	ev	[3.310]	1993	
CoNiPt alloy	rfs	[3.311]	1990	

Table 3.A.3. Rare earth, rare earth–transition metal and other films

System	Prep.	Reference	Year	Comment
Gd/sapphire	MBE	[3.312]	1992	Determination of stiffness constant and anisotropy
$La(FeAl)_{13}$ (Invar)	rfms	[3.313]	1994	Determination of stiffness constant, elastic constants
TbFeCo/Si	sp	[3.314]	1997	
GdTbFe	ev	[3.315, 3.316]	1994	Determination of stiffness constant, comparison to domain wall model
GdCo	ev	[3.315, 3.316]	1994	ditto
GdFe	ev	[3.315]	1994	ditto
GdNdFe	ev	[3.316]	1994	ditto
$(YLu)(FeSc)_2Fe_3O_{12}$	LPE	[3.317]	1987	Spin wave stiffness constant as function of Sc substitution
Ga: YIG(111)	LPE	[3.318]	1997	Spin wave stiffness constant as function of Ga substitution

Table 3.A.4. Fe bilayers

System	Prep.	Reference	Year	Comment
$Fe/Ag/Fe/SiO_2$	ev	[3.319]	1987	
Fe/Ag/Fe/sapphire	ev	[3.320]	1985	
Fe/Au/Fe(001)/Cr(001)/sapphire	MBE	[3.321, 3.206]	1989	A_{12} positive
$Fe/Au/Fe/SiO_2$	MBE	[3.319]	1987	
Fe/Au/Fe/sapphire	MBE	[3.320, 3.26]	1985–86	
Fe/Cr/Fe(211)/Cr(211)/MgO(110), Fe/Cr/Fe/Ag/GaAs(001)	MBE	[3.322]	1995	Determination of A_{12} and B_{12}, comparison of BLS, MOKE
Fe/Cr/Fe(001)whisker	MBE	[3.323, 3.324]	1993	Determination of A_{12} and B_{12}, short period oscillations
	MBE	[3.325]	1994	Determination of A_{12} and B_{12} as a function of temperature
	MBE	[3.326]	1995	Determination of A_{12} and B_{12}
	MBE	[3.210]	1995	Determination of A_{12} and B_{12}, comparison of BLS, MOKE, theoretical and experimental study of spin wave properties including surface canting of magnetization
	MBE	[3.264]	1995	ditto
Fe/Cr/Fe(001)/GaAs(001)	MBE	[3.327]	1994	Experimental and theoretical study of splitting of Stokes/anti-Stokes spin wave frequencies in the canted magnetization state
Fe/Cr/Fe/GaAs	MBE	[3.328]	1997	
Fe/Cr/Fe(001)/Cr(001)/sapphire	MBE	[3.321, 3.206]	1989	
Fe/Cr/Fe(001)/Cr(001)/MgO(001)	dcms	[3.211]	1996	Determination of A_{12} and B_{12}, comparison of BLS, MOKE, MR

Table 3.A.4. (Continued)

System	Prep.	Reference	Year	Comment
Fe/Cr/Fe	ev	[3.320]	1985	
	MBE	[3.26]	1986	First finding of antiferromagnetic coupling
Fe/Cu/Fe/Ag(001)	MBE	[3.189, 3.190]	1990	Oscillating coupling through bcc Cu(001)
	MBE	[3.256]	1991	Combined FMR and BLS study
	MBE	[3.324, 3.326, 3.329]	1993–95	Determination of A_{12} and B_{12}, test of loose spin model
Fe/Cu/Fe/SiO_2	ev	[3.319]	1987	
Fe/Cu/Fe/sapphire	ev	[3.320]	1985	
Fe/Pd/Fe/MgO(001)	MBE	[3.265]	1994	Comparison of BLS, SQUID, MOKE experiments, $A_{12} > 0$
Fe/Pd/Fe/Ag(001)	MBE	[3.330, 3.331]	1990–91	Combined FMR and BLS study, coupling through lattice-expanded ferromagnetic Pd
	MBE	[3.332]	1991	Combined FMR and BLS study, oscillatory coupling on ferromagnetic background
Fe/Pd/Fe/SiO_2	ev	[3.319]	1987	
Fe/Pd/Fe/sapphire	ev	[3.320]	1985	
Fe/V/Fe/sapphire	ev	[3.320]	1985	
FeSi/ZnO/FeSi, FeSi/SiO_2/FeSi	sp	[3.333]	1990	Coupling through 28Å SiO_2 interlayer observed

A_{12}, B_{12}: bilinear and biquadratic coupling constants
MR: magnetoresistance
FMR: ferromagnetic resonance

Table 3.7. Co and other bilayer systems

System	Prep.	Reference	Year	Comment
Co/Cr/Co	sp	[3.334]	1991	Ferromagnetic interlayer exchange coupling
Co/Cr/Co		[3.335]	1996	
Co/Cu/Co/glass or Si	MBE	[3.336, 3.337]	1995	Oscillatory interlayer exchange coupling
Co/Au/Co/glass or Si	MBE	[3.336, 3.337]	1995	ditto
Co/Au/Co(111)	MBE	[3.292]	1998	
CoFe/Mn/CoFe	MBE	[3.209]	1997	Non-trigonometric biquadratic coupling, exp. and theoretical study of spin wave dispersion
$Ni_{80}Fe_{20}$/Pd/$Ni_{80}Fe_{20}$	MBE	[3.207]	1989	Combined theoretical and experimental study, ferromagnetic interlayer coupling
$Ni_{80}Fe_{20}$/Pd/$Ni_{80}Fe_{20}$/SiO_2	MBE	[3.319]	1987	
$Ni_{80}Fe_{20}$/Cr/Co	MBE	[3.207]	1989	Combined theoretical and experimental study, ferromagnetic interlayer coupling
$Ni_{80}Fe_{20}$/Bi/$Ni_{80}Fe_{20}$/SiO_2	MBE	[3.319]	1987	
$Ni_{80}Fe_{20}$/Cu/$Ni_{80}Fe_{20}$/SiO_2	MBE	[3.319]	1987	
$Ni_{80}Fe_{20}$/Ag/$Ni_{80}Fe_{20}$/SiO_2	MBE	[3.319]	1987	
$Ni_{80}Fe_{20}$/Ge/$Ni_{80}Fe_{20}$/SiO_2	MBE	[3.319]	1987	
$Ni_{80}Fe_{20}$/Au/$Ni_{80}Fe_{20}$/SiO_2	MBE	[3.319]	1987	
$Ni_{80}Fe_{20}$/Gd/$Ni_{80}Fe_{20}$/SiO_2	MBE	[3.319]	1987	
$Ni_{80}Fe_{20}$/V/$Ni_{80}Fe_{20}$/SiO_2	MBE	[3.319]	1987	
$Ni_{80}Fe_{20}$/Si/$Ni_{80}Fe_{20}$/SiO_2	MBE	[3.319]	1987	
$Ni_{80}Fe_{20}$/Cu/Co	MBE	[3.228]	1996	

Table 3.A.6. Multilayers

System	Prep.	Reference	Year	Comment
(Fe/Cr(211))/MgO(110)	MBE	[3.338]	1996	Combined theoretical and experimental study of low-field antiferromagnetic state, surface spin flop regime, symmetric spin flop regime and high field state
(Fe/Cr)	sp	[3.339]	1993	Antiferromagnetically coupled multilayer
		[3.340]	1993	Two bulk spin wave bands in antiferromagnetically coupled multilayer
(Fe/Cu)/sapphire, (Fe/Cu)/Si	dcms	[3.341]	1995	Combined static magnetization, FMR and BLS study, interface anisotropy
(Fe/Mo)	sp	[3.339]	1993	Antiferromagnetically coupled multilayer
(Fe/Ni)/sapphire	MBE	[3.254, 3.342]	1992,93	Effective Fe-Fe interlayer exchange coupling
(Fe/Pd)/sapphire	dcms	[3.173, 3.343]	1986,87	First observation of collective spin wave band using a TFP-interferometer
	dcms	[3.29, 3.264]	1988,89	Collective band of spin waves, comparison of static and BLS measurements, intrinsic magnetic moment
	dcms	[3.344, 3.264]	1988	Multilayer with two repeat periods
(Fe/W)/sapphire	dcms	[3.173, 3.343]	1986,87	First observation of collective spin wave band using a TFP-interferometer
(Co/Nb)	MBE	[3.345, 3.346, 3.347]	1984-86	Theoretical and expermental investigation of spin wave properties
(Co/Cu(111))/Nb/sapphire	MBE	[3.348]	1996	
(Co/Cu)/sapphire	rf	[3.349]	1997	
(Co/Ru)/Si(111)	dcms	[3.215, 3.214]	1992,93	First application of BLS for the determination of oscillatory coupling

Table 3.A.6. (Continued)

System	Prep.	Reference	Year	Comment
(Co/Pt(110))	MBE	[3.181]	1993	In-plane magnetoelastic anisotropy, comparison with strains determined by x-ray diffraction
(Co/Pt(111))/Si	ev	[3.218]	1994	Line width broadening due to spatially varying anisotropies
(Co/Pt(111))/Si	ev	[3.219]	1992	Observation of collective exchange modes
(Co/Pt(001))/GaAs(001)	MBE	[3.219]	1992	
(Co/Pt)	ev	[3.308]	1991	
(Co/Au)/sapphire	ibsp	[3.186]	1991	Characterisation of interface quality by BLS
(Co/Ti)/sapphire	dcms	[3.350]	1990	Determination of dead layer thickness
(Co/Pd(110))	MBE	[3.181]	1993	In-plane anisotropy, line width broadening due to spatially varying anisotropies
(Co/Pd)/sapphire	dcms	[3.213, 3.351]	1990,91	First observation of collective exchange modes
	dcms	[3.91]	1991	Combined BLS-SQUID study, determination of Pd polarization
(Co/Gd)		[3.352]	1993	Change in spin wave spectrum near T_C of Gd
(Ni/Mo)		[3.28, 3.27]	1983,84	First observation of spin waves in multilayers
(Ni/Pt)/Si	sp	[3.353]	1996	
($Fe_{80}Ni_{20}$/Ru)/Si(111)	dcms	[3.215]	1993	First application of BLS for the determination of oscillatory coupling
(FeCo/Ag)/GaAs(001)	rfms	[3.354]	1997	
(FeCo/Cu)/GaAs(001)	rfms	[3.354]	1997	
(FeCo/FeSiBC)/GaAs(001)	rfms	[3.354]	1997	

Table 3.A.7. BLS from microwave excited spin waves (including bulk samples)

System	Reference	Year	Comment
$FeBO_3$ (bulk)	[3.126, 3.355]	1979	BLS from parametrically excited magnons in bulk material
$CoCO_3$ (bulk)	[3.356, 3.357]	1979-80	ditto
YIG (bulk)	[3.233, 3.358]	1977-79	ditto
K_2CuF_4 (bulk)	[3.234]	1980	ditto
YIG	[3.93]	1983	BLS from parametrically excited magnons
YIG, $FeBO_3$	[3.126]	1984	First combined BLS-FMR experiment on planar structures, spatically resolved experiment
YIG	[3.94]	1984	Wavevector selective BLS setup
YIG	[3.127, 3.128], [3.359]	1985-87	BLS from signal-to-noise-enhancer, observation of MSSW and parametric magnons, wave vector distribution
YIG	[3.129]	1987	BLS from decay line and signal-to-noise-enhancer structures; observation of BVMSW
YIG	[3.130]	1988	Combined BLS-FMR experiment, quantitative determination of light scattering intensity
YIG	[3.360]	1988	Anomalous parametric spin wave character in subsidiary absorption
YIG	[3.361]	1988	YIG film strip line, new type of non-propagating surface spin waves observed
YIG	[3.134, 3.135]	1991–92	Combined BLS-FMR spectrometer, determination of mode profile in cavity, discussion of dipolar narrowing
Fe/Cr/Fe	[3.131]	1993	Combined BLS-FMR spectrometer for low temperatures
YIG	[3.97]	1994	Distribution of critical mode wavevectors above instability threshold
YIG	[3.362]	1998	Investigation of wavevector distribution for magnetic envelope solitons
YIG	[3.100]	1998	Spatio-temporal self-focusing, spin wave "bullets"
YIG	[3.101]	1999	Collisions of spinwave envelope solitons and self-focused spin wave packets ("bullets")
YIG	[3.98]	1996	Observation of self channeling, spin wave beam profiles
$Bi_{0.96}Lu_{2.04}Fe_5O_{12}$	[3.99]	1997	BLS from a wave guide structure; two-dimensional mapping of spin wave intensity, first observation of self-focusing
$Lu_{2.04}Bi_{0.96}Fe_5O_{12}$	[3.132]	1998	Mode beating

MSW: magnetostatic wave
BVMSW: backward volume magnetostatic wave
MSSW: magnetostatic surface wave

References

3.1 see review chapters in *Ultrathin Magnetic Structures Vol I and II*, J.A.C. Bland, B. Heinrich (eds.) Springer Verlag (Heidelberg, Berlin, London, New York, Tokyo) (1992), and references therein
3.2 J.H.E. Griffiths: Nature **158**, 670 (1946)
3.3 C. Kittel: Interpretation of anomalous Larmor frequencies in ferromagnetic resonance experiment, Phys. Rev. **71**, 270 (1947)
3.4 C. Kittel: On the theory of ferromagnetic resonance absorption, Phys. Rev. **73**, 155 (1948)
3.5 W.S. Ament, G.T. Rado: Electromagnetic effects of spin wave resonance in ferromagnetic metals, Phys. Rev. **97**, 1558 (1955)
3.6 G.T. Rado, J.R. Weertman: Spin-wave resonance in a ferromagnetic metal, J. Phys. Chem. Solids **11**, 315 (1959)
3.7 C. Kittel: Excitation of spin waves in a ferromagnet by a uniform rf field, Phys. Rev. **110**, 1295 (1958)
3.8 M.H. Seavey, P.E. Tannenwald: Direct observation of spin-wave resonance, Phys. Rev. Lett. **1**, 168 (1958)
3.9 L. Brillouin: Ann. Phys. (Paris) **17**, 88 (1922)
3.10 L.I. Mandelstam: Zh. Russ. Fiz.-Khim. Ova **58**, 381 (1926)
3.11 L.I. Mandelstam, G.S. Landsberg, M. Leontovitsch: Über die Theorie der molekularen Lichtzerstreuung in Kristallen (klassische Theorie), Zs. Physik **60**, 334 (1930)
3.12 E.F. Gross: Nature **126**, 201 (1930)
3.13 P.A. Fleury, S.P.C. Porto, L.E. Cheesman, H.J. Guggenheim: Light scattering by spin waves in FeF_2, Phys. Rev. Lett. **17**, 84 (1966)
3.14 P.A. Fleury, S.P.C. Porto, R. Loudon: Two-magnon light scattering in antiferromagnetic MnF_2, Phys. Rev. Lett. **18**, 658 (1967)
3.15 J. R. Sandercock, M. Balkanski: The design and use of a stabilised multipassed interferometer of high contrast ratio, Proceedings of the 2nd Intern. Conf. on Light Scattering in Solids, Flammarion, Paris, 9, (1971)
3.16 J.R. Sandercock: Brillouin scattering study of SbSi using a double-passed, stabilised scanning interferometer, Optics Commun. **2**, 73 (1970)
3.17 J.R. Sandercock: Brillouin-Scattering Measurements on Silicon and Germanium, Phys. Rev. Lett. **28**, 237 (1972)
3.18 J.R. Sandercock: A light scattering study of the ferromagnet $CrBr_3$, Solid State Commun. **15**, 1715 (1974)
3.19 J.R. Sandercock, W. Wettling: Light scattering from thermal acoustic magnons in Yttrium iron garnet, Solid State Commun. **13**, 1729 (1973)
3.20 P. Grünberg, F. Metawe: Light scattering from bulk and surface spin waves in EuO, Phys. Rev. Lett. **39**, 1561 (1977)
3.21 J.R. Sandercock, W. Wettling: Light scattering from thermal magnons in iron and nickel, IEEE Trans. Mag. **14**, 442 (1978)
3.22 J.R. Sandercock, W. Wettling: Light scattering from surface and bulk thermal magnons in iron and nickel, J. Appl. Phys. **50**, 7784 (1979)
3.23 A.P. Malozemoff, M. Grimsditch, J. Aboaf, A. Brunsch: Brillouin-scattering studies of polycrystalline and amorphous sputtered films of $Fe_{1-x}B_x$ and $Co_{1-x}B_x$, J. Appl. Phys. **50**, 4821 (1979)

3.24 P. Grünberg, M.G. Cottam, W. Vach, C. Mayr, R.E. Camley: Brillouin scattering of light by spin waves in thin ferromagnetic films, J. Appl. Phys. **53**, 2078 (1982)
3.25 P. Grünberg: Light scattering from magnetic surface excitations, Progr. Surf. Sci. **18**, 1 (1985)
3.26 P. Grünberg, R. Schreiber, Y. Pang, M.B. Brodsky, H. Sowers: Layered magnetic structures: evidence for antiferromagnetic coupling of Fe layers across Cr interlayers, Phys. Rev. Lett. **57**, 2442 (1986)
3.27 M. Grimsditch, M.R. Khan, A. Kueny, I.K. Schuller: Collective behavior of magnons in superlattices, Phys. Rev. Lett. **51**, 498 (1983)
3.28 A. Kueny, M.R. Khan, I.K. Schuller, M. Grimsditch: Magnons in superlattices: a light scattering study, Phys. Rev. B **29**, 2879 (1984)
3.29 B. Hillebrands, A. Boufelfel, C.M. Falco, P. Baumgart, G. Güntherodt, E. Zirngiebl, J.D. Thompson: Brillouin scattering from collective spin waves in magnetic superlattices, J. Appl. Phys. **63**, 3880 (1988)
3.30 T. Wolfram, R.E. DeWames: Surface dynamics of magnetic materials, Progr. Surf. Sci. **2**, 233 (1972)
3.31 D.L. Mills, K.R. Subbaswamy: Surface and size effects on the light scattering spectra of solids, Progress in Optics XIX, E. Wolf (ed.) North Holland, 47 (1981)
3.32 J.R. Sandercock: Trends in Brillouin light scattering: studies of opaque materials, supported films and central modes, *Light Scattering in Solids III*, M. Cardona, G. Güntherodt (eds.) Springer Series Topics Appl. Physics **51**, 173 (1982)
3.33 A.S. Borovik-Romanov, N.M. Kreines: Brillouin-Mandelstam scattering from thermal and excited magnons, Physics Reports **81**, 351 (1982)
3.34 C.E. Patton: Magnetic excitations in solids, Physics Reports **103**, 251 (1984)
3.35 D.L. Mills: Surface spin waves on magnetic crystals, Surface excitations, V. M. Agranovich, R. Loudon (eds.) Elsevier Science Publishers B. V. (1984)
3.36 P. Grünberg: Light scattering from spin waves in thin films and layered magnetic structures, *Light Scattering in Solids V*, M. Cardona, G. Güntherodt (eds.) Springer Series Topics Appl. Physics **66**, 303 (1989)
3.37 M. Grimsditch: Brillouin light scattering from metallic superlattices, *Light Scattering in Solids V*, M. Cardona, G. Güntherodt (eds.) Springer Series Topics Appl. Physics **66**, 285 (1989)
3.38 B. Heinrich, J.F. Cochran: Ultrathin metallic magnetic films: magnetic anisotropies and exchange interactions, Advances in Physics **42**, 523 (1993)
3.39 J.F. Cochran: Light scattering from ultrathin magnetic layers and bilayers, *Ultrathin Magnetic Structures II*, B. Heinrich, J. A. C. Bland (eds.) Springer Verlag (Heidelberg, Berlin, London, New York, Tokyo) (1994), and references therein.
3.40 B. Hillebrands, G. Güntherodt: Brillouin light scattering in magnetic superlattices, *Ultrathin Magnetic Structures II*, B. Heinrich, J. A. C. Bland (eds.) Springer Verlag (Heidelberg, Berlin, London, New York, Tokyo) (1994)
3.41 J.R. Dutcher: Light scattering and microwave resonance studies of spin-waves in metallic films and multilayers, *Linear and Nonlinear Spin Waves in Magnetic Films and Superlattices*, M. G. Cottam (ed.) World Scientific, (Singapore) (1994)

3.42 B. Hillebrands, P. Krams, J. Fassbender, C. Mathieu, G. Güntherodt, R. Jungblut, M.T. Johnson: Light scattering investigations of magnetic anisotropies in ultrathin epitaxial Co films, Acta Phys. Pol. A **85**, 179 (1994)
3.43 S.O. Demokritov, E. Tsymbal: Light scattering from spin waves in thin films and layered systems, J. Phys. C **6**, 7145 (1994)
3.44 M.G. Cottam, I.V. Rojdestvenski, A.N. Slavin: Brillouin light scattering from dipole-exchange microwave spin waves in magnetic films, *High Frequency Processes in Magnetic Materials*, G. Srinivasan, A. N. Slavin (eds.) World Scientific, Singapore (1995)
3.45 B. Hillebrands, J. Fassbender, P. Krams, C. Mathieu, G. Güntherodt, R. Jungblut: Magnetic properties of ultrathin epitaxial Co films studied by Brillouin light scattering, Il Vuoto, Scienza e Tecnologia **25**, 24 (1996)
3.46 *Ultrathin Magnetic Structures Vol I and II*, J. A. C. Bland, B. Heinrich (eds.) Springer Verlag (Heidelberg, Berlin, London, New York, Tokyo) (1994)
3.47 U. Gradmann: *Magnetism in ultrathin transition metal films*, Handbook of magnetic materials, Vol. 7, K. H. J. Buschow (ed.) Elsevier Publishers B. V., (Amsterdam) (1993)
3.48 G.T. Rado, R.J. Hicken: Theory of magnetic surface anisotropy and exchange effects in the Brillouin scattering of light by magnetostatic spin waves, J. Appl. Phys. **63**, 3885 (1988)
3.49 J.F. Cochran, J.R. Dutcher: Light scattering from thermal magnons in thin metallic ferromagnetic films, J. Appl. Phys. **63**, 3814 (1988)
3.50 B. Hillebrands: Calculation of spin waves in multilayered structures including interface anisotropies and exchange contributions, Phys. Rev. B **37**, 9885 (1988)
3.51 B. Hillebrands: Spin-wave calculations for multilayered structures, Phys. Rev. B **41**, 530 (1990)
3.52 R.L. Stamps, B. Hillebrands: Long-wave-length spin waves in structures with large out-of-plane anisotropies, J. Appl. Phys. **69**, 5718 (1991)
3.53 R.L. Stamps, B. Hillebrands: Dipole-exchange modes in multilayers with out-of-plane anisotropies, Phys. Rev. B **44**, 5095 (1991)
3.54 R.L. Stamps, B. Hillebrands: Dipole-exchange modes in single thin films and multilayers with large out-of-plane anisotropies, J. Magn. Magn. Mater. **93**, 616 (1991)
3.55 R.L. Stamps, B. Hillebrands: Dipolar interactions and the magnetic behavior of two-dimensional ferromagnetic systems, Phys. Rev. B **44**, 12417 (1991)
3.56 M.J. Hurben, C.E. Patton: Theory of magnetostatic waves for in-plane magnetized isotropic films, J. Magn. Magn. Mater. **139**, 263 (1995)
3.57 This requirement is not fulfilled, if the effective internal field, i.e., the sum of the external field, the demagnetizing field and the anisotropy field, is close to zero
3.58 B. Hillebrands, P. Baumgart, G. Güntherodt: In-situ Brillouin scattering from surface-anisotropy-dominated Damon-Eshbach modes in ultrathin epitaxial Fe(110) layers, Phys. Rev. B **36**, 2450 (1987)
3.59 J.O. Artman: Microwave resonance relations in anisotropic single-crystal ferrites, Phys. Rev. **105**, 62 (1957)
3.60 E.P. Valstyn, I.P. Hanton, A.M. Morish: Ferromagnetic resonance of single-domain particles, Phys. Rev. **128**, 2078 (1962)
3.61 G. Rupp, W. Wettling, R.S. Smith, W. Janitz: Surface magnons in anisotropic ferromagnetic films, J. Magn. Magn. Mater. **45**, 404 (1984)

3.62 F. Hoffmann, A. Stankoff, H. Pascard: Evidence for an exchange coupling at the interface between two ferromagnetic films, J. Appl. Phys. **41**, 1022 (1970)
3.63 F. Hoffmann: Dynamic pinning induced by nickel layers on permalloy films, Phys. Status Solidi **41**, 807 (1970)
3.64 We have corrected the Hoffman boundary condition by adding the term $(a_{n'} \partial \boldsymbol{M}_{n'} / \partial n_{n'})$ to the square bracket in the second term, in order to obtain a consistent form for the strong interlayer coupling case
3.65 L. Néel: Anisotropic magnétique superficielle et surstructured d'orientation, J. Phys. Radium **15**, 225 (1954)
3.66 P. Bruno: Magnetic surface anisotropy of cobalt and surface roughness effects within Néel's model, J. Phys. F **18**, 1291 (1988)
3.67 For simplicity we here have assumed, that the $\hat{y}$-axis is the symmetry axis for the two-fold in-plane anisotropy. This can be easily generalized, see, e.g., Eq. (3.80)
3.68 G.T. Rado: Conditional replaceability of magnetic surface anisotropies by effective volume anisotropies in the ferromagnetic resonance of ultrathin films, J. Appl. Phys. **61**, 4262 (1987)
3.69 U. Gradmann, J. Korecki, G. Waller: In-plane magnetic surface anisotropies in Fe(110), Appl. Phys. A **39**, 101 (1986)
3.70 B. Hillebrands, J.R. Dutcher: Origin of very large in-plane anisotropies in (110)-oriented Co/Pd and Co/Pt coherent superlattices, Phys. Rev. B **47**, 6126 (1993)
3.71 Contrary to Refs 3.57, 3.68, 3.69 we here use a different sign convention for K_p for better consistency
3.72 R. F. Pearson: *Experimental Magnetism*, Vol. 1, G. M. Kalvius, R. S. Tebble (eds.) John Wiley and Sons, (Chichester, New York, Brisbane, Toronto) (1979)
3.73 I. Berenbaum, R. Rosenberg: Surface topology changes during electromigration in metallic thin films stripes, Thin Solid Films **5**, 187 (1970)
3.74 W.A. Jesser, J.W. Matthews: Pseudomorphic deposits of cobalt and copper, Phil. Mag. **17**, 461 (1968)
3.75 A.I. Fedorenko, R. Vincent: The epitaxial growth of cobalt and copper, Phil. Mag. **24**, 55 (1971)
3.76 W.A. Jesser, D. Kuhlmann-Wilsdorf: On the theory of interfacial energy and elastic strain of epitaxial overgrowths in parallel alignment on single crystal substrates, Phys. Stat. Solidi **19**, 95 (1967)
3.77 C. Chappert, P. Bruno: Magnetic anisotropy in metallic ultrathin films and related experiments on cobalt films, J. Appl. Phys. **64**, 5736 (1988)
3.78 P. Bruno, J. Seiden: Theoretical investigations on magnetic surface anisotropy, J. Physique Colloque 49 C **8**, 1645 (1988)
3.79 R.W. Damon, J.R. Eshbach: Magnetostatic modes of a ferromagnet slab, J. Phys. Chem. Solids **19**, 308 (1961)
3.80 B. Heinrich, S.T. Purcell, J.R. Dutcher, K.B. Urquhart, J.F. Cochran, A.S. Arrott: Structural and magnetic properties of ultrathin Ni/Fe bilayers grown epitaxially on Ag(001), Phys. Rev. B **38**, 12879 (1988)
3.81 J.F. Cochran, B. Heinrich, A.S. Arrott: Ferromagnetic resonance in a system composed of a ferromagnetic substrate and an exchange-coupled thin ferromagnetic overlayer, Phys. Rev. B **34**, 7788 (1986)

3.82 R.L. Stamps, B. Hillebrands: Dipole-exchange modes in thin ferromagnetic films with strong out-of-plane anisotropies, Phys. Rev. B **43**, 3532 (1991)
3.83 R.E. de Wames, T. Wolfram: Dipole-exchange spin waves in ferromagnetic films, J. Appl. Phys. **41**, 987 (1970)
3.84 R.L. Stamps: Spin configurations and spin-wave excitations in exchange-coupled bilayers, Phys. Rev. B **49**, 339 (1994)
3.85 N.S. Almeida, D.L. Mills: Effective-medium theory of long-wavelength spin waves in magnetic superlattices, Phys. Rev. B **38**, 6698 (1988)
3.86 F.C. Nörtemann, R.L. Stamps, R.E. Camley, B. Hillebrands, G. Güntherodt: Effective-medium theory for finite magnetic multilayers: effect of anisotropy on dipolar modes, Phys. Rev. B **47**, 3225 (1993)
3.87 W.B. Zeper, F.J.A.M. Greidanus, P.F. Carcia, C.R. Fincher: Perpendicular magnetic anisotropy and magneto-optical Kerr effect of vapor-deposited Co/Pt-layered structures, J. Appl. Phys. **65**, 4971 (1989)
3.88 The light scattering cross section is proportional to the net part of the fluctuating magnetization, which is very small for exchange-type modes. Only dipolar terms contribute to the cross section
3.89 C. Demeangeat, D.L. Mills: Evidence for magnetic surface reconstruction from ferromagnetic resonance studies of EuS films, Phys. Rev. B **16**, 2321 (1977)
3.90 J.G. LePage, R.E. Camley: Surface phase transitions and spin-wave modes in semi-infinite magnetic superlattices with antiferromagnetic interfacial coupling, Phys. Rev. Lett. **65**, 1152 (1990)
3.91 J.V. Harzer, B. Hillebrands, R.L. Stamps, G. Güntherodt, C.D. England, C.M. Falco: Magnetic properties of Co/Pd multilayers determined by Brillouin light scattering and SQUID magnetometry, J. Appl. Phys. **69**, 2448 (1991)
3.92 A possible enhancement of the Co moment for ultrathin Co layers is not included since here large interface anisotropies dominate over the saturation magnetization
3.93 W. Wettling, W.D. Wilber, P. Kabos, C.E. Patton: Light scattering from parallel-pump instabilities in yttrium iron garnet, Phys. Rev. Lett. **51**, 1680 (1983)
3.94 W.D. Wilber, W. Wettling, P. Kabos, C.E. Patton, W. Jantz: A wave-vector selective light scattering magnon spectrometer, J. Appl. Phys. **55**, 2533 (1984)
3.95 V.G. Khotikov, N.M. Kreines: Scattering of light in weakly ferromagnetic $CoCO_3$ following excitation of the spin system by large microwave power, JETP Letters **26**, 360 (1977)
3.96 W.D. Wilber, J.G. Booth, C.E. Patton, G. Srinivasan, R.W. Cross: Light-scattering observation of anomalous parametric spin-wave character in subsidiary absorption, J. Appl. Phys. **64**, 5477 (1988)
3.97 P. Kabos, G. Wiese, C.E. Patton: Measurement of spin wave instability magnon distributions for subsidiary absorption in yttrium iron garnet films by Brillouin light scattering, Phys. Rev. Lett. **72**, 2093 (1994)
3.98 J.W. Boyle, S.A. Nikitov, A.D. Boardman, J.G. Booth, K. Booth: Nonlinear self-channeling and beam shaping of magnetostatic waves in ferromagnetic films, Phys. Rev. B **53**, 12173 (1996)

3.99 M. Bauer, C. Mathieu, S.O. Demokritov, B. Hillebrands, P.A. Kolodin, S. Sure, H. Dötsch, A.N. Slavin: Direct observation of two-dimensional self-focusing and initial stages of a wave collapse for spin waves in magnetic films, Phys. Rev. B **56**, 8483 (1997)
3.100 M. Bauer, O. Büttner, S.O. Demokritov, B. Hillebrands, Y. Grimalsky, Y. Rapoport, A.N. Slavin: Observation of spatiotemporal self-focusing of spin waves in magnetic films, Phys. Rev. Lett. **81**, 3769 (1998)
3.101 O. Büttner, M. Bauer, S.O. Demokritov, B. Hillebrands, M.P. Kostylev, B.A. Kalinikos, A.N. Slavin: Collisions of spin wave envelope solitons and self-focused spin wave packets in magnetic films, Phys. Rev. Lett., **82**, 4320 (1999)
3.102 M.J. Lighthill: J. Inst. Math. Appl. **1**, 269 (1965)
3.103 A.S. Borovik-Romanov, S.O. Demokritov, N.M. Kreines, V.I. Kudinov: Observation of new inelastic light scattering mechanism in antiferromagnet in AFMR studies, J. Magn. Magn. Mater. **54–57**, 1181 (1986)
3.104 Y.R. Shen, N. Bloembergen: Interaction between light waves and spin waves, Phys. Rev. **143**, 372 (1966)
3.105 P.A. Fleury, R. Loudon: Scattering of light by one- and two-magnon excitations, Phys. Rev. **166**, 514 (1968)
3.106 T. Moriya: J. Phys. Soc. Japan **23**, 163 (1967)
3.107 H. LeGall, J.P. Jamet: Theory of the elastic and inelastic scattering of light by magnetic crystals, Phys. Status Solidi **46**, 467 (1971)
3.108 W. Wettling, M.G. Cottam, J.R. Sandercock: The relation between one-magnon light scattering and the complex magneto-optic effects in YIG, J. Phys. C **8**, 211 (1975)
3.109 R.E. Camley, D.L. Mills: Surface response of exchange- and dipolar-coupled ferromagnets: application to light scattering from magnetic surfaces, Phys. Rev. B **18**, 4821 (1978)
3.110 R.E. Camley, D.L. Mills: Theory of light scattering by spin waves, J. Appl. Phys. **50**, 7779 (1979)
3.111 R.E. Camley, T.S. Rahman, D.L. Mills: Theory of light scattering by the spin-wave excitations of thin ferromagnetic films, Phys. Rev. B **23**, 1226 (1981)
3.112 R.E. Camley, T.S. Rahman, D.L. Mills: Magnetic excitations in layered media: spin waves and the light-scattering spectrum, Phys. Rev. B **27**, 261 (1983)
3.113 R.E. Camley, M. Grimsditch: Brillouin scattering from magnons in ferromagnetic thin films of polycrystalline iron, Phys. Rev. B **22**, 5420 (1980)
3.114 M.G. Cottam: Magnetostatic theory for Brillouin scattering of light from surface spin waves in thin ferromagnetic films, J. Phys. C **16**, 1573 (1983)
3.115 M.G. Cottam, A.N. Slavin: Theory for Brillouin light scattering from dipole-exchange surface spin waves in ferromagnetic films, IEEE Trans. Mag. **27**, 5489 (1991)
3.116 I.V. Rojdestvenski, M.G. Cottam: A dipole-exchange theory for Brillouin light scattering from ferromagnetic thin films, J. Appl. Phys. **73**, 7001 (1993)
3.117 I.V. Rojdestvenski, M.G. Cottam, A.N. Slavin: Dipole-exchange theory for Brillouin light scattering from hybridized spin waves in ferromagnetic thin films, Phys. Rev. B **48**, 12768 (1993)

3.118 J.F. Cochran, J.R. Dutcher: Calculation of Brillouin light scattering intensities from pairs of exchange-coupled thin films, J. Appl. Phys. **64**, 6092 (1988)
3.119 J.R. Dutcher, J.F. Cochran, I. Jacob, W.F. Egelhoff: Brillouin light-scattering intensities for thin magnetic films with large perpendicular anisotropies, Phys. Rev. B **39**, 10430 (1989)
3.120 J.F. Cochran, J.R. Dutcher: Calculation of the intensity of light scattered from magnons in thin films, J. Magn. Magn. Mater. **73**, 299 (1988)
3.121 J.F. Cochran: Brillouin light scattering intensities for a thin film exchange coupled to a bulk magnetic substrate, J. Magn. Magn. Mater. **169**, 1 (1997)
3.122 L.D. Landau, E.M. Lifshitz: Electrodynamics of continuous media, Pergamon Press (Oxford) (1960)
3.123 H. Moosmüller, J.R. Truedson, C.E. Patton: Oscillations in the Stokes-anti-Stokes ratio in Brillouin scattering from magnons in thin permalloy films, J. Appl. Phys. **69**, 5721 (1991)
3.124 R. Mock, B. Hillebrands, J.R. Sandercock: Construction and performance of a Brillouin scattering set-up using a triple-pass tandem Fabry-Perot interferometer, J. Phys.E **20**, 656 (1987)
3.125 B. Hillebrands: Progress in multipass tandem Fabry-Perot interferometry: I. A fully automated, easy to use, self-aligning spectrometer with increased stability and flexibility, Rev. Sci. Instr., **70**, 1589 (1999)
3.126 W. Wettling, W. Jantz: Investigation of ferromagnetic resonance by the inelastic light scattering technique, J. Magn. Magn. Mater. **45**, 364 (1984)
3.127 G. Srinivasan, C.E. Patton: Direct observation of surface wave excitations in magnetostatic wave devices, IEEE Trans. Mag. **21**, 1797 (1985)
3.128 G. Srinivasan, C.E. Patton, P.R. Emtage: Brillouin light scattering on yttrium iron garnet films in a magnetostatic wave device structure, J. Appl. Phys. **61**, 2318 (1987)
3.129 G. Srinivasan, J.G. Booth, C.E. Patton: Characterization of magnetostatic wave devices by Brillouin light scattering, IEEE Trans. Mag. **23**, 3718 (1987)
3.130 G. Srinivasan, C.E. Patton, J.G. Booth: Brillouin light scattering detection of ferromagnetic resonance in thin films, J. Appl. Phys. **63**, 3344 (1988)
3.131 S.O. Demokritov: Combined light scattering-FMR technique for the investigation of magnetic coupling in layered systems, J. Magn. Magn. Mater. **126**, 291 (1993)
3.132 O. Büttner, M. Bauer, C. Mathieu, S.O. Demokritov, B. Hillebrands, P.A. Kolodin, M.P. Kostylev, S. Sure, H. Dötsch, V. Grimalsky, Y. Rapoport, A.N. Slavin: Mode beating of spin wave beams in ferrimagnetic $Lu_{2.04}Bi_{0.96}Fe_5O_{12}$ films, IEEE Trans. Magn. **34**, 1381 (1998)
3.133 P. Kabos, W.D. Wilber, C.E. Patton, P. Grünberg: Brillouin light scattering study of magnon branch crossover in thin iron films, Phys. Rev. B **29**, 6396 (1984)
3.134 S.M. Rezende, A. Azevedo: Dipolar narrowing of ferromagnetic resonance lines, Phys. Rev. B **44**, 7062 (1991)
3.135 A. Azevedo, S.M. Rezende: Spatial distribution of magnetostatic modes in a thin YIG slab, J. Magn. Magn. Mater. **104-107**, 1039 (1992)
3.136 P. Baumgart, B. Hillebrands, G. Güntherodt: In-situ determination of magnetic anisotropies of thin epitaxial Fe(110) and Co(0001) films on W(110) using Brillouin light scattering, J. Magn. Magn. Mater. **93**, 225 (1991)

3.137 P. Baumgart, B. Hillebrands, J.V. Harzer, G. Güntherodt: In-situ investigation of structure and magnetic properties of the epitaxial system Pd(111)/Fe(110) on W(110) by means of Brillouin spectroscopy, Mat. Res. Soc. Symp. Proc. **151**, 199 (1989)
3.138 P. Baumgart, Ph.D. thesis, Aachen, 1989, Baumgart uses a different choice of anisotropy definitions, which expressed as a function of those used in this chapter are: $k_{ip} = 2(k_p^{(2)} - k_p^{(4)})$, $k_{ipp} = 2k_p^{(4)}$, $K_u = K_p$.
3.139 B. Hillebrands, P. Baumgart, G. Güntherodt: Brillouin light scattering from spin waves in magnetic layers and multilayers, Appl. Phys. A **49**, 589 (1989)
3.140 G.A. Prinz, G.T. Rado, J.J. Krebs: Magnetic properties of single-crystal (110) iron films grown on GaAs by molecular beam epitaxy, J. Appl. Phys. **53**, 2087 (1982)
3.141 R. Kurzawa, K.P. Kämper, W. Schmitt, G. Güntherodt: Spin-resolved photoemission study of in-situ grown epitaxial Fe layers on W(110), Solid State Commun. **60**, 777 (1986)
3.142 H.J. Elmers, U. Gradmann: Magnetic anisotropies in Fe(110) films on W(110), Appl. Phys. A **51**, 255 (1990)
3.143 N.D. Mermin, H. Wagner: Absence of ferromagnetism or antiferromagnetism in one- or two-dimensional isotropic Heisenberg models, Phys. Rev. Lett. **17**, 1133 (1966)
3.144 Y. Yafet, J. Kwo, E.M. Gyorgy: Dipole-dipole interactions and two-dimensional magnetism, Phys. Rev. B **33**, 6519 (1986)
3.145 P. Krams, F. Lauks, R.L. Stamps, B. Hillebrands, G. Güntherodt: Magnetic anisotropies of ultrathin Co(001) films on Cu(001), Phys. Rev. Lett. **69**, 3674 (1992)
3.146 P. Krams, F. Lauks, R.L. Stamps, B. Hillebrands, G. Güntherodt, H.P. Oepen: Magnetic anisotropies of ultrathin Co films on Cu(001) and Cu(1 1 13) substrates, J. Magn. Magn. Mater. **121**, 483 (1993)
3.147 D.S. Wang, R. Wu, A.J. Freemann: Magnetocrystalline anisotropy of interfaces: first-principles theory for Co-Cu interface and interpretation by an effective ligand interaction model, J. Magn. Magn. Mater. **129**, 237 (1994)
3.148 B. Heinrich, J.F. Cochran, M. Kowalewski, J. Kirschner, Z. Celinski, C.M. Schneider, K. Myrtle: Magnetic anisotropies and exchange coupling in ultrathin fcc Co(001) structures, Phys. Rev. B **44**, 9348 (1991)
3.149 M. Kowalewski, C.M. Schneider, B. Heinrich: Thickness and temperature dependence of magnetic anisotropies in ultrathin fcc Co(001) structures, Phys. Rev. B **47**, 8748 (1993)
3.150 P. Krams, B. Hillebrands, G. Güntherodt, H.P. Oepen: Magnetic anisotropies of ultrathin Co films on Cu(1 1 13) substrates, Phys. Rev. B **49**, 3633 (1994)
3.151 A. Berger, U. Linke, H.P. Oepen: Symmetry-induced uniaxial anisotropy in ultrathin epitaxial Cobalt films given on Cu(1 1 13), Phys. Rev. Lett. **68**, 839 (1992)
3.152 H.P. Oepen, C. Schneider, D.S. Chuang, C.A. Ballentine, R.C. O'Handley: Magnetic anisotropy in epitaxial fcc Co/Cu(1 1 13), J. Appl. Phys. **73**, 6186 (1993)
3.153 H.P. Oepen, A. Berger, C.M. Schneider, T. Reul, J. Kirschner: Uniaxial anisotropy in epitaxial cobalt films grown on Cu(1 1 13), J. Magn. Magn. Mater. **121**, 490 (1992)

3.154 H. Fujiwara, H. Kadomatsu, T. Tokunaga: Magnetocrystalline anisotropy and magnetostriction of Pd-Co alloys, J. Magn. Magn. Mater. **31-34**, 809 (1983)
3.155 Elastic, Piezoelectric, Pyroelectric, Piezooptic, Electrooptic Constants and Nonlinear Dielectric Susceptibilities of Crystals, K.-H. Hellwege, A. M. Hellwege (eds.) Landolt-Börnstein, New Series, Group 3, Vol. **18**, Springer Verlag (Berlin) (1984)
3.156 J. Fassbender, C. Mathieu, B. Hillebrands, G. Güntherodt, R. Jungblut, M.T. Johnson: Identification of magnetoelastic and magnetocrystalline anisotropy contributions in ultrathin epitaxial Co(110) films, J. Appl. Phys. **76**, 6100 (1994)
3.157 J. Fassbender, C. Mathieu, B. Hillebrands, G. Güntherodt: Identification of magnetocrystalline and magnetoelastic anisotropy contributions in ultrathin epitaxial Co(110) films, J. Magn. Magn. Mater. **148**, 156 (1995)
3.158 B. Hillebrands, J. Fassbender, R. Jungblut, G. Güntherodt, D.J. Roberts, G.A. Gehring: Suppression of the magnetocrystalline bulk anisotropy in thin epitaxial Co(110) films on Cu(110), Phys. Rev. B **53**, 10548 (1996)
3.159 J. Fassbender, G. Güntherodt, C. Mathieu, B. Hillebrands, R. Jungblut, J. Kohlhepp, M.T. Johnson, D.J. Roberts, G.A. Gehring: Correlation between structure and magnetic anisotropies of Co on Cu(110), Phys. Rev. B **57**, 5870 (1998)
3.160 W. Sucksmith, J.E. Thompson: The magnetic anisotropy of cobalt, Proc. R. Soc. London Ser. A **225**, 362 (1954)
3.161 D.J. Roberts, G.A. Gehring: Surface magnetic anisotropies in ultrathin magnetic films, J. Magn. Magn. Mater. **156**, 293 (1996)
3.162 M. Cinal, D.M. Edwards, J. Mathon: Magnetocrystalline anisotropy in ferromagnetic films, Phys. Rev. B **50**, 3754 (1994)
3.163 R. Lorenz, J. Hafner, J. Phys. Chem. Matter **7**, 253 (1995)
3.164 K.W.H. Stevens, C.A. Bates: *Magnetic Oxides*, D.J. Craik (ed.) Wiley, London, New York (1975)
3.165 D.J. Craik: *Magnetism, Principles and Applications*, Wiley, Chichester (1995)
3.166 F.J.A. den Broeder, W. Hoving, P.J. Bloemen: Magnetic anisotropy of multilayers, J. Magn. Magn. Mater. **93**, 562 (1991)
3.167 J.R. Dutcher, B. Heinrich, J.F. Cochran, D.A. Steigerwald, W.F. Egelhoff: Magnetic properties of sandwiches and superlattices of FCC Fe(001) grown on Cu(001) substrates, J. Appl. Phys. **63**, 3464 (1988)
3.168 D.A. Steigerwald, W.F. Egelhoff Jr.: Observation of intensity oscillations in *RHEED* during the epitaxial growth of Cu and Fe on Cu(100), Surf. Sci. **192**, 887 (1987)
3.169 B. Heinrich, K.B. Urquhart, J.R. Dutcher, S.T. Purcell, J.F. Cochran, A.S. Arrott, D.A. Steigerwald, W.F. Egelhoff: Large surface anisotropies in ultrathin films of BCC and FCC Fe(001), J. Appl. Phys. **63**, 3863 (1988)
3.170 In Refs. [3.167, 3.119] the authors include a Gilbert damping term and the sample conductivity into their theory, which, however, cause only very slight modifications of the spin wave frequencies as calculated in Sect. 2.1.
3.171 An even better fit to the theory discussed in Sect. 2.1 can be achieved on assuming that the direction of the applied field is slightly tilted out of plane by 2 deg. (R. L. Stamps, private communication), which is within the error margin of the setup.

3.172 C.M. Falco: Structural and electronic properties of artificial metallic superlattices, J. Phys. (Paris) Colloq. **45**, 499 (1984)
3.173 B. Hillebrands, P. Baumgart, R. Mock, G. Güntherodt, A. Boufelfel, C.M. Falco: Collective spin waves in Fe-Pd and Fe-W multilayer structures, Phys. Rev. B **34**, 9000 (1986)
3.174 This holds for zero anisotropies. A general rule for the existence of a distinct stack surface mode is given in [3.53].
3.175 F.J.A. den Broeder, H.C. Donkersloot, H.J.G. Draaisma, W.J.M. de Jonge: Magnetic properties and structure of Pd/Co and Pd/Fe multilayers, J. Appl. Phys. **61**, 4317 (1987)
3.176 R. Van Leeuwen, C.D. England, J.R. Dutcher, C.M. Falco, W.R. Bennett, B. Hillebrands: Structural and magnetic properties of Ti/Co multilayers, J. Appl. Phys. **67**, 4910 (1990)
3.177 C.D. England, B.N. Engel, C.M. Falco: Preparation and structural characterization of epitaxial Co/Pd(111) superlattices, J. Appl. Phys. **69**, 5310 (1991)
3.178 B.N. Engel, M.H. Wiedmann, R.A. Van Leeuwen, C.M. Falco, L. Wun, N. Nakayama, T. Shinjo: Influence of structure on the magnetic anisotropy of Co/Pd(001) epitaxial superlattices, Appl. Surf. Sci. **60**, 776 (1992)
3.179 B.N. Engel, C.D. England, R.A. Van Leeuwen, M.H. Wiedmann, C.M. Falco: Interface magnetic anisotropy in epitaxial superlattices, Phys. Rev. Lett. **67**, 1910 (1991)
3.180 B.N. Engel, C.D. England, R.A. Van Leeuwen, M.H. Wiedmann, C.M. Falco: Magnetocrystalline and magnetoelastic anisotropy in epitaxial Co/Pd superlattices, J. Appl. Phys. **70**, 5873 (1991)
3.181 B. Hillebrands, J.V. Harzer, G. Güntherodt, J.R. Dutcher: Study of Co-based multilayers by Brillouin light scattering, J. Magn. Soc. Jpn. **17**, 17 (1993)
3.182 I. Pogosova, J.V. Harzer, B. Hillebrands, G. Güntherodt, D. Guggi, D. Weller, R.F.C. Farrow, C.H. Lee: Interface analysis of (110)-oriented Co/Pt superlattices with large magnetic anisotropies using x-ray diffraction, J. Appl. Phys. **76**, 908 (1994)
3.183 F.J.A. den Broeder, D. Kuiper, A.P. van de Mosselaer, W. Hoving: Perpendicular magnetic anisotropy of Co-Au multilayers induced by interface sharpening, Phys. Rev. Lett. **60**, 2769 (1988)
3.184 T.B. Massalski: *Binary Alloy Phase Diagrams*, American Society for Metals, Metals Park (Ohio) (1986)
3.185 K. Spörl, D. Weller, G. Rupp: Interface quality in ion beam sputtered multilayers: Au-Co and Tb-Fe, J. Magn. Magn. Mater. **83**, 91 (1990)
3.186 P. Krams, B. Hillebrands, G. Güntherodt, K. Spörl, D. Weller: Interface quality and magnetic properties of Co/Au multilayers investigated by Brillouin light scattering, J. Appl. Phys. **69**, 5307 (1991)
3.187 S.S.P. Parkin, N. More, K.P. Roche: Oscillations in exchange coupling and magnetoresistance in metallic superlattice structures: Co/Ru, Co/Cr, and Fe/Cr, Phys. Rev. Lett. **64**, 2304 (1990)
3.188 G. Binasch, P. Grünberg, F. Saurenbach, W. Zinn: Enhanced magnetoresistance in layered magnetic structures with antiferromagnetic interlayer exchange, Phys. Rev. B **39**, 4828 (1989)
3.189 B. Heinrich, Z. Celinski, J.F. Cochran, W.B. Muir, J. Rudd, Q.M. Zhong, A.S. Arrott, K. Myrtle, J. Kirschner: Ferromagnetic and antiferromagnetic exchange coupling in bcc epitaxial ultrathin Fe(001)/Cu(001)/Fe(001) trilayers, Phys. Rev. Lett. **64**, 673 (1990)

3.190 J.F. Cochran, J. Rudd, W.B. Muir, B. Heinrich, Z. Celinski: Brillouin light-scattering experiments on exchange-coupled ultrathin bilayers of iron separated by epitaxial copper (001), Phys. Rev. B **42**, 508 (1990)
3.191 D.H. Mosca, F. Petroff, A. Fert, P.A. Schroeder, W.P. Pratt Jr., R. Laloee: Oscillatory interlayer coupling and giant magnetoresistance in Co/Cu multilayers, J. Magn. Magn. Mater. **94**, 1 (1991)
3.192 S.S.P. Parkin: Systematic variation of the strength and oscillation period of indirect magnetic exchange coupling through the 3d, 4d, and 5d transition metals, Phys. Rev. Lett. **67**, 3598 (1991)
3.193 S.S.P. Parkin, R. Bhadra, K.P. Roche: Oscillatory magnetic exchange coupling through thin copper layers, Phys. Rev. Lett. **66**, 2152 (1991)
3.194 F. Petroff, A. Barthelemy, D.H. Mosca, D.K. Lottis, A. Fert, P.A. Schroeder, W.P. Pratt Jr., R. Laloee, S. Lequien: Oscillatory interlayer exchange and magnetoresistance in Fe/Cu multilayers, Phys. Rev. B **44**, 5355 (1991)
3.195 S.S.P. Parkin, A. Mansour, G.P. Felcher: Antiferromagnetic interlayer exchange coupling in sputtered Fe/Cr multilayers: dependence on number of Fe layers, Appl. Phys. Lett. **58**, 1473 (1991)
3.196 S.O. Demokritov, J.A. Wolf, P. Grünberg: Evidence for oscillations in the interlayer coupling of Fe films across Cr films from spin waves and M(H) curves, Europhys. Lett. **15**, 881 (1991)
3.197 J. Unguris, R.J. Celotta, D.T. Pierce: Observation of two different periods in the exchange coupling of Fe/Cr/Fe(100), Phys. Rev. Lett. **67**, 140 (1991)
3.198 S.T. Purcell, W. Folkerts, M.T. Johnson, N.W.E. McGee, K. Jager, J. aan de Stegge, W.B. Zeper, W. Hoving, P. Grünberg: Oscillations with a period of two Cr monolayers in the antiferromagnetic exchange coupling in a (001) Fe/Cr/Fe sandwich structure, Phys. Rev. Lett. **67**, 903 (1991)
3.199 P. Grünberg, S.O. Demokritov, A. Fuss, R. Schreiber, J.A. Wolf, S.T. Purcell: Interlayer exchange, magnetotransport and magnetic domains in Fe/Cr layered structures, J. Magn. Magn. Mater. **106**, 1734 (1992)
3.200 A. Fuß, S.O. Demokritov, P. Grünberg, W. Zinn: Short- and long period oscillations in the exchange coupling of Fe across epitaxially grown Al- and Au-interlayers, J. Magn. Magn. Mater. **103**, 221 (1992)
3.201 M.N. Baibich, J.M. Broto, A. Fert, F.N. Van Dau, F. Petroff, P. Eitenne, G. Creuzet, A. Friedrich, J. Chazelas: Giant magnetoresistance of (001)Fe/(001)Cr magnetic superlattices, Phys. Rev. Lett. **61**, 2472 (1988)
3.202 M.D. Stiles: Exchange coupling in magnetic heterostructures, Phys. Rev. B **48**, 7238 (1993)
3.203 J.C. Slonczewski: Overview of interlayer exchange coupling, J. Magn. Magn. Mater. **150**, 13 (1995)
3.204 B. Dieny, V.S. Speriosu, S. Metin, S.S.P. Parkin, B.A. Gurney, P. Baumgart, D.R. Wilhoit: Magnetotransport properties of magnetically soft spin-valve structures, J. Appl. Phys. **69**, 4774 (1991)
3.205 S.S.P. Parkin, D. Mauri: Spin engineering: direct determination of the Ruderman-Kittel-Kasuya-Yosida far-field range function in ruthenium, Phys. Rev. B **44**, 7131 (1991)
3.206 J. Barnaś, P. Grünberg: Spin waves in exchange-coupled epitaxial double-layers, J. Magn. Magn. Mater. **82**, 186 (1989)
3.207 M. Vohl, J. Barnas, P. Grünberg: Effect of interlayer exchange coupling on spin-wave spectra in magnetic double layers: Theory and experiment, Phys. Rev. B **39**, 12003 (1989)

3.208 M. Macciò, M.G. Pini, P. Politi, A. Rettori: Spin-wave study of the magnetic excitations in sandwich structures coupled by bilinear and biquadratic interlayer exchange, Phys. Rev. B **49**, 3283 (1994)
3.209 S. Tschopp, G. Robins, R.L. Stamps, R. Sooryakumar, M.E. Filipkowski, C.J. Gutierrez, G.A. Prinz: Observation of magnons by light scattering in epitaxial CoFe/Mn/CoFe trilayers, J. Appl. Phys. **81**, 3785 (1997)
3.210 J.F. Cochran: Magnetization reversal in a thin ferromagnetic film weakly exchange coupled to a bulk ferromagnet, J. Magn. Magn. Mater. **147**, 101 (1995)
3.211 M. Grimsditch, S. Kumar, E.F. Fullerton: Brillouin light scattering study of Fe/Cr/Fe (211) and (100) trilayers, Phys. Rev. B **54**, 3385 (1996)
3.212 B. Hillebrands, J.V. Harzer, G. Güntherodt, C.D. England, C.M. Falco: Experimental evidence for the existence of exchange-dominated collective spin-wave excitations in multilayers, Phys. Rev. B **42**, 6839 (1990)
3.213 B. Hillebrands, J.V. Harzer, R.L. Stamps, G. Güntherodt, C.D. England, C.M. Falco: Evidence of collective exchange modes in Co/Pd multilayers observed by Brillouin light scattering, J. Magn. Magn. Mater. **93**, 211 (1991)
3.214 J. Fassbender, F.C. Nörtemann, R.L. Stamps, R.E. Camley, B. Hillebrands, G. Güntherodt, S.S.P. Parkin: Oscillatory interlayer exchange coupling of Co/Ru multilayers investigated by Brillouin light scattering, Phys. Rev. B **46**, 5810 (1992)
3.215 J. Fassbender, F.C. Nörtemann, R.L. Stamps, R.E. Camley, B. Hillebrands, G. Güntherodt, S.S.P. Parkin: Oscillatory interlayer exchange coupling of Co/Ru and permalloy/Ru multilayers investigated by Brillouin light scattering, J. Magn. Magn. Mater. **121**, 270 (1993)
3.216 Experimentally this band has not been observed so far, however light scattering cross section calculations show that the modes of this band would only contribute weakly to the cross section. (R. L. Stamps, private commun.)
3.217 R.L. Stamps, R.E. Camley, B. Hillebrands, G. Güntherodt: Spin-wave propagation on imperfect ultrathin ferromagnetic films, Phys. Rev. B **47**, 5072 (1993)
3.218 G. Güntherodt, B. Hillebrands, P. Krams, J.V. Harzer, F. Lauks, R.L. Stamps, W. Weber, D. Hartmann, D.A. Wesner, A. Rampe, U.A. Effner, H.P. Oepen, D. Weller, R.F.C. Farrow, B.N. Engel, C.M. Falco: Effects of interfaces on magnetic properties of Co-based multilayers, Phil. Mag. B **70**, 767 (1994)
3.219 J.V. Harzer, B. Hillebrands, R.L. Stamps, G. Güntherodt, D. Weller, C. Lee, R.F.C. Farrow, E.E. Marinero: Characterization of large magnetic anisotropies in (100)- and (111)-oriented Co/Pt multilayers by Brillouin light scattering, J. Magn. Magn. Mater. **104-107**, 1863 (1992)
3.220 B. Hillebrands, P. Krams, K. Spörl, D. Weller: Influence of the interface quality on the elastic properties of Co/Au multilayers investigated by Brillouin light scattering, J. Appl. Phys. **69**, 938 (1991)
3.221 F. Scheurer, R. Allenspach, P. Xhonneux, E. Courtens: Magnetic coupling of structural microdomains in bcc Fe on Cu(001), Phys. Rev. B **48**, 9890 (1993)
3.222 W.H. Meiklejohn, C.P. Bean: New magnetic anisotropy, Phys. Rev. B **102**, 1413 (1956)
3.223 W. Stoecklein, S.S.P. Parkin, J.C. Scott: Ferromagnetic resonance studies of exchange-biased permalloy thin films, Phys. Rev. B **38**, 6847 (1988)

3.224 M.J. Carey, A.E. Berkowitz: CoO-NiO superlattices: Interlayer interactions and exchange anisotropy with $Ni_{81}Fe_{19}$, J. Appl. Phys. **73**, 6892 (1993)

3.225 R. Jungblut, R. Coehoorn, M.T. Johnson, J. aan de Stegge, A. Reinders: Orientational dependence of the exchange biasing in molecular-beam-epitaxy-grown $Ni_{80}Fe_{20}/Fe_{50}Mn_{50}$ bilayers, J. Appl. Phys. **75**, 6659 (1994)

3.226 R. Jungblut, R. Coehoorn, M.T. Johnson, C. Sauer, P. J. van der Zaag, A.R. Ball, T.G. S.M. Rijks, J. aan de Stegge, A. Reinders: Exchange biasing in MBE-grown $Ni_{80}Fe_{20}/Fe_{50}Mn_{50}$ bilayers, J. Magn. Magn. Mater. **148**, 300 (1995)

3.227 A.P. Malozemoff: Random-field model of exchange anisotropy at rough ferromagnetic-antiferromagnetic interfaces, Phys. Rev. B **35**, 3679 (1987)

3.228 A. Ercole, T. Fujimoto, M. Patel, C. Daboo, R.J. Hicken, J.A.C. Bland: Direct measurement of magnetic anisotropies in epitaxial FeNi/Cu/Co spin-valve structures by Brillouin light scattering, J. Magn. Magn. Mater. **156**, 121 (1996)

3.229 C. Mathieu, M. Bauer, B. Hillebrands, J. Fassbender, G. Güntherodt, R. Jungblut, J. Kohlhepp, A. Reinders: Brillouin light scattering investigations of exchange biased (110)-oriented NiFe/FeMn bilayers, J. Appl. Phys. **83**, 2863 (1998)

3.230 R. Jungblut, M.J. Decker, J.T. Kohlhepp, J. J. de Fries, K. Ramstöck, A. Reinders, R. Coehoorn: to be published

3.231 T.C. Schulthess, W.H. Butler: Consequences of spin-flop coupling in exchange biased films, Phys. Rev. Lett. **81**, 4516 (1998)

3.232 A.S. Borovik-Romanov, V.G. Jotikov, N.M. Kreines, A.A. Pankow: Microwave modulation of light by antiferromagnetic resonance in $CoCO_3$, JETP Letters **23**, 649 (1976)

3.233 V.N. Venitskii, V.V. Eremenko, E.V. Matyushkin: Light scattering from parametric spin waves in $Y_3Fe_5O_{12}$ under longitudinal pumping, JETP Lett. **27**, 222 (1978)

3.234 A.S. Borovik-Romanov, N.M. Kreines, R. Laiho, T. Levola, V.G. Zhotikov: Optical detection of ferromagnetic resonance in K_2CuF_4, J. Physics C **13**, 879 (1980)

3.235 C. Elachi: Magnetic wave propagation in a periodic medium, IEEE Trans. Mag. **11**, 36 (1975)

3.236 M. Tsutsumi, Y. Sakaguchi, N. Kumagai: Behavior of the magnetostatic wave in a periodically corrugated YIG slab, IEEE Transactions on Microwave Theory and Techniques, 224 (1976)

3.237 M. Tsutsumi, Y. Sakaguchi, N. Kumagai: The magnetostatic surface-wave propagation in a corrugated YIG slab, Appl. Phys. Lett. **31**, 779 (1977)

3.238 S.R. Seshadri: Magnetic wave interactions in a periodically corrugated YIG film, IEEE Transactions on Microwave Theory and Techniques **27**, 199 (1979)

3.239 N.S. Chang, Y. Matsuo: Magnetostatic surface wave propagation on a periodic YIG film layer, Appl. Phys. Lett. **35**, 352 (1979)

3.240 Y.V. Gulyaev, S.A. Nikitov, V.P. Plesskii: The propagation of magnetostatic waves in a normally magnetized ferrite plate with periodically uneven surfaces, Sov. Phys. Solid State **22**, 1651 (1980)

3.241 Y.V. Gulyaev, S.A. Nikitov, V.P. Plesskii: Reflection of surface magnetostatic waves from a periodically uneven ferrite surface, Sov. Phys. Solid State **23**, 724 (1981)

3.242 Y.V. Gulyaev, S.A. Nikitov: Bragg reflection of an obliquely incident surface magnetostatic wave from a periodic part of the surface of a ferrite, Sov. Phys. Solid State **23**, 2138 (1981)
3.243 P.A. Kolodin, B. Hillebrands: Spin-wave propagation across periodically corrugated thin metallic ferromagnetic films, J. Magn. Magn. Mater. **161**, 199 (1996)
3.244 B.A. Gurney, P. Baumgart, V. Speriosu, R. Fontana, A. Patlac, T. Logan, P. Humbert: Finite size effects in ferromagnetic patterned submicron structures observed by Brillouin light scattering, Digest of the International Conference on Magnetic Films and Surfaces, Glasgow (1991)
3.245 C. Mathieu, J. Jorzick, A. Frank, S.O. Demokritov, B. Hillebrands, A.N. Slavin, B. Bartenlian, C. Chappert, D. Decanini, F. Rousseaux, E. Cambril: Lateral quantization of spin waves in micron size magnetic wires, Phys. Rev. Lett. **81**, 3968 (1998)
3.246 B. Hillebrands, S.O. Demokritov, C. Mathieu, S. Riedling, O. Büttner, A. Frank, B. Roos, J.Jorzick, A.N. Slavin, B. Bartenlian, C. Chappert, D. Decanini, A. Müller, U. Hartmann: Arrays of interacting magnetic dots and wires: static and dynamic properties, Jap. J. Appl. Phys, in press
3.247 A. Ercole, A.O. Adeyeye, C. Daboo, J.A.C. Bland, D.G. Hasko: Finite size effects in the static and dynamic magnetic properties of FeNi wire array structures, J. Appl. Phys. **81**, 5452 (1997)
3.248 A. Ercole, A.O. Adeyeye, J.A.C. Bland, D.G. Hasko: Size-dependent spin-wave frequency in ferromagnetic wire-array structures, Phys. Rev. B **58**, 345 (1998)
3.249 S.M. Cherif, Y. Roussigne, P. Moch, J. Hennequin, M. Labrune: Brillouin investigation of the dynamic properties and of cobalt wire arrays, J. Appl. Phys., **85**, 5477 (1999)
3.250 B. Hillebrands, C. Mathieu, M. Bauer, S.O. Demokritov, B. Bartenlian, C. Chappert, D. Decanini, F. Rousseaux, F. Carcenac: Brillouin light scattering investigations of structured permalloy films, J. Appl. Phys. **81**, 4993 (1997)
3.251 C. Mathieu, C. Hartmann, M. Bauer, O. Büttner, B. Roos, S.O. Demokritov, B. Hillebrands, B. Bartenlian, C. Chappert, D. Decanini, F. Rousseaux, E. Cambril, A. Müller, B. Hoffmann, U. Hartmann: Anisotropic magnetic coupling of permalloy micron dots forming a square lattice, Appl. Phys. Lett. **70**, 2912 (1997)
3.252 C. Mathieu, C. Hartmann, M. Bauer, O. Büttner, S. Riedling, B. Roos, S.O. Demokritov, B. Hillebrands, B. Bartenlian, C. Chappert, D. Decanini, F. Rousseaux, E. Cambril, A. Müller, B. Hoffmann, U. Hartmann: Static and dynamic properties of patterned magnetic permalloy films, J. Magn. Magn. Mater. **175**, 10 (1997)
3.253 S.M. Cherif, C. Dugautier, J. Hennequin, P. Moch: Brillouin light scattering by magnetic surface waves in dot-strucured permalloy layers, J. Magn. Magn. Mater. **175**, 228 (1997)
3.254 H. Litschke, M. Schilberg, T. Kleinefeld, B. Hillebrands: Spin wave dynamics in Fe-Ni double layers and multilayers, J. Magn. Magn. Mater. **104-107**, 1807 (1992)
3.255 B. Heinrich, J.F. Cochran, A.S. Arrott, S.T. Purcell, K.B. Urquhart, J.R. Dutcher, W.F. Egelhoff: Development of magnetic anisotropies in ultrathin epitaxial films of Fe(001) and Ni(001), Appl. Phys. A **49**, 473 (1989)

3.256 J.F. Cochran, W.B. Muir, J.M. Rudd, B. Heinrich, Z. Celinski, T.-T. Le-Tran, W. Schwarzacher, W. Bennett, W.F. Egelhoff: Magnetic anisotropies in ultrathin FCC Fe(001) films grown on Cu(001) substrates, J. Appl. Phys. **69**, 5206 (1991)
3.257 J.F. Cochran, J.M. Rudd, M. From, B. Heinrich, W. Bennett, W. Schwarzacher, W.F. Egelhoff Jr.: Magnetic anisotropies in ultrathin fcc Fe(001) films grown on Cu(001) substrates, Phys. Rev. B **45**, 4676 (1992)
3.258 J.F. Cochran, B. Heinrich, A.S. Arrott, K.B. Urquhart, J.R. Dutcher, S.T. Purcell: Anisotropies in ultrathin films of iron grown on silver, Journal de Physique Colloque **49**, 1671 (1988)
3.259 M. From, J.F. Cochran, B. Heinrich, Z. Celinski: Investigation of the dependence of BLS frequencies on angle of incidence for thin iron films, J. Appl. Phys. **73**, 6754 (1993)
3.260 R.J. Hicken, A. Ercole, S.J. Gray, C. Daboo, J.A.C. Bland: In-situ Brillouin light scattering from ultrathin epitaxial Fe/Ag(100) films with Cr and Ag overlayers, J. Appl. Phys. **79**, 4987 (1996)
3.261 A. Yoshihara, Y. Suzuki, T. Katayama: Collective spin waves in an ultrathin Fe wedge: a Brillouin scattering study, J. Phys. Soc. Japan **65**, 1469 (1996)
3.262 S.O. Demokritov, C. Mathieu, M. Bauer, S. Riedling, O. Büttner, H. de Gronckel, B. Hillebrands: Interface anisotropy and magnetization reversal in epitaxial Fe(001)/Pd double layers, J. Appl. Phys. **81**, 4466 (1997)
3.263 B. Heinrich, A.S. Arrott, J.F. Cochran, K.B. Urquhart, K. Myrtle, Z. Celinski, Q.M. Zhong: In-situ techniques for studying epitaxially grown layers and determining their magnetic properties, Mat. Res. Soc. Symp. Proc. **151**, 177 (1989)
3.264 J.F. Cochran, K. Totland, B. Heinrich, D. Venus, S. Govorkov: BLS studies of exchange coupling in the iron whisker/Cr/Fe system, Mat. Res. Soc. Symp. Proc. **384**, 131 (1995)
3.265 R.J. Hicken, A.J.R. Ives, D.E.P. Eley, C. Daboo, J.A.C. Bland, J.R. Childress, A. Schuhl: Interlayer exchange coupling and weakened surface exchange in ultrathin epitaxial Fe(100)/Pd(100)/MgO(100) structures, Phys. Rev. B **50**, 6143 (1994)
3.266 G.W. Anderson, M.C. Hanf, P.R. Norton, C. Gigault, J.R. Dutcher: The structure and magnetic properties of Fe overlayers on Al(100), Proceedings of the SPIE - The International Society for Optical Engineering **2140**, 112 (1994)
3.267 C. Daboo, R.J. Hicken, D.E.P. Fley, M. Gester, S.J. Gray, A.J.R. Ives, J.A.C. Bland: Magnetization reversal processes in epitaxial Fe/GaAs(001) films,
J. Appl. Phys. **75**, 5586 (1994)
3.268 C. Daboo, M. Gester, S.J. Gray, J.A.C. Bland, R.J. Hicken, E. Gu: Magnetic anisotropy and magnetization reversal in Fe films grown on GaAs (001) and (110), J. Magn. Magn. Mater. **148**, 262 (1995)
3.269 R.J. Hicken, D.E.P. Eley, M. Gester, S.J. Gray, C. Daboo, A.J.R. Ives, J.A.C. Bland: Brillouin light scattering studies of magnetic anisotropy in epitaxial Fe/GaAs films, J. Magn. Magn. Mater. **145**, 278 (1995)
3.270 W. Jantz, G. Rupp, R.S. Smith, W. Wettling, G. Bayreuther: Investigation of single crystal Fe films grown by MBE on GaAs substrates, IEEE Trans. Mag. **19**, 1859 (1983)

3.271 W. Wettling, W. Jantz, R.S. Smith: *Brillouin scattering study of surface magnons in single crystal iron films epitaxially grown on GaAs*, Proceedings of the international conference on surface studies with lasers, Mauterndorf, Fr. Mussenegg, A. Leitner, M.E. Lippitsch (eds.) Springer, Berlin, Heidelberg, New York, Tokyo (1983), p. 135.
3.272 P. Grünberg, C.M. Mayr, W. Vach: Determination of magnetic parameters by means of Brillouin scattering. Examples: Fe, Ni, $Ni_{0.8}Fe_{0.2}$, J. Magn. Magn. Mater. **28**, 319 (1982)
3.273 P. Kabos, C.E. Patton, W.D. Wilber: Anomalous angle dependence of the surface-magnon frequency in thin films, Phys. Rev. Lett. **53**, 1962 (1984)
3.274 P.X. Zhang, Y.Z. Pang, W. Zinn: Brillouin scattering from ultra thin iron films, Solid State Commun. **60**, 449 (1986)
3.275 P.H. Chang, A.P. Malozemoff, M. Grimsditch, W. Senn, G. Winterling: Brillouin scattering from ferromagnetic metallic glasses, Solid State Commun. **27**, 617 (1978)
3.276 M. Grimsditch, A. Malozemoff, A. Brunsch: Standing spin waves observed by Brillouin scattering in amorphous metallic $Fe_{80}B_{20}$ films, Phys. Rev. Lett. **43**, 711 (1979)
3.277 A.P. Malozemoff, P.H. Chang, M. Grimsditch: Brillouin scattering from rapidly quenched ferromagnetic metallic glasses, J. Appl. Phys. **50**, 5896 (1979)
3.278 M. Mendik, Z. Frait, P. Wachter: A comparative study of FMR and Brillouin light scattering on FeNiB alloys, Solid State Commun. **84**, 951 (1992)
3.279 X. Liu, R. Sooryakumar, C.J. Gutierrez, G.A. Prinz: Exchange stiffness and magnetic anisotropies in BCC $Fe_{1-x}Co_x$ alloys, J. Appl. Phys. **75**, 7021 (1994)
3.280 A. Yoshihara, Y. Shimada, T. Maro, O. Kitakami, K. Mizushima: Brillouin scattering in Fe-polyethylene co-evaporated films, J. Magn. Magn. Mater. **104-107**, 1835 (1992)
3.281 A. Yoshihara, Y. Haneda, Y. Shimada, T. Fujimura: Magnon Brillouin scattering in FeSi sputtering films, J. Appl. Phys. **65**, 4968 (1989)
3.282 G. Gubbiotti, G. Carlotti, G. Socino, F. D'Orazio, F. Lucari, R. Bernardini, M.D. Crescenzi: Perpendicular and in-plane magnetic anisotropy in epitaxial Cu/Ni/Cu/Si(111) ultrathin films, Phys. Rev. B **56**, 11073 (1997)
3.283 G. Gubbiotti, G. Carlotti, G. Socino: Determination of the magnetic parameters of thin Ni films by Brillouin light scattering, Materials Science Forum **195**, 215 (1995)
3.284 G. Gubbiotti, G. Carlotti, G. Socino: Inelastic light scattering by surface spin waves in thin ferromagnetic films, Proceedings of the national school: new developments and magnetism's applications (1995)
3.285 G. Gubbiotti, G. Carlotti, G. Socino: Brillouin light scattering investigations of the magnetic properties of Ni films deposited on different substrates, Il Vuoto, Scienza e Technologia **25**, 75 (1996)
3.286 P. Djemia, F. Ganot, P. Moch: Brillouin scattering in ultra thin permalloy films: monolayers and multilayers with alumina interfaces, J. Magn. Magn. Mater. **165**, 428 (1997)
3.287 G. Srinivasan, C.E. Patton: Light scattering study on surface magnons in permalloy films, J. Appl. Phys. **61**, 4120 (1987)

3.288 D. Kerkmann, J.A. Wolf, D. Pescia, T. Wolke, P. Grünberg: Spin waves and two-dimensional magnetism in the Co-monolayer on Cu(100), Solid State Commun. **72**, 963 (1989)

3.289 M. G. Pini, A. Rettori, D. Pescia, N. Majlis, S. Selzer: Microscopic spin wave study of the Co/Cu(100) epitaxial monolayer, *Microscopic Aspects of Nonlinearity in Condensed Matter*, A.R. Bishop et al. (eds.) Plenum Press, New York (1991) p. 349

3.290 K. Hyomi, A. Murayama, J. Eickmann, C.M. Falco: Perpendicular magnetic anisotropy in Au/Co/Au(111) films: interface anisotropy and effect of strain, J. Magn. Magn. Mater., in press

3.291 A. Murayama, K. Hyomi, J. Eickmann, C.M. Falco: Low frequency excitations of spin-wave Brillouin scattering in ultrathin Co/Au(111) films, J. Magn. Magn. Mater., **198-199**, 372 (1999)

3.292 A. Murayama, K. Hyomi, J. Eickmann, C.M. Falco: Underlayer-induced perpendicular magnetic anisotropy in ultrathin Co/Au/Cu(111) films: A spin-wave Brillouin-scattering study, Phys. Rev. B **58**, 8596 (1998)

3.293 K. Hyomi, A. Murayama, J. Eickmann, C.M. Falco: Magnetic inhomogeneity in ultrathin Co films studied by spin-wave Brillouin scattering, J. Magn. Magn. Mater., **198-199**, 528 (1999)

3.294 A. Moghadam, J.G. Booth, D.G. Lord, J. Boyle, A.D. Boardman: Brillouin light scattering studies of Co/GaAs(001) films, IEEE Trans. Mag. **30**, 769 (1994)

3.295 E. Gu, M. Gester, R.J. Hicken, C. Daboo, M. Tselepi, S.J. Gray, J.A.C. Bland, L.M. Brown: Fourfold anisotropy and structural behaviour of epitaxial hcp Co/GaAs(001) thin films, Phys. Rev. B **52**, 14704 (1995)

3.296 X. Liu, M. Steiner, R. Sooryakumar, G.A. Prinz, R.F.C. Farrow, G. Harp: Exchange stiffness, magnetization, and spin waves in cubic and hexagonal phases of cobalt, Phys. Rev. B **53**, (1996)

3.297 S. Subramanian, X. Liu, R.L. Stamps, R. Sooryakumar, G.A. Prinz: Magnetic anisotropies in body-centered-cubic cobalt films, Phys. Rev. B **52**, 10194 (1995)

3.298 J.M. Karanikas, R. Sooryakumar, G.A. Prinz, B.T. Jonker: Thermal magnons in BCC cobalt-itinerancy and exchange stiffness, J. Appl. Phys. **69**, 6120 (1991)

3.299 X. Liu, R.L. Stamps, R. Sooryakumar, G.A. Prinz: Magnetic anisotropies in thick bcc Co, J. Appl. Phys. **79**, (1996)

3.300 M. Grimsditch, E.E. Fullerton, R.L. Stamps: Exchange and anisotropy effects on spin waves in epitaxial Co films, Phys. Rev. B **56**, 2617 (1997)

3.301 R.L. Stamps, R.E. Camley, R.J. Hicken: Influence of exchange-coupled anisotropies on spin-wave frequencies in magnetic layered systems: application to Co/CoO, Phys. Rev. B **54**, 4159 (1996)

3.302 R.L. Stamps, R.E. Camley, R.J. Hicken: Spin wave frequency shifts in exchange coupled ferromagnet/antiferromagnet structures: Application to Co/CoO, J. Appl. Phys. **81**, 4485 (1997)

3.303 S.P. Vernon, S.M. Lindsay, M.B. Stearns: Brillouin scattering from thermal magnons in a thin Co film, Phys. Rev. B **29**, 4439 (1984)

3.304 A. Yoshihara, Y. Haneda, Y. Shimada, T. Fujimura: Brillouin scattering from spin waves in sputtered Co-Zr films, J. Appl. Phys. **66**, 328 (1989)

3.305 A. Yoshihara, K. Takanashi, M. Shimoda, O. Kitakami, Y. Shimada: Magnon Brillouin scattering from an amorphous CoNbZr thin film, Japanese Journal of Appl. Phys. **33**, 3927 (1994)
3.306 G. Srinivasan, C.E. Patton: Brillouin light scattering on cobalt-chromium films, IEEE Trans. Mag. **22**, 996 (1986)
3.307 A. Yoshihara, Y. Haneda, Y. Shimada, K. Ouchi: Magnetostatic spin waves in sputtered ferromagnetic films with perpendicular anisotropy: CoCr, Japanese Journal of Appl. Phys. **30**, 2010 (1991)
3.308 A. Murayama, M. Miyamura, K. Nishiyama, K. Miyata, Y. Oka: Brillouin spectroscopy of spin waves in sputtered CoPt alloy films and Co/Pt/Co multilayered films, J. Appl. Phys. **69**, 5661 (1991)
3.309 A. Murayama, M. Miyamura, K. Nishiyama, Y. Oka: Spin-wave Brillouin scattering in CoPt alloy films with perpendicular c-axis orientation of hcp-Co, J. Mag. Soc. Japan **18**, 265 (1994)
3.310 J.V. Harzer, B. Hillebrands, I.S. Pogosova, M. Herrmann, G. Güntherodt, D. Weller: Structural and magnetic properties of e–beam prepared Co_xPt_{1-x} alloy films, Mat. Res. Soc. Symp. Proc. **313**, 387 (1993)
3.311 A. Murayama, M. Miyamura, S. Ishikawa, Y. Oka: Brillouin study of spin waves in sputtered CoNiPt alloy films, J. Appl. Phys. **67**, 410 (1990)
3.312 S.H. Kong, M.V. Klein, F. Tsui, C.P. Flynn: Brillouin-light scattering study of long-wavelength spin waves in a single-crystal 300 Å gadolinium film, Phys. Rev. B **45**, 12297 (1992)
3.313 A. Yoshihara, Y. Shimada, T.-H. Chiang, K. Fukamichi: Brillouin scattering determination of magnetic and elastic constants in Invar-type $La(Fe_xAl_{1-x})_{13}$ amorphous alloy films, J. Appl. Phys. **75**, 1733 (1994)
3.314 J.W. Boyle, A.J.A. Cowen, K.M. Yasseen, J.G. Booth, A.D. Boardman, K.M. Booth, D.G. Lord: Brillouin light scattering study of TbFeCo thin films, J. Appl. Phys. **82**, 4453 (1997)
3.315 D. Raasch, J. Reck, C. Mathieu, B. Hillebrands: Exchange stiffness constant and wall energy density of amorphous GdTb-FeCo thin films, J. Appl. Phys. **76**, 1145 (1994)
3.316 C. Mathieu, B. Hillebrands, D. Raasch: Exchange stiffness constant and effective gyromagnetic factor of Gd, Tb and Nd containing, amorphous rare earth-transition metal films, IEEE Trans. Mag. **30**, 4434 (1994)
3.317 J.G. Booth, G. Srinivasan, C.E. Patton, P. de Gasperis: Brillouin light scattering study of the spin wave stiffness parameter in Sc-substituted lutetium-yttrium iron garnet, IEEE Trans. Mag. **23**, 3494 (1987)
3.318 J.W. Boyle, J.G. Booth, A.D. Boardman, I. Zavislyak, V. Bobkov, V. Romanyuk: Investigations of epitaxial Ga:YIG (111) films by Brillouin light scattering and microwave spectroscopy, Proc. ICF7, J. de Physique IV (Colloque) **7**, 497 (1997)
3.319 P. Swiatek, F. Saurenbach, Y. Pang, P. Grünberg, W. Zinn: Exchange coupling of ferromagnetic films across nonmagnetic interlayers, J. Appl. Phys. **61**, 3753 (1987)
3.320 P. Grünberg: Some ways to modify the spin-wave mode spectra of magnetic multilayers, J. Appl. Phys. **57**, 3673 (1985)
3.321 F. Saurenbach, J. Barnaś, G. Binasch, M. Vohl, P. Grünberg, W. Zinn: Spin waves and magnetoresistance in exchange coupled layered magnetic structures, Thin Solid Films **175**, 317 (1989)

3.322 R.J. Hicken, C. Daboo, M. Gester, A.J.R. Ives, S.J. Gray, J.A.C. Bland: Interlayer exchange coupling in epitaxial Fe/Cr/Fe/Ag/GaAs(100) structures, J. Appl. Phys. **78**, 6670 (1995)

3.323 B. Heinrich, M. From, J.F. Cochran, L.X. Liao, Z. Celinski, C.M. Schneider, K. Myrtle: Studies of exchange coupling in Fe(001) whisker/Cr/Fe structures using BLS and RHEED techniques, Mat. Res. Soc. Symp. Proc. **313**, 119 (1993)

3.324 B. Heinrich, Z. Celinski, J.F. Cochran, M. From: *Bilinear and biquadratic exchange coupling in bcc Fe/Cu/Fe trilayers. Exchange coupling in Fe whisker/Cr/Fe(001) structures*, Proceedings of the NATO Advanced Studies Institute Series B, **309**, Magnetism and Structure in Systems of Reduced Dimension, R.F.C Farrow, B. Dieny, M. Donath, A. Fert, B.D. Hermsmeier (eds.) Plenum Press (New York, London 1993) p. 175

3.325 M. From, L.X. Liao, J.F. Cochran, B. Heinrich: Temperature dependence of the exchange coupling in the Fe(001) whisker/11 ML Cr/20 ML Fe structure, J. Appl. Phys. **75**, 6181 (1994)

3.326 B. Heinrich, M. From, J.F. Cochran, M. Kowalewski, D. Atlan, Z. Celinski, K. Myrtle: MBE growth and FMR, BLS and MOKE studies of exchange coupling in Fe whisker/Cr/Fe(001) and in Fe/Cu/Fe(001) 'loose spin' structures, J. Magn. Magn. Mater. **140–144**, 545 (1995)

3.327 P. Kabos, C.E. Patton, M.O. Dima, D.B. Church, R.L. Stamps, R.E. Camley: Brillouin-light scattering on Fe/Cr/Fe thin-film sandwiches, J. Appl. Phys. **75**, 3553 (1994)

3.328 S.M. Rezende, M.A. Lucena, F.M.d. Aguiar, A. Azevedo, C. Chesman, P. Kabos, C.E. Patton: High-resolution Brillouin light scattering and angle-dependent 9.4 GHz ferromagnetic resonance in MBE-grown Fe/Cr/Fe on GaAs, Phys. Rev. B **55**, 8071 (1997)

3.329 M. Kowalewski, B. Heinrich, K. Totland, J.F. Cochran, S. Govorkov, D. Atlan: Studies of exchange coupling in Fe/Cu/Fe(001) "Loose Spin" structures, Mat. Res. Soc. Symp. Proc. **384**, 171 (1995)

3.330 Z. Celinski, B. Heinrich, J.F. Cochran, W.B. Muir, A.S. Arrott, J. Kirschner: Growth and magnetic studies of lattice expanded Pd in ultrathin Fe(001)/Pd(001)/Fe(001) structures, Phys. Rev. Lett. **65**, 1156 (1990)

3.331 W.B. Muir, J.F. Cochran, J.M. Rudd, B. Heinrich, Z. Celinski: Observation of the exchange-coupled modes in Fe(001)/Pd/Fe(001) ultrathin trilayers, J. Magn. Magn. Mater. **93**, 229 (1991)

3.332 Z. Celinski, B. Heinrich, J.F. Cochran: Ferromagnetic and antiferrromagnetic exchange coupling in ultrathin Fe(001)/Pd(001)/Fe(001) structures, J. Appl. Phys. **70**, 5870 (1991)

3.333 T. Otake, Y. Haneda, A. Yoshihara, Y. Shimada, T. Fujimura: Brillouin scattering in FeSi/X/FeSi double-layer films, X=ZnO and SiO_2, J. Appl. Phys. **67**, 3456 (1990)

3.334 A. Murayama, M. Miyamura, K. Miyata, Y. Oka: Interlayer exchange coupling in Co/Cr/Co double-layered recording films studied by spin-wave Brillouin scattering, IEEE Trans. Mag. **27**, 5064 (1991)

3.335 H. Niedoba, B. Mirecki, M. Jackson, S. Jordan, S. Thompson, J.S.S. Whiting, P. Djemia, F. Ganot, P. Moch, T.P. Hase, I. Pape, B.K. Tanner: Magnetization process and magnetic properties of Co/Cr/Co trilayers, Phys. Stat. Sol. (a) **158**, 259 (1996)

3.336 Y. Roussigne, F. Ganot, C. Dugautier, P. Moch: Brillouin scattering in Co/Cu/Co and Co/Au/Co trilayers: anisotropy fields and interlayer magnetic exchange, J. Magn. Magn. Mater. **148**, 213 (1995)
3.337 Y. Roussigne, F. Ganot, C. Dugautier, P. Moch: Brillouin scattering in Co/Cu/Co and Co/Au/Co trilayers: anisotropy fields and interlayer magnetic exchange, Phys. Rev. B **52**, 350 (1995)
3.338 R.W. Wang, D.L. Mills, E.E. Fullerton, S. Kumar, M. Grimsditch: Magnons in antiferromagnetically coupled superlattices, Phys. Rev. B **53**, 2627 (1996)
3.339 M. Grimsditch, S. Kumar, E.E. Fullerton: Magnetic excitations in antiferromagnetically coupled superlattices: Fe/Mo and Fe/Cr, Journal of Physics **5**, 369 (1993)
3.340 A. Yoshihara, K. Takanashi, Y. Obi, H. Fujimori: Brillouin scattering from spin waves in Fe/Cr multilayer films, J. Magn. Magn. Mater. **126**, 333 (1993)
3.341 M.J. Pechan, E.E. Fullerton, W. Robertson, M. Grimsditch, I.K. Schuller: Determination of magnetic anisotropy in Fe/Cu multilayers: equivalence of dynamic and static measurements, Phys. Rev. B **52**, 3045 (1995)
3.342 M. Schilberg, T. Kleinefeld, B. Hillebrands: Spin wave dynamics and domain structure in exchange coupled FeNi multilayers, J. Magn. Magn. Mater. **121**, 303 (1993)
3.343 B. Hillebrands, P. Baumgart, R. Mock, G. Güntherodt, A. Boufelfel, C.M. Falco: Collective spin waves in Fe/Pd and Fe/W multilayers, J. Appl. Phys. **61**, 4308 (1987)
3.344 A. Boufelfel, B. Hillebrands, G.I. Stegemann, C.M. Falco: Fe/Pd second-order superlattices, Solid State Commun. **68**, 201 (1988)
3.345 R.K.W. Wettling, G. Rupp, W. Jantz: Surface spin waves in magnetic multilayers. *Magnetic Thin Films*, R. Krishnan, Les Ulis (eds.) Editions Phys. 69 (1986)
3.346 G. Rupp, W. Wettling, W. Jantz, R. Krishnan: Brillouin scattering study of multilayer cobalt-niobium films, Appl. Phys. A **37**, 73 (1985)
3.347 R. Krishnan, W. Jantz, W. Wettling, G. Rupp: Investigation of compositionally modulated and amorphous cobalt-niobium films, IEEE Trans. Mag. **20**, 1264 (1984)
3.348 J.A. Cowen, J.G. Booth, J. Boyle, A.D. Boardman, K.M. Booth, F. Schreiber: CoCu(111) superlattices investigated by Brillouin light scattering, J. Magn. Magn. Mater. **156**, 163 (1995)
3.349 G. Carlotti, G. Gubbiotti, L. Pareti, G. Socino, G. Turilli: Elastic and magnetic properties of Co/Cu multilayers studied by Brillouin spectroscopy, J. Magn. Magn. Mater. **165**, 424 (1997)
3.350 R.V. Leeuwen, C.D. England, J.R. Dutcher, C.M. Falco, W.R. Bennett, B. Hillebrands: Structural and magnetic properties of Ti/Co multilayers, J. Appl. Phys. **67**, 4910 (1990)
3.351 B. Hillebrands, J.V. Harzer, G. Güntherodt, C.D. England, C.M. Falco: Experimental evidence for the existence of exchange-dominated collective spin-wave excitation in multilayers, Phys. Rev. B **42**, 6839 (1990)
3.352 G.S. Bains, A. Yoshihara, K. Takanashi, H. Fujimori: Brillouin scattering form spin waves in Co/Gd multilayer films, J. Magn. Magn. Mater. **126**, 329 (1993)
3.353 J. Boyle, J.G. Booth, J.A. Cowen, A.D. Boardman, K.M. Booth, R. Krishnan: Brillouin light scattering study of Ni/Pt multilayers, J. Magn. Magn. Mater. **156**, 33 (1996)

3.354 J.A. Cowen, J.G. Booth, A.D. Boardman, K.M. Booth, M.R.J. Gibbs, C. Shearwood: Brillouin light scattering studies of iron-cobalt multilayers, J. Magn. Magn. Mater. **165**, 383 (1997)
3.355 W. Wettling, W. Jantz, C.E. Patton: Light scattering study of phonons parametrically excited in the weak ferromagnet $FeBO_3$, J. Appl. Phys. **50**, 2030 (1979)
3.356 V.G. Zhotikov, N.M. Kreines: Scattering of light by parametric magnons and phonons in $CoCO_3$, Sov. Phys.-JETP **50**, 1202 (1979)
3.357 V.G. Jotikov, N.M. Kreines: Light scattering from magnons and phonons excited by microwave pumping in antiferromagnets, J. Magn. Magn. Mater. **15**, 809 (1980)
3.358 V.N. Venitskii, V.V. Eremenko, E.V. Matyushkin: Optical investigation of spin and magnetoelastic waves in a single crystal of yttrium iron garnet under longitudinal magnetic pumping, Sov. Phys. JETP **50**, 934 (1979)
3.359 G. Srinivasan, C.E. Patton: Direct detection of magnetostatic wave excitations in magnetostatic wave device structures by Brillouin light scattering, Appl. Phys. Lett. **47**, 759 (1985)
3.360 W.D. Wilber, J.G. Booth, C.E. Patton, G. Srinivasan, R.W. Cross: Light-scattering observation of anomalous parametric spin-wave character in subsidiary absorption, J. Appl. Phys. **64**, 5477 (1988)
3.361 G. Srinivasan, P.R. Emtage, J.G. Booth, C.E. Patton: Optical observation of evanescent surface magnons in thin magnetic films, J. Appl. Phys. **63**, 3817 (1988)
3.362 H. Xia, P. Kabos, H.Y. Zhang, P. Kolodin, C.E. Patton: Brillouin light scattering and magnon wave vector distributions for microwave magnetic envelope solitons in yttrium iron garnet thin films, Phys. Rev. Lett. **81**, 449 (1998)

Index

Topics in Applied Physics Founded by Helmut K. V. Lotsch

Topics in Applied Physics [illegible]

[illegible]

Interactions on Metal Surfaces [illegible]

[illegible]

Picture Processing and Digital Filtering 2nd Edition. Editor: T. S. Huang

Integrated Optics 2nd Ed. Editor: T. Tamir

[illegible]

Hydrogen in Metals I [illegible] Editors: G. Alefeld and J. Völkl

[illegible]

Solar Energy Conversion. Solid-State Physics Aspects. Editor: B. O. Seraphin

[illegible]

X-Ray Optics [illegible]

[illegible]

Acoustic Surface Waves Editor: A. A. Oliner

Laser Beam Propagation in the Atmosphere. Editor: J. W. Strohbehn

Photoemission in Solids I. General Principles. Editors: M. Cardona and L. Ley

[illegible]

Glassy Metals I Editors: H.-J. Güntherodt and H. Beck

Sputtering by Particle Bombardment I Editor: R. Behrisch

Optical Information Processing Fundamentals. Editor: S. H. Lee

MIX
Papier aus verantwortungsvollen Quellen
Paper from responsible sources
FSC® C105338

If you have any concerns about our products,
you can contact us on
ProductSafety@springernature.com

In case Publisher is established outside the EU,
the EU authorized representative is:
Springer Nature Customer Service Center GmbH
Europaplatz 3, 69115 Heidelberg, Germany

Printed by Libri Plureos GmbH
in Hamburg, Germany